INTEGRATED NATURAL RESOURCE MANAGEMENT IN THE HIGHLANDS OF EASTERN AFRICA

This book documents a decade of research, methodological innovation, and lessons learned in an eco-regional research-for-development program operating in the eastern African highlands, the African Highlands Initiative (AHI). It does this through reflections of the protagonists themselves—AHI site teams and partners applying action research to development innovation as a means to enhance the impact of their research. This book summarizes the experiences of farmers, research and development workers, policy and decision-makers who have interacted within an innovation system with the common goal of implementing an integrated approach to natural resource management (NRM) in the humid highlands.

This book demonstrates the crucial importance of "approach" in shaping the outcomes of research and development, and distils lessons learned on what works, where and why. It is enriched with examples and case studies from five benchmark sites in Ethiopia, Uganda, Kenya, and Tanzania, whose variability provides the reader with an in-depth knowledge of the complexities of integrated NRM in agro-ecosystems that play an important role in the rural economy of the region. It is shown that the struggle to achieve sustainable agricultural development in challenging environments is a complex one, and can only be effectively achieved through combined efforts and commitment of individuals and institutions with complementary roles.

Laura German is Senior Scientist at the Center for International Forestry Research in Bogor, Indonesia.

Jeremias Mowo is Regional Coordinator and Senior Scientist at the World Agroforestry Centre in Nairobi, Kenya.

Tilahun Amede is Nile Basin Leader for the Challenge Programme on Water for Food and Scientist at the International Livestock Research Institute and the International Water Management Institute in Addis Ababa, Ethiopia.

Kenneth Masuki is Knowledge Management Specialist for the African Highlands Initiative at the World Agroforestry Centre in Nairobi, Kenya.

INTEGRATED NATURAL RESOURCE MANAGEMENT IN THE HIGHLANDS OF EASTERN AFRICA

From concept to practice

Edited by Laura German,
Jeremias Mowo, Tilahun Amede,
and Kenneth Masuki

International Development Research Centre
Ottawa • Cairo • Dakar • Montevideo • Nairobi • New Delhi • Singapore

earthscan
publishing for a sustainable future
LONDON AND NEW YORK

World Agroforestry Centre
TRANSFORMING LIVES AND LANDSCAPES

First published 2012
by Earthscan
2 Park Square, Milton Park, Abingdon, Oxon OX14 4RN

Simultaneously published in the USA and Canada
by Earthscan
711 Third Avenue, New York, NY 10017

Earthscan is an imprint of the Taylor & Francis Group, an informa business

Co-published with the
International Development Research Centre
PO Box 8500, Ottawa, ON K1G 3H9, Canada
info@idrc.ca / www.idrc.ca
(IDRC publishes an ebook edition of this book, ISBN 978-1-55250-530-4)

British Library Cataloguing in Publication Data
A catalogue record for this book is available from the British Library

Library of Congress Cataloging in Publication Data

Integrated natural resource management in the highlands of eastern Africa : from concept to practice / edited by Laura German ... [et al.].
p. cm.
Includes bibliographical references and index.
1. African Highlands Initiative. 2. Natural resources--Africa, Eastern--Management. 3. Upland conservation--Africa, Eastern. 4. Ecosystem management--Africa, Eastern. 5. Watershed management--Africa, Eastern. 6. Conservation of natural resources--Africa, Eastern. 7. Sustainable development--Africa, Eastern. I. German, Laura A.
HC840.Z65I58 2011
333.709676--dc22
2011028391

ISBN: 978-1-84971-424-2 (hbk)
ISBN: 978-0-203-13918-9 (ebk)

Typeset in Bembo
by GreenGate Publishing Services, Tonbridge, Kent

This book is dedicated to Dr. Ann Stroud—colleague, mentor, and friend. Ann served as Regional Coordinator of the African Highlands Initiative (AHI) system-wide program from 1998 to 2006. Ann approached her work as a visionary, with a remarkable level of personal commitment and a strong philosophical underpinning that gave form to both the program and those of us within its expansive conceptual and geographical reach. This book is a symbolic gesture of gratitude for the sacrifices she made so that we all might have a space in which to "create." By buffering us from the exigencies of ever-shifting institutional demands of host institutions and the ongoing challenge to keep the program resourced, we were free to rally behind her vision and unlock creative energies in an effort to advance understanding and impact among the poor—whose needs and perspectives have so often been bypassed by externally imposed development schemes. We recognize you as the intellectual force behind this volume, and are grateful for the opportunity to have learned so much at your side.

Ann Stroud was born in California in 1952 and obtained degrees from the State University of New York at Geneseo (BSc) and Cornell University (MSc, PhD). After working as a pioneering organic farmer in the USA, she moved to Africa in 1982, where she lived and worked for 24 years (in Kenya, Ethiopia, Tanzania, and Uganda). Before working with AHI, her jobs included the design of a sustainable gardens project in Kenya, employment with FAO as a weed management adviser to Ethiopia's Ministry of Agriculture, and national adviser to the Tanzanian farming systems research program with support provided by DGIS Netherlands. Ann passed away in May 2007. She leaves behind two children, Katie and Andrew, husband Roger Kirkby, and her mother, Frances Stroud.

I worked with Ann in the Western Kenya Benchmark site for a long time and admired her vision and passion for success in whatever she put her fingers on. She became a great inspiration to us in the benchmark site and to date we are proud of her for the legacy she bequeathed on us.

Dr. Kenneth Otieno

My colleagues in IDRC appreciated your contributions and all of them are with me in this acknowledgement and celebration of your life and contributions to development thinking and application in the moment of your departure. You inspired many who will become your followers and call on you for inspiration. I know I will, as long as I can stay on this track of the ideals and objectives we shared and valued so much.

Dr. Luis Navarro

In her short lifetime, Ann made a tremendous contribution to people-centred research and development—and her enthusiasm and humour and innovative ideas inspired countless people.

Ann Waters-Bayer

All of us who had the privilege of knowing and working with Ann will recall her professionalism, her instinctive understanding of the values of inclusiveness and participation and the pleasure she gave and took in working with anybody and everybody. Her illness robbed Africa of an outstanding scientist and a true friend with a commitment to improving the lives of resource-poor rural people.

Dr. Monty Jones

Ann was a pioneering, innovative and caring scientist, willing to take risks, adopting participatory and integrated research as fundamental, and strongly committed to capacity building. She received a great deal of recognition, donor and partner support for the work of AHI yet was down to earth, not seeking honors for herself but constantly promoting others. She always strove to fully participate, to add value to any meeting or workshop, to bring her insights and years of experience to enrich any discussion. Ann was the best colleague one could have.

Diane Russell

In all the spectrum of people I worked with in Africa Ann was one that I most admired. She had the unique combination of being a good scientist, having a deep understanding of the needs and aspirations of the farmers and very effective management skills. But above all she was always a delight to be with. I have many fond memories of discussions and arguments in workshops and over lunch or dinner, which always left me with something new to take away. The world is a lesser place without her; but … her work will remain as a memorial; she will remain in the thoughts of the countless people she helped.

Dr. Mike Swift

CONTENTS

FIGURES

Colour plate section (between pages 146 and 147)

TABLES

BOXES

CONTRIBUTORS

Zenebe Adimassu is a PhD student at Wageningen University and Research Centre, the Netherlands. Zenebe served as AHI site coordinator for the Galessa site before he initiated his PhD studies. He co-edited a book entitled *Working with Rural Communities on Integrated Natural Resources Management (INRM)*, which profiles AHI research and development outputs from Galessa site, Ethiopia.

Diana Akullo works with the Association for Strengthening Agricultural Research in eastern and central Africa (ASARECA) Secretariat in Entebbe, Uganda, under the Staple Crops Programme. She previously worked as a Senior Research Officer in Quality Assurance with the Secretariat of Uganda's National Agricultural Research Organization (NARO). Her work has been to facilitate learning in the agricultural research domain in eastern and central Africa. She holds a PhD in social science, with a focus on institutional reforms in response to national and regional development policies.

Tilahun Amede is currently a leader of the Nile Basin Challenge Programme on Water for Food (CPWF), based in Addis Ababa, Ethiopia. Before assuming his current position, he served as a Systems Agronomist for the African Highlands Initiative and the Tropical Soils Biology and Fertility Institute with the International Center for Tropical Agriculture (CIAT). Tilahun has served as Guest Editor for special issues in the *Rangeland Journal* (2009) and *Experimental Agriculture* (2011).

Shenkut Ayele works with the Catholic Relief Service in Ethiopia as Early Warning Assessment and Response Manager. He previously served as a Socio-Economics and Extension Researcher and Head of the Research and Extension Department at the Ethiopian Institute of Agricultural Research. Shenkut holds an MSc in Agricultural Extension Education from the University of Agriculture Dharwad (UAS-Dharwad), India.

Leulseged Begashaw is a plant pathologist with the Southern Agricultural Research Institute, based at the Areka Agricultural Research Center,

Ethiopia. He led AHI's research on vertebrate pest control in Areka benchmark site.

Berhanu Bekele is a wheat breeder at Debre Zeit Agricultural Research Center, Ethiopia. He was one of the researchers who introduced different improved barley varieties in Galessa and the surrounding areas.

Francis Byekwaso is Planning and Monitoring Manager for the National Agricultural Advisory Services (NAADS) in Uganda. He holds a PhD in Nutrition Planning and Decentralization. Past work has included action research on pro-poor decentralized service delivery; community-based agricultural extension services; farmer empowerment and demand for agricultural extension services; and natural resource management.

Mulugeta Diro holds a PhD in Plant Physiology and currently works as an independent consultant on a collaborative NRM project with Wageningen University. Before assuming this responsibility, Mulugeta was the Director of the Crop Directorate, Southern Regional Agricultural Research Institute (SARI), Awassa, Ethiopia. He also served as a National Coordinator for enset research in Ethiopia.

Laura German is Senior Scientist at the Center for International Forestry Research (CIFOR) in Bogor, Indonesia, where she contributes to CIFOR's research on managing the impacts of globalized trade and investment on forests and forest communities. Past work has included action research on integrated natural resource management and landscape governance with AHI; human ecological research in the Brazilian Amazon; and agricultural development in Honduras. She holds a PhD in Cultural and Ecological Anthropology. Her recent publications include two edited volumes, *Governing Africa's Forests in a Globalized World* (Earthscan, 2009) and *Beyond the Biophysical* (Springer, 2010), and a special issue in *Ecology and Society* on the social and environmental impacts of biofuels.

Girma Hailu holds a PhD in Agroecology from Kenyatta University, Kenya. He is an agricultural scientist with over 12 years' experience in designing, conducting, and managing research projects. He has worked as Program Coordinator at Farm Radio International (Canada), Project Manager at African Highlands Initiative (AHI), and Research Scientist at World Agroforestry Centre. He currently works as an independent consultant.

Rick Kamugisha is a social scientist with the World Agroforestry Centre (ICRAF), based in Kampala, Uganda. He holds an MA in Development Studies from Mbarara University of Science and Technology, Uganda. Rick has been involved in facilitating local communities in natural resource management and governance through efforts to strengthen local level policies and grassroots institutions under Landcare and AHI. He has vast experience working with policy taskforces, innovation platforms and conflict management in NRM. Rick has worked with AHI for nine years.

Bekele Kassa is a Senior Pathologist with the Ethiopian Institute of Agricultural Research, based at Holetta Agricultural Research Center (HARC). Bekele served as a site coordinator in the second phase of AHI and was one of the researchers who facilitated Farmers' Field Schools on potato late blight in the Galessa and Jeldu areas.

Charles Lyamchai is a principal agricultural research officer at Selian Agricultural Research Institute, Arusha, Tanzania. His current work focuses on improving natural resource governance at community level and mainstreaming Integrated Natural Resource Management in research and development institutions to foster sustainable management of natural resources, increased income and poverty alleviation. He holds an MSc in Agro-meteorology.

Hussein Mansoor holds a PhD in Plant and Soil Sciences from the University of Aberdeen, Scotland. He is currently an assistant director for crop research in the Division of Research and Development, Ministry of Agriculture, Food and Cooperatives, Tanzania. Before assuming his current post he was involved in numerous research activities in land management, integrated natural resource management, integrated soil fertility management and soil and water conservation at Selian Agricultural Research Institute, Arusha, Tanzania.

Kenneth Masuki is a knowledge management specialist with the African Highlands Initiative, based at the World Agroforestry Centre, Nairobi, Kenya. His professional experience includes research in agriculture, soil and water management, integrated natural resource management, and knowledge management. He holds a PhD in Agricultural Education and Extension from Sokoine University of Agriculture, Morogoro, Tanzania.

Waga Mazengia is a senior researcher in agronomy at the Southern Agricultural Research Institute, based as the Hawassa Agricultural Research Center. Waga coordinated AHI project activities at Areka benchmark site for many years.

Joel Meliyo holds an MSc in Soil Science and Land Management from Sokoine University of Agriculture (SUA). He has extensive experience in soil and land resources inventories, mapping, land evaluation and suitability assessment for land use planning, using tools such as remote sensing, geographical information systems and research on integrated soil fertility management. He is currently pursuing his PhD at SUA.

Jeremias Mowo holds a PhD in Soil Science from Wageningen University, the Netherlands. He has over 30 years of research and development experience in natural resources management in eastern Africa. Currently, he is the Regional Coordinator for ICRAF eastern Africa. His areas of interest include integrated natural resource management, farmer participatory research and research on methods and approaches for technology transfer.

Simon Nyangas is a coordinating secretary for the Landcare Chapter in Kapchorwa, Uganda and serves as the NAADS Coordinator in the town council overseeing Agricultural Technology and Agribusiness Advisory Services. Previous responsibilities included agricultural extension services in the Ministry of Agriculture, Animal Husbandry and Fisheries (MAAIF) in Kapchorwa local government, where he worked on sustainable conservation and development in Mt. Elgon. He is an agronomist by training.

Chris Opondo, sociologist by profession, joined AHI in 2002 as a regional research fellow with an emphasis on monitoring and evaluation. In 2008 he joined United Nations Office on Drugs and Crime (UNODC) as a monitoring and evaluation manager, based in Pretoria, South Africa. In May 2011 Chris passed away after an acute asthmatic attack. All members of the AHI and partners who worked with him will continue to remember him for the contribution he made to science and practice, as illustrated in this volume.

Pascal Sanginga is a senior programme specialist for Agriculture and Environment at the International Development Research Centre's regional office for eastern and southern Africa. Pascal joined the African Highlands Initiative in 1999 as a regional research fellow, and went on to work for the International Center for Tropical Agriculture (CIAT) as Senior Social Scientist. He holds a PhD in Rural Sociology and has conducted and managed interdisciplinary and participatory action research for agriculture and natural resource management in eastern, southern, western and central Africa. His recent publications include two edited volumes, *Innovation Africa: Enriching Farmers' Livelihoods* (Earthscan, 2009) and *Managing Natural Resources for Development in Africa: A Resource Book* (University of Nairobi Press, 2010).

Joseph Tanui is an agricultural economist and coordinates the Landcare program based at the World Agroforestry Centre, Nairobi, Kenya. Joseph has extensive experience on smallholder collective action and grassroots institutional development, from a research, policy and extension perspective. He is currently undertaking PhD research at Wageningen University, the Netherlands, with a focus on the institutional economics of sustainable land management.

Wilberforce Tirwomwe holds a BA in Social Sciences with a specialization in Development Studies and has extensive experience in facilitating local communities in natural resource management and governance in Kabale District, Uganda. He worked as a consultant for both AHI and ICRAF-Uganda's Landcare program. He currently works for CARE International in Kabale District, Uganda.

Adugna Wakjira holds a PhD in Crop Breeding, with a focus on oil crops. He is a senior researcher with the Ethiopian Agricultural Research Institute, based at Holetta Agricultural Research Center.

Juma Wickama holds an MSc in Soil and Land Management from Sokoine University of Agriculture in Tanzania. He currently works at the Agricultural Research Institute of Tanzania as a principal researcher in soils and soil management consultant. The bulk of his work has focused on researching and promoting efficient resource use in the West Usambara mountains and reclamation of salt affected fields on the lower slopes of Mt. Kilimanjaro in northern Tanzania. He is currently a PhD student at Wageningen University, the Netherlands.

Gebremedhin Woldegiorgis is a senior researcher at the Ethiopian Institute of Agricultural Research, based at the Holetta Agricultural Research Center (HARC). With expertise in potato breeding and production, he is currently seconded by the International Potato Center in Addis Ababa, Ethiopia.

FOREWORD

As this book goes to press, the Horn of Africa is experiencing one of the worst food crises and droughts in five decades, with millions of lives in danger. In times of such crises, attention is naturally focused on the urgent short-term responses needed to keep people alive. Well-intentioned efforts often fail to recognize the complex, multidimensional, and long-term causes of hunger, poverty, and environmental degradation. Experience and lessons from over 40 years of international development and research suggest that there are no easy solutions to avoiding such crises. Rather, it requires integrated, multilevel and multidimensional approaches that have the support of the people affected, and a sustained commitment by society at large to build resilience for rural Africans and their food systems in the face of increasing uncertainty.

Since its establishment in 1970, the International Development Research Centre (IDRC) has supported networks and partnerships of researchers in developing countries to contribute to the operationalization of participatory action research and institutional innovation for community-based natural resources management, in "hot spots" of poverty and environmental degradation. The highlands of eastern and central Africa are home to several millions of people whose livelihoods depend on natural resources and agriculture. The African Highlands Initiative is one of the networks and innovative programs that has received consistent and continued support from IDRC and other donors, since 1992, to advance the conceptualization and the practice of integrated natural resources management (INRM) as a modest but effective approach to support inclusive decision making, access to resources, rehabilitation of degraded natural resources, and the development of agriculture in a sustainable manner.

This book describes a compelling success story of an ambitious program with an inspiring narrative. Early work of the African Highlands Initiative focused on developing and facilitating the uptake of techniques and practices to intensify and diversify farming systems. AHI then embarked on innovative but untested concepts of participatory integrated watershed management, collective action, integrated agricultural research for development, and institutional change processes, under the umbrella of Integrated Natural Resource Management (INRM). This marked a significant departure from most agricultural research initiatives in the region that concentrated on commodity-based and uni-disciplinary research. While AHI experienced several challenges in its organizational structures, capacities, leadership, and institutional mandates, as this book demonstrates, it has remained focused, nimble, flexible, and willing to experiment with new approaches at the forefront of INRM in eastern Africa.

After over a decade of support to AHI, IDRC commissioned an external review and impact assessment of the whole initiative. The findings of this evaluation were summarized by the evaluators in a simple phrase: "AHI has achieved unique success in implementing INRM that works." A crucial feature of this success is the production of an impressive number of peer reviewed articles in mainstream international journals, edited books, method guides, and briefs, vital contributions to international and national public goods. This book consolidates those contributions to the science and practice of INRM in an easy-to-read volume. As the six chapters demonstrate, AHI has developed innovative tools and approaches and practical methods for putting INRM principles into action, with tangible results that have bettered the lives of resource-poor farmers in five eastern African countries. Examples of impacts are discussed in the book with honesty, recognizing challenges and limitations and, more importantly, reflecting on the lessons learned to provide some practical "how to" tips for practitioners.

However, as the different chapters of the book demonstrate, INRM remains a work in progress. The challenging conjunction of poverty, environmental degradation, and food and personal insecurity remains daunting. A vital next step, conceptual and practical, strategic and tactical, methodological and outcomes-oriented, is to "learn our way" toward innovating on a much broader scale. We need to bring together people, institutions, and ideas, to experiment in a very explicit and systematic manner, and on a society-wide scale, with the concepts of social learning, adaptive management, and socio-ecological resilience that are emerging from the work of AHI and others.

This book and the evolution of AHI illustrate a key principle of IDRC's mission: long-term strategic investment in building the capacity of thinkers, innovators, and practitioners in the global South. It is clear from the experiences of AHI that long-term support and commitment are needed for INRM; natural resources management research requires patience and a significant commitment to investment in resources and people, over periods of ten years or longer. At IDRC we have been privileged to see the fruits of consistent support, intellectual mentoring, practical advice, and a willingness to take risks in uncertain ventures. IDRC staff placed great emphasis on engaging with AHI researchers and their partners, helping to open spaces for intellectual leadership in a spirit of mutual learning. It is a rare opportunity and privilege to be involved with an organization such as AHI from inception and program establishment to "closing the loop." We have been privileged to be involved in this initiative, and have learned much ourselves.

As much as this book is an institutional memory of over a decade of "muddling through complexity" of INRM research and development, it is also a strong tribute to the late Luis Navarro, an IDRC colleague and friend. Luis championed and channeled IDRC's support to AHI from its inception to its maturity in 2006 when he fell ill. Luis was instrumental in the conceptualization of INRM and in its practical application with farming communities and other stakeholders. This book represents a scientific and professional testimony to Luis' pioneering efforts to further the application of participatory approaches and systems thinking in integrated natural resources management in eastern Africa. The application of these principles, which he spent his career thinking about, in farming-systems intensification and diversification, in landscape innovations, in district-level initiatives linking local communities to decentralized government structures and institutions, and in support of institutional change in research and development organizations, has inspired scientists in this region and elsewhere, and makes compelling reading. I think he would feel gratified that the lessons are now being shared.

Simon Carter
Director, Regional Office for Eastern and Southern Africa
International Development Research Centre (IDRC)

PREFACE

This book represents an ambitious effort to document over ten years of research, methodological innovation, and lessons learning in AHI through reflections by the protagonists themselves—AHI site teams and partners involved in innovating to enhance the impact orientation of research. It attempts to summarize the experiences of farmers, researchers, and development workers, and policy and decision-makers who have interacted with AHI on the theme of Integrated Natural Resource Management in the humid highlands of eastern Africa. It aims both to raise awareness of the crucial importance of methods or "approaches" in the outcomes derived from research and development work, and to distill lessons learned on "what works, where, and why." The book therefore is enriched with examples and case studies from five benchmark sites whose variability provides the reader with an in-depth knowledge of the complexities of natural resource management (NRM) in agro-ecosystems that play an important role in the rural economy of the region. The struggle to achieve sustainable agricultural development in challenging environments is a difficult one, and can only be effectively achieved through combined efforts and commitment of individuals and institutions with complementary roles.

The book is organized into six chapters. Chapter 1 gives an overview of INRM as a concept and the birth and evolution of AHI, including the methodological framework through which innovations were developed and tested and its results. Chapter 2 provides an overview of farm-level methodological innovations oriented towards participatory intensification and diversification of smallholder farming systems for optimal system productivity (economic, social, and ecological). Chapter 3 summarizes AHI experiences with a set of approaches employed to operationalize participatory watershed management through an integrated lens which looks not only at soil and water but at a wider set of system components and interations. Chapter 4 explores lessons learned to date on methods and approaches for participatory landscape governance, exploring how processes that cut across farm boundaries, involve trade-offs between different land users or require collective action may be addressed effectively and equitably.

Chapter 5 explores the role of district level institutions and cross-scale linkages in supporting grassroots development and conservation initiatives, including improved coordination and better support to local livelihood priorities and bottom-up governance reforms. Chapter 6 explores methods and approaches for scaling up and institutionalizing integrated natural resource management innovations (e.g., those presented in earlier chapters), as well as approaches for self-led institutional change that can institutionalize the *process* of methodological innovation and impact-oriented research.

ACKNOWLEDGMENTS

Many people and organizations contributed to the work of AHI in the decade of work profiled in this volume and in the effort to compile it, and it would be an impossible task to list them all. The authors would like to recognize, first and foremost, the commitment of the farmers of the humid highlands of eastern Africa whose enthusiasm, dedication, and patience enabled the program to engage in a host of previously untested innovations in order to accomplish its mandate. The dedication of AHI site teams and their courage to explore the unknown were key factors in the success the program has achieved, for which AHI staff, partners, and donors are grateful. Without the leadership and support of the managers of national agricultural research institutes (NARIs) in Ethiopia, Kenya, Madagascar, Rwanda, Tanzania, and Uganda, this work would not have been successful; we bank on their continued support as we move into the next decade. Other partners who served instrumental roles in the evolution of AHI include sister CGIAR programs and centers (CAPRi, CIAT, ICRAF, IFPRI, ILRI, CIMMYT, CIP, and ICRISAT); non-governmental organizations (CARE International—Uganda, Africare, Africa 2000 Network, Action Aid, Farm Africa, and SOS-Sahel); the National Agricultural Advisory Services of Uganda (NAADS); extension organizations in AHI benchmark sites; the Kapchorwa District Landcare Chapter (KADLACC); and a host of farmers' organizations in AHI benchmark sites.

The ability of AHI to sustain innovation and learning over more than a decade is in large part attributable to the generous financial support AHI has received from a few committed donors who have shown total commitment to the piloting of new forms of professionalism and its potential to unlock change in rural livelihoods and environmental management—namely, the Swiss Development Cooperation (SDC); the International Development Research Centre (IDRC); the European Commission; Cooperazione Italiana; and the Ministerie van Buitenlandse Zaken (the Netherlands). We are also grateful to other donors who funded AHI for specific initiatives that have gone a long way in furthering our learning on topics of instrumental importance to the program and the region. Among these donors, we would like to acknowledge the Collective Action and Property Rights program of the CGIAR (CAPRi);

the Department for International Development (UK); Aus-AID (Australia); the Rockefeller Foundation; and the Norwegian government.

A number of individuals provided periodic mentoring and facilitation to support the conceptualization of the AHI approach over the years. While many individuals have made significant contributions in this regard, Jürgen Hagmann and Rajiv Khandelwal deserve special mention. We would also like to acknowledge the support of those who have helped to usher us through the publication process. Those deserving special mention include Dina Hubudin at CIFOR for her patient assistance in formatting the manuscript to Earthscan style guides, Paul Stapleton and Rose Onyango at ICRAF for facilitating various communications with Earthscan, and Tim Hardwick and Ashley Irons at Earthscan, and Bill Carman of IDRC who patiently guided the book to its completion.

We are also grateful to our institutional hosts, ASARECA and ICRAF, for their technical, financial and administrative support to AHI activities and guidance during times of transition. We give special thanks to Dr. Seyfu Ketema, Dr. Denis Garrity, Dr. Eldad Tukahirwa, Dr. Jan Laarman, Dr. Mohamed Bakarr, and the ICRAF finance and human resource departments. We would like to also express our sincere appreciation to Dr. Roger Kirkby, who assisted in steering AHI through a challenging transition away from Dr. Ann Stroud's competent leadership.

And last but not least, we would like to acknowledge the intellectual guidance and commitment of the late Dr. Luis Navarro. From the very beginning, Luis was much more than a donor to AHI – participating in steering committee meetings, engaging in regular dialogue to shape the program's vision and evolution, and challenging the AHI team to step beyond conventional boundaries. We are most grateful for the creative tensions inspired by your mentorship.

ABBREVIATIONS

AAR	after action review
AHI	African Highlands Initiative
ASARECA	Association for Strengthening Agricultural Research in Eastern and Central Africa
BMS	benchmark sites
CA	collective action
CAP	community action plan
CAPRi	CGIAR Systemwide Program on Collective Action and Property Rights
CBF	community-based facilitator
CBO	community-based organization
CEED	Coalition for Effective Extension Delivery
CIAT	Centro Internacional de Agricultura Tropical
CIP	International Potato Center
CGIAR	Consultative Group on International Agricultural Research
DEM	digital elevation model
DfID	Department for International Development (UK)
DRC	Democratic Republic of the Congo
DRD	Department of Research and Development (Tanzania)
DU	developmental unit
ECA	Eastern and Central Africa
EIAR	Ethiopian Institute of Agricultural Research
EU	European Union
FAO	Food and Agriculture Organization of the United Nations
FFS	farmer field school
FPR	farmer participatory research
FRG	farmer research group
FSR	farming systems research
GPS	global positioning systems
HARC	Holetta Agricultural Research Center
IAR4D	integrated agricultural research for development
IARCs	International Agricultural Research Centers

ICRAF	World Agroforestry Centre
ICTs	information and communication technologies
IDRC	International Development Research Centre
IFPRI	International Food Policy Research Institute
ILAC	institutional learning and change
INRM	integrated natural resource management
ISAR	Institut des Sciences Agronomiques du Rwanda
KADLACC	Kapchorwa District Landcare Chapter
LCC	legume cover crops
NAADS	National Agricultural Advisory Services (Uganda)
NARES	national agricultural research and extension systems
NARIs	national agricultural research institutes
NARO	National Agricultural Research Organization (Uganda)
NARS	national agricultural research system
NGO	non-governmental organization
NRM	natural resource management
PA	peasant association
PAR	participatory action research
PCC	parish coordination committee
PIWM	participatory integrated watershed management
PM&E	participatory monitoring and evaluation
PTF	policy task force
PTD	participatory technology development
QAC	Quality Assurance Committee
R&D	research and development
SACCOS	savings and credit cooperative society
SDC	Swiss Development Cooperation
Tsh	Tanzanian shilling
ToT	transfer of technology
UgSh	Ugandan shilling
USD	United States dollar
VICE	village information center
VWC	village watershed committee
WHO	World Health Organization

1

INTEGRATED NATURAL RESOURCE MANAGEMENT

From theory to practice

Laura German, Jeremias Mowo, and Chris Opondo

Introduction

Highlands worldwide are important repositories of biodiversity as well as water towers for vast lowland and urban populations. The highlands of eastern Africa are no exception, with the Eastern Arc mountains home to a host of endemic species and a globally renowned biodiversity hotspot (Burgess et al., 1998). Yet while the ecological importance of highland areas cannot be underestimated, neither should their cultural and economic importance. Surrounded in most eastern African countries by semi-arid lowlands, the highlands have historically been home to disproportionately large human populations attracted by good rainfall, relatively good soils, and—in some locations—the potential to develop vast irrigation networks (SCRP, 1996).

In the past four decades, the eastern African highlands have seen rapid population growth and unprecedented land-use changes (Zhou et al., 2004), heightening the challenge of sustaining the resource base while providing for a growing population heavily dependent on natural resources for their livelihoods. Population growth and inheritance practices have contributed to very small household landholdings, reducing incomes and food security and in turn undermining farmers' capacity to invest in conservation activities, often characterized by delayed returns.

Population pressure has also caused people to expand into marginal hilly areas, increasing soil and water loss, and destroying unique habitats (Amede et al., 2001; Stroud, 2002). The erosion in collective action traditions and traditional authority structures at a time when interactions among adjacent landscape units and users are ever more tightly coupled has undermined cooperation in natural resource management and led to increased incidence of conflict (German et al., 2009; Sanginga et al., 2007, Sanginga et al., 2010).

Historical factors have also established powerful path dependencies in local attitudes and behavior that continue to undermine local natural resource management investments. Colonial era agricultural policies—externally imposed, coercive and often brutal—led to such an active resistance to soil conservation practices that they played a key role in the growth of organized resistance to colonial rule (Anderson, 1984; Throup, 1988). Fortress conservation policies in the colonial and post-colonial era have served as a similar disincentive to sustainable forest management (Borrini-Feyerabend et al., 2002; Colchester, 2004; Western, 1994). Shifts in political regimes and related land reform policies have also resulted in significant ambiguity in resource ownership and control, contributing in some cases to resource mining behaviors (Bekele, 2003; Omiti et al., 1999).

Yet if the colonial state "failed to show the farmer what tangible benefits the conservation effort would bring to the land … [or to] provide an adequate incentive for this effort" (Anderson, 1984: 321), to what extent have contemporary natural resource management interventions done any better? Within the conservation establishment, fortress conservation policies and approaches have slowly given way to a host of decentralized approaches—variously labeled Joint Forest Management, Co-Management, and Integrated Conservation and Development (Blomley et al., 2010; Brown, 2003; Hobley, 1996). Yet with the bottom line almost always one of natural resource conservation, some authors have begun to question whether such approaches have shifted the burden of conservation to local people without corresponding shifts in authority and benefits (Nsibambi, 1998; Wells, 1992). Others question whether these more participatory approaches are even suited to biodiversity conservation (Oates, 1999; Terborgh, 1999). Furthermore, these efforts have focused almost exclusively on production and protection of forests, leaving what happens in surrounding "anthropized" landscapes either beyond the scope of concern or of interest only to the extent that it furthers conservation objectives within protection forests (Hughes and Flintan, 2001). Where real powers for forest management have been devolved to local communities, it has often been where resources are already degraded and therefore of limited economic or conservation value (Blomley and Ramadani, 2006; Oyono et al., 2006).

Within the agricultural establishment, on the other hand, natural resource management concerns are squarely focused on landscapes where the human influence on landscape structure and function is dominant. Early emphasis, still prominent among agricultural research and extension institutions, was placed on soil erosion and its effect on soil fertility decline—with smaller communities of researchers focused on crop and livestock pests, agro-biodiversity, and rangeland management. With the vast majority of agricultural scientists and practitioners emanating from biophysical disciplines, problem definition has focused almost exclusively on biophysical constraints, and solutions put forward to address these have been largely technological (German et al., 2010). Early approaches, still evident today in the structure and functions of agricultural research and extension agencies, stressed a unilinear "transfer of technology" (ToT) approach in which research diagnoses problems and generates technologies to address these and passes them along to extension agents, who in turn disseminate them to farmers (Hagmann, 1999). Criticized for limited adoption levels and inability of this approach to catalyze effective responses to local needs, the rhetoric—and to some extent research practice as well—has slowly given way to a focus on more participatory forms of research. This has led to the proliferation of new approaches to address deficiencies in the old model, from "on-farm research" (advocated to adjust technologies to local conditions), "farmer participatory research" and "participatory technology development" (seeking to integrate farmers' criteria into technology testing and evaluation), and "farming systems research" (to take a more holistic look at farms as systems and explicitly address component interactions and the allocation of finite resources among multiple production objectives) (Byerlee and Tripp, 1988; Farrington and Martin, 1987; Haverkort et al., 1991; Walters-Bayer, 1989). While these approaches went a long way in adapting agricultural research and extension to local concerns and priorities, their institutionalization has been partial at best; the focus has remained largely technological and exclusive attention to the farm level and individualized decision-making has left many natural resource management problems unaddressed.

A host of newer approaches to conservation and natural resource management[1] hold great promise in placing the nexus of ownership and control squarely with local institutions, taking a wider view on natural resource management (beyond the farm, beyond the biophysical) and linking local users with outside actors and institutions. Yet, with a few notable exceptions (e.g., Colfer, 2005), the proliferation of jargon and rhetoric far outpaces efforts to operationalize them in practice (Rhoades, 2000; Sayer, 2001). This volume focuses on one such approach, Integrated Natural Resource Management (INRM), and tries to address this gap by profiling efforts to conceptualize the concept, develop and test approaches to operationalize it, and distil lessons learned. This chapter seeks to set the stage for the rest of the book by providing an overview of the INRM approach and how it is framed and interpreted in this volume, and by introducing and defining key concepts that form the "conceptual core" of the approaches profiled in the chapters to follow. The second half of the chapter is dedicated to an introduction to an eco-regional program operating in the eastern African highlands where the INRM concept was defined, piloted, and evaluated.

Integrated natural resource management

Key aims

Integrated natural resource management is a scientific and resource management paradigm uniquely suited to managing complex natural resource management challenges in densely settled landscapes where people are highly dependent on local resources for their livelihoods, thus heightening the tension between livelihood and conservation aims. The explicit effort to bridge productivity enhancement, environmental protection, and social well-being (Sayer and Campbell, 2003b) therefore makes INRM strategically relevant in such situations. The CGIAR (2001) defines INRM as "an approach to research that aims at improving livelihoods, agro-ecosystem resilience, agricultural productivity and environmental services. It does this by helping solve complex real-world problems affecting natural resources in agro-ecosystems."[2]

Yet, what does this mean in practice? A wide variety of research, development and conservation actors would already claim to be working towards such aims without employing the INRM label. So what aims and features set INRM apart from other approaches designed to address complex agricultural and natural resource management challenges? The CGIAR Task Force on INRM identified a number of success factors in managing an effective INRM process (CGIAR, 2002). Grouped by organizational level, these include:

Research and development teams:

- Employment of a participatory action research, learning process approach by all.
- Partnerships built on mutual trust, respect, and ownership by all.
- Multi-institutional arrangements with clear roles and commitments.
- Effective facilitation and coordination of interactive processes.
- Cross-disciplinary adaptive learning of research teams and development agents.
- Explicit scaling up/out strategy, building on "to-be" successes and strategic entry points.
- Effective communication strategy.

Partner and target communities/institutions:

- Application of a participatory action research, learning process approach by all.
- Shared problem and opportunity-driven focus.
- Short-term gains through the process itself (rather than via "handouts").
- Local organizational capacity for INRM.
- Access to knowledge, technological, policy, and institutional options.

Thus, the concept as it has evolved within the CGIAR emphasizes the process through which NRM innovations evolve. As stated by Hagmann et al. (2003), INRM is grounded in a learning paradigm, premised upon a social constructivist approach[3] to development and grounded in learning process approaches. Yet what is the substance of INRM? Key proponents emphasize the following core aims (Campbell, 2001; Sayer, 2001; Voss, 2001):

- Fostering sustainable agriculture, forestry and fisheries.
- Enhancing local adaptive capacity (in agriculture, forestry and fisheries), while supporting adaptive management beyond community level (e.g., the

evolution of NARS, government agencies, and international organizations into learning organizations).

- Acknowledging and addressing trade-offs in NRM through negotiation support.
- Emphasizing sustainable livelihoods through a client-centered approach.
- Solving real-world problems with partners through the integration of system components, disciplines, stakeholders, and scales.

Given the complexity of aims and the arbitrariness of "system boundaries" within multi-scale NRM initiatives, it is essential that these boundaries be set in some clear way (Campbell, 2001). Aside from bounding the "system" spatially, it is important to clearly define the nature of challenges to be addressed. While INRM could encompass efforts to reconcile local livelihood needs and NRM concerns with societal and global interests in environmental services emanating from rural areas, this volume makes an explicit effort to focus on the NRM concerns of local land users. It therefore focuses on natural resource management within landscapes managed by local resource users to meet their own livelihood goals, addressing issues related to protected areas only to the extent that these more "exclusionary" conservation efforts are of concern to adjacent land users. The scope of issues encompassed in subsequent chapters therefore includes the following:

- Stimulating farmer investment in natural resource management and adoption of land management innovations through innovative efforts to package and deliver technologies which address the livelihood and NRM concerns of farmers in an integrated manner.
- Addressing the social, economic, and cultural factors influencing NRM at farm and landscapes scales, including the pervasive tension between individual and collective goods.
- Improving farmer feedback to research, extension and development agencies within a social learning process, so as to exploit the complementary knowledge, skills, and mandates of different sets of actors in addressing pressing development and NRM problems.
- Achieving synergies between local technological, institutional, market, and policy innovations in NRM.
- Enabling higher-level innovations within research and development institutions to support local resource users, foster synergies in knowledge and skills, and institutionalize lessons learned.

Conceptual overview

This section provides an introduction to some of the key concepts utilized to frame this book and the methodological innovations that underlie it. It therefore sets the conceptual foundations to the chapters that follow.

Integration

AHI has worked with the concept of integration in its efforts to pursue integrated research and development innovations since its inception. In the first two phases, this concept was advanced by the work done to foster synergies among diverse system goals at farm level. In many benchmark sites, for example, teams experimented with "linked technologies," defined as a set of technologies whose benefits are best manifested when applied as an integrated whole rather than in isolation. For example, farmers experimented with soil conservation structures (bench terraces, *fanya juu*) stabilized with fodder, which was in turn fed to zero-grazed livestock in improved sheds, which in turn helped to make more efficient use of dung for fertilizing high-value crops on these structures. The integration concept is seen in the functional linkages established between system components (crops, livestock, soil, trees, water), which may be either ecological or economic. Ecologically, tighter nutrient recycling between crop and livestock components is designed to enhance the productivity of both crop and livestock components. From an economic standpoint, improved income from high-value crops on conservation structures may give farmers an incentive to invest in soil and water conservation, as well as additional disposable income. In Phase II the concept continued to be employed for achieving multiple and linked production objectives at farm level, but was further expanded to consider how integrated forms of support (e.g., technological, organizational, credit) to farm-level production could generate synergies and unlock change. Several years of systems modeling and experimentation in Areka, Ethiopia, also led to methodological innovations for understanding farms as systems—namely, integrated production units where multiple aims are pursued simultaneously in a context of limited financial, nutrient, and labor resources. It also led to strategies for enhancing component contributions to the wider farming system (as opposed to research efforts seeking to simply maximize returns to the component itself) and to participatory approaches to systems intensification.

During Phase 3, AHI began to experiment with integration concepts at the watershed level. These innovations helped to consolidate our understanding not only about what integration means at this level, but also overall. A typology of three forms of integration was developed during this phase to concretize current understanding of the integration concept (German, 2006). The first form, "component integration," involves understanding and managing the impacts of any given component (or component innovations) on other parts of the system. Farm-level components include trees, crops, livestock, and soil, while landscape components include these same components plus common property resources (including water, both for productive and domestic use). "Integration" in this sense implies moving beyond component-specific objectives (i.e., maximizing the yield of edible plant products) and outcomes. Integration generally implies an optimizing logic, ensuring balanced returns to diverse system components (yields from tree, crop, and livestock components) or increasing biomass yield without depleting system nutrients or water. At times, optimization requires sacrificing yield gains in one component of the system so as to balance returns to other system components or goals. An example of this would be the selection of a crop variety that does not exhibit the highest grain yield, but yields an optimal return to crop and livestock components in the form of grain and plant biomass (for fodder). Yet the integration concept may also cater to the logic of maximization, common to market-oriented production systems, by elucidating the consequences to other system components (synergies and/or trade-offs) when maximizing outputs from one component. For example, research might quantify the effects of fast-growing tree species—chosen to maximize timber yield and tree income—on crop yield and income within the landholding and on adjacent farms, and on spring discharge. Understanding such trade-offs provides information on what is gained and lost to different system goals and land users, which may provide important inputs into development practice or policy.

The second form, "constructivist integration," is more socio-political in nature—aiming to integrate the needs and priorities of different interest groups into research. "Constructivism" acknowledges that there is not one 'correct' view of reality but rather multiple, socially constructed realities (Chambers et al., 1992). In systems innovation, the priorities of these different social actors are actively solicited and integrated into the design of innovations. One form of constructivist integration is participatory research, in which farmers articulate research priorities and variables to be maximized or optimized. Variables that will often enter into research through participatory processes (and which would otherwise be absent) include those associated with risk; those exposing trade-offs related to the allocation of limited resources (land, labor, organic nutrient resources, capital) to different system components; and cultural variables such as those relating to local culinary practices and preferences (German, 2006). A second form of constructivist integration acknowledges the social trade-offs of current and alternative land-use scenarios by making explicit who gains and who loses from diverse types of innovations. By making social trade-offs explicit during the planning stage, alternative solutions or means of implementation can be considered that aim to optimize gains to diverse social actors while minimizing losses to any given one. By monitoring who wins and loses during an implementation process, creative strategies can be developed to ameliorate losses suffered by any given land user and to enable more equitable access to the benefits stream.

A third form of integration involves efforts to foster positive synergies among diverse types of innovations—for example, linking biophysical innovations to the social, policy, and institutional processes required to bring far-reaching change. This "sectoral integration" concept helps to frame scientific inquiry around the synergies among technological and other forms of innovation (social, organizational, policy, economic). The latter might include negotiation processes, participatory policy reforms and strategies to enhance market access so as to foster multiple goals simultaneously (i.e., income generation, equity, good governance, sustainable NRM).

Participation

"Participation" means different things to different people. All too often, however, it is taken to mean mere numbers of people present in community fora. AHI has been experimenting with ways to understand participation in more meaningful terms, by exploring the mechanisms and processes through which equitable development strategies and local empowerment may be achieved. Empowerment means enhancing people's ability—individually or collectively—to address their own concerns by leveraging existing resources and capacities and capturing opportunities. Equity, on the other hand, is about the fairness and social justice in the distribution of resources, opportunities, and benefits within a society. It is also about how approaches used by external facilitators or local change agents structure patterns of benefits capture.

Fostering these two goals requires experimenting with different approaches at different stages of farm and watershed-level natural resource management. It may involve attention to gendered participation and outcomes; strategies to mobilize communities around a common cause; mechanisms to ensure adequate representation of the many "voices" in large villages or watersheds through representational democracy (e.g., watershed structures and decision-making processes, socially targeted consultations); strategies to "level the playing field" between more and less powerful actors (e.g., stakeholder analysis, negotiation support); or instruments to hold people accountable to negotiated agreements (e.g., by-law reforms). Attention to participation is often concentrated at the planning stage of community interventions. Yet attention to equity and empowerment is needed at all stages—from problem diagnosis and prioritization to planning, implementation, and monitoring. At the planning stage, attention must be given to adequately capturing the diversity of "voices" in rural communities who may have different interests and goals. For farm-level innovations, it involves identification of variables of importance to male and female household members—not just to researchers or elite farmers. For watershed-level innovations, it involves instruments to explicitly capture a diversity of opinions when diagnosing problems, prioritizing, and planning. Similarly, during implementation and monitoring, it involves consulting diverse local interest groups (including participants and non-participants) on their views of how things are evolving to ensure diverse interests and concerns are considered when exploring how to improve upon ongoing change processes. Importantly, each of these phases of farm and watershed innovation requires attention to divergent opinions within communities, and means to reconcile these.

Collective action

Collective action may be defined as action taken by a group, either directly or on its behalf through an organization, in pursuit of members' perceived shared interests (Marshall, 1998). This pursuit of common goals may go well beyond formal social structures (farmers' groups) or direct activities carried out by such groups. In the context of AHI, we have experimented with a set of approaches to foster collective action in watershed management, leading us to identify a number of different forms of collective action (see German and Taye, 2008 for a related discussion). The first, and by far the most widely used, refers to direct actions carried out by groups of people working towards common goals (Lubell et al., 2002; Swallow et al., 2001; Tanner, 1995). This may range from two neighboring resource users managing a common boundary to the mobilization of large groups to work towards common interests. German et al. (2006) have called this the "social movement" dimension of collective action.

Another form of collective action involves collective regulation of individual actions (Meinzen-Dick et al., 2002; Pender and Scherr, 2002; Gebremedhin et al., 2002; Scott and Silva-Ochoa, 2001). In other words, rather than involving direct actions by groups of people pursuing common goals, this dimension of collective action refers to collectively agreed-upon rules to govern individual behavior—often proscribing what "not to do" or individual responsibilities towards the group. Such rules are generally formulated to minimize the negative impacts of one person's behavior on another person or on an environmental service of public concern, or to bolster individual commitments to group activities. Such rule-making is often one element of other forms of collective action which enable them to work owing to the prior agreement on "rules of the game."

Mechanisms for group representation in decision-making may also be considered another form of collective action. Given the sheer number of resource users in watersheds, equal levels of direct participation in decision-making on natural resource management or interaction with outside actors is seldom possible. Mechanisms for effective representation of all watershed users in decision-making and benefits sharing are therefore essential to minimize the tendency for elite capture of decision authority and benefits. This role of collective action has been included in collective action definitions of some authors (Meinzen-Dick et al., 2002), but features little in actual case studies.

A final form of collective action includes mechanisms for addressing power relations so as to achieve political equality. This dimension of collective action involves acknowledgment of diverse political interests around any given resource or management decision, and their effective integration into more equitable decision-making processes (Sultana et al., 2002). Issues of political equality among stakeholders have largely been addressed in the literature through case studies illustrating the "winners and losers" of development and conservation interventions owing to the frequent failure to establish mechanisms for equitable outcomes (Munk Ravnborg and Ashby, 1996; Rocheleau and Edmunds, 1997; Schroeder, 1993).

Watershed

The standard definition of watershed refers to a region of land drained by a watercourse and its tributaries to a common confluence point (outlet) (Pattanayak,

2004). However, given the AHI emphasis on *participatory* watershed management and an *integrated* approach to NRM, the spatial delineation of hydrological watershed boundaries was taken as only a tentative unit of analysis and engagement. These units were to be adjusted as the landscape-level natural resource management priorities of local users, and the spatial dimensions of these problems and related solutions, came to light. Following the participatory diagnosis of watershed problems, it was found that some "watershed problems" conformed to hydrological boundaries but many others did not. Problems related to the declining quality and quantity of water and the destruction of property from excess run-off, and the land-use practices contributing to this resource degradation, had clear hydrological boundaries. Yet many other landscape-level natural resource management problems did not conform to hydrological boundaries. These included damage caused by free grazing, incompatible trees on farm boundaries, conflict surrounding protected areas, and pests and diseases. Yet even when problems may be defined by hydrological boundaries, solutions may be more readily found through the use of administrative boundaries. For example, spring rehabilitation may require village-level organizing and the support of government institutions whose mandate covers larger administrative units (e.g., districts), in addition to collective action among land users within catchment areas. In these cases, a flexible approach to defining watershed processes and boundaries was used. Use of the term "watershed" in this book is often, therefore, used interchangeably with the word "landscape." Similarly, "watershed management" often encompasses problems and solutions whose dimensions extend beyond the biophysical realm altogether.

Institutional innovations

Addressing farm and landscape-level natural resource management problems—and capturing related opportunities—often requires innovation in the institutions that structure patterns of interaction among land users and other entities. Institutions may be defined as "rules of the game in society, … the humanly devised constraints that shape human interaction" (North, 1990) or "decision structures" (Ostrom, 1994). Institutional innovations may therefore be defined as changes in the standard set of rules governing social behavior and in the social structures and processes through which decisions are made. AHI has experimented with each of these forms of institutional innovation. Innovations in organizational structure have included the testing of diverse forms of farmer organization for farm-level technological innovation, demand-driven technology and information provision, and policy innovation. It has also included the testing of novel organizational structures (platforms) to foster district-level collaboration in natural resource management. AHI experimentation with organizational processes has been even more extensive, as exhibited throughout this book. It has included processes for planning, for negotiating rules to govern collective action processes or natural resource management, for monitoring, and for enforcing agreed-upon rules. It has also included organizational processes for improving the effectiveness of innovations at farm, landscape, district, and national level (e.g. within national agricultural research systems). Finally, AHI has experimented with rules for governing how external resources (technologies, training, credit) will be shared within communities; for governing collective action processes (contributions to be made, benefits accruing to different members, and sanctions to be applied when contributions are not made); and for governing individual behavior at farm and landscape level (for example, to curtail certain practices having negative effects on other resource users or to negotiate and incentivize actions that individuals must take in addressing a common problem).

The birth and evolution of the African Highlands Initiative

The African Highlands Initiative (AHI) is an eco-regional research program working to improve livelihoods and reduce natural resource degradation in the densely settled highlands of eastern Africa. To this end, AHI has been developing and pilot testing an integrated natural resource management approach in selected highland areas of Ethiopia, Kenya, Tanzania, and Uganda and institutionalizing its use in key partner organizations. AHI work targets the poor in degraded highland watersheds where environmental and related livelihood problems are widely visible on farms and landscapes and of concern to local residents due to their effects on livelihoods. It is this awareness and concern of natural resource management issues by local land users, rather than external conservation concerns, that has framed the scope of innovations tested by AHI.

The idea of a highland eco-regional program was tabled in 1992 at a regional meeting of the National Agricultural Research Institute (NARI) directors and International Agricultural Centers (IARCs) out of the concern that sustainable use of natural resources was given insufficient attention in agricultural research in the region. AHI was conceived as a NARI–IARC collaborative initiative aimed at improving farmer livelihoods while improving natural resource management so as to sustain rural livelihoods into the future. For most of its history, AHI has therefore operated as an eco-regional program of the Consultative Group for International Agricultural Research (CGIAR) and a regional network of the Association for Strengthening Agricultural Research in East and Central Africa (ASARECA), convened by the World Agroforestry Centre. While AHI's hosting arrangements have shifted over the years, and the focus of core innovations has evolved to build on lessons learned and address new challenges, its core vision of developing an integrated approach to improved livelihoods and better management of natural resources has remained unchanged.

The main impetus behind AHI's conception was a growing concern that the absence of a coordinated, inter-institutional effort had contributed to the limited adoption by farmers and communities of technologies and practices that improve and sustain natural resources. Although previous, independent research efforts had generated technologies to improve soil fertility and conserve water and other natural resources, they were not necessarily suited to the diverse socio-economic and biophysical circumstances of farmers living in the humid tropical highlands. Nor were farmer decision frames—the key considerations driving behavior, the "bottom lines" (e.g., food sufficiency and income generation in the near term) and the timeframe over which these are manifest—very often taken into consideration. Yet the ideas leading to the program's birth had less to do with the deficiency of research "inputs" to development and natural resource management (e.g., technologies, knowledge, decision tools) than with the limitations in the approaches through which the research–development interface was structured and these inputs brought to bear on the everyday challenges lived by rural households.

Implementing a regional research-for-development initiative involving multiple stakeholders at multiple levels, and accountable to different actors (farmers, national, regional, and international agricultural research institutes), is no simple task. It requires careful thought regarding institutional aims and design, key concepts that will help to anchor program evolution, and effective program governance (see Annex I). It also requires periodic evaluations to adjust program directions and governance as needed to effectively position the program to make unique contributions or to address challenges emerging through implementation. Adding to this complexity is the emphasis on the development and testing of new methodologies and approaches for integrated natural resource management (INRM) at different scales. This requires a strong methodological backbone to operationalize a social learning process at village, district, and higher levels and to link actors at different levels in a research and innovation process.

With these considerations as a background, the program was born with a mandate to do the following (Stroud, 2001):

- Develop a participatory approach to foster farmer innovation and adaptation.
- Employ an integrated systems approach rather than a commodity-based approach, so as to solve multiple and linked problems and make an impact on livelihoods and the environment.
- Develop a more integrated approach among research and development (R&D) actors in solving land degradation and related poverty issues.
- Give attention to social dimensions of natural resource management, such as local institutional arrangements for managing communal resources or issues of mutual concern.
- Consider how the short-term concerns of smallholders, which often override other considerations and lead to an inability or unwillingness to make investments with medium- to long-term returns, could be taken on board in efforts to support improved natural resource management.
- Explore mechanisms to identify and address external circumstances that act as disincentives to technology adoption—such as lack of market outlets, credit and input supplies.
- Interface with local and national policies that shape local natural resource management and the forms of institutional support to rural development.

Key phases in AHI's evolution

Since its inception in 1995, the AHI has been operationalized through discrete conceptual and funding phases of approximately three years each. This book reports on the first four phases of program evolution. In Phase I (1995–1997), a competitive grant scheme was employed to foster partnerships for multidisciplinary research in Ethiopia, Kenya, Madagascar, and Uganda. Yet achieving changes in mindset and practices among those accustomed to working in isolation proved challenging in practice, leading to a reconceptualization of modes of operation. In Phase II (1998–2000), the program shifted away from the competitive grant approach to the use of benchmark sites as a means of operationalizing multidisciplinary approaches and teamwork for farm-level innovations. Eight benchmark sites were established and the geographical coverage was expanded to include Tanzania.

Following a favorable review of Phase 2, it was suggested that the program continue with the benchmark site approach in Phase III (2002–2004). It was also suggested that the number of sites be reduced, resulting in a reduction from the original eight sites to six (Figure 1.1). The program was also encouraged to shift focus from the farm to the watershed level, so as to address NRM issues that cannot be effectively addressed at farm level. Watershed and landscape-level innovations continued in Phase IV (2005–2007), but greater emphasis was placed on institutional innovations to expand the reach of INRM in benchmark sites and to institutionalize lessons and approaches within the region. A more detailed description of each phase of work is presented in Annex II.

Operationalizing "approach development"

An important question underlying all of this work is the "how" of methodological innovation. How are new ways of managing natural resources identified? How are they tested in practice? And how are they evaluated for their effectiveness? There are two answers to these questions, one looking at the "big picture" of how collaboration is structured within benchmark sites and regionally to foster a culture of methodological innovation—and the other looking at the methodological framework through which innovation was fostered and lessons captured.

The regional "infrastructure" for methodological innovation

The organizational structure and functions of AHI are in many ways explicitly designed for enhancing methodological innovation. The most crucial ingredient has been the presence of functional research and development (R&D) teams in AHI benchmark sites, consisting of representatives of diverse disciplines and institutions with different mandates and organizational competencies (including, minimally, those working in the realms of "research" and "development") and a well-facilitated process for collective deliberation and experiential learning. This was often operationalized through smaller theme-based teams and periodic

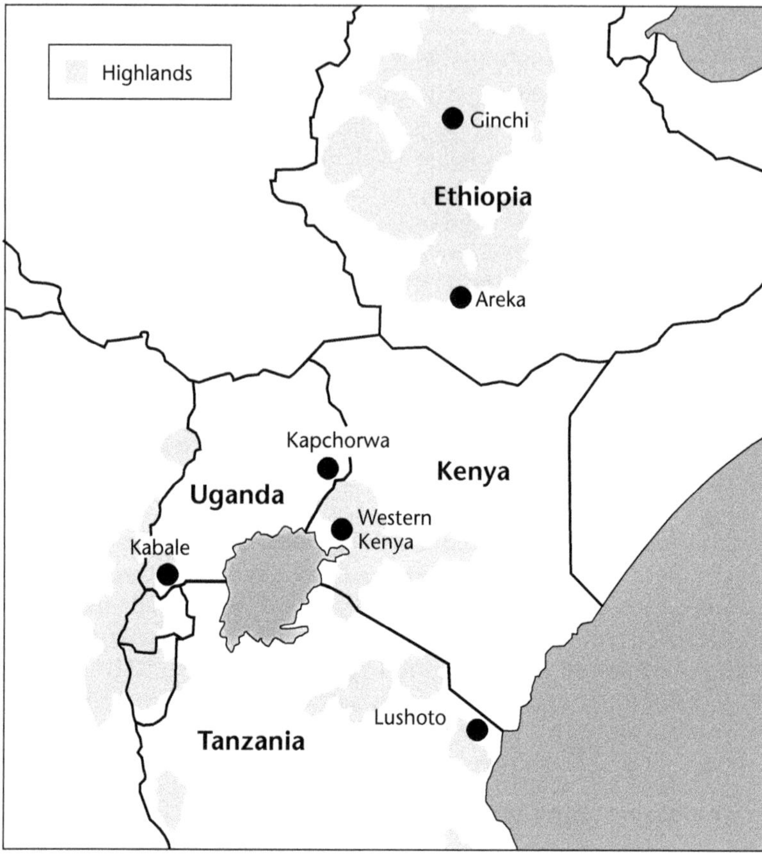

FIGURE 1.1 Map of eastern Africa showing AHI mandated areas and the benchmark sites

Note: Two Phase II benchmark sites in Madagascar are not shown here.

meetings for cross-team reflection and replanning. A diverse institutional and disciplinary composition has helped ensure that efforts to conceptualize the "system" and the approaches to be tested are holistic and integrative. It has also helped to instill a more critical perspective on approaches under development, for example to ensure that unfounded assumptions are questioned. For example, the common misperception of communities as homogeneous entities for which interactions with or benefits flowing to one or more members will automatically constitute communication with or benefits to all can be regularly questioned by bringing experienced development practitioners into planning. This diversity of voices has also helped to ensure that work being done on diverse sub-components (e.g., soil and water, animal and crop husbandry) or themes (e.g., technological innovation, watershed management) harmonize their engagements with communities and one another.

Regional research team members, hired to fill gaps in disciplines and perspectives represented in site teams, have also played a critical role in periodic reflection and re-planning sessions. Their engagement with multiple teams at a time has enabled them to bring in unique observations from other sites, which may be at different stages in the implementation cycle or experimenting with different approaches. It also provides a unique "birds-eye" perspective that enables patterns and lessons to be captured across sites, thereby grounding the development of regional synthesis products within different thematic areas. Assisting site teams to distil lessons learned into Methods Guides and other public goods has been a fundamental step in coming "full circle" in the innovation and learning process, and in clarifying the overall role and institutional niche of AHI in the region.

The thematic focus of learning and innovation has been structured through external phase reviews, where broad targets for the next phase are set, and regional phase planning meetings, where representatives of all site teams and research managers come together to agree how to operationalize new concepts and first steps of methodological innovation. Key regional themes are distilled and used to structure learning by site teams, as well as regional team members who specialize in one or more of the themes. By Phase III, the key themes structuring learning and innovation across the program were consolidated into the following:

- approach for integrated natural resource management for watersheds;
- local organizational capacity for collective action;
- innovation systems through partnerships and institutional arrangements (alternatively called, "district institutional and policy innovations");
- scaling up and institutionalization.

Different donors[4] have historically funded different pieces of the whole (thematic and geographical), depending on their thematic and country priorities. This has resulted in a complex matrix of sites, projects and to some extent thematic thrusts, from which the methodological innovations and lessons in this book are derived. Having a set of cross-cutting thematic priorities and coordination functions at regional, national, and site levels has therefore been instrumental in ensuring the coherence of the program as a whole.

AHI's benchmark sites have also played a fundamental role in building and consolidating expertise in interdisciplinary teamwork, methodological innovation, and demand-driven research and development over time, and in linking levels of innovation. By having multidisciplinary and multi-institutional teams in place and a specific location where new ideas could be tested, an opportunity was provided for the mindsets and practices of individual professionals to evolve over time as well as for lessons to be more systematically learned and accumulated. At the time of writing, AHI had five active benchmark sites (BMS): Areka, located near Soddo town in south-central Ethiopia; Ginchi, located in the

Galessa Highlands near Ginchi town, in central Ethiopia; Kabale, located in the Kigezi highlands of south-western Uganda; Kapchorwa, located on the foothills of Mount Elgon in eastern Uganda; and Lushoto, located in the West Usambara Mountains of Tanzania (Table 1.1). Each of these sites and the wider eco-regions in which they are embedded are characterized by high population density, natural resource degradation, and declining agricultural productivity—posing significant challenges to farmers to provide for a growing population while maintaining the productivity of basic resources (water, food, fuel, fodder). Benchmark sites are delineated by topographical boundaries (micro-watersheds), encompass from six to nine villages, and lie within larger administrative units (districts or woredas) where some of the activities take place. A more detailed description of these benchmark sites may be found in Annex III.

Methodological "nuts and bolts"

Once the institutional "infrastructure" for learning and innovation are in place, how are new methods actually conceived and tested in practice to derive broader lessons about the "approaches that matter"? Perhaps the most important component of this is that the innovations tested by different site teams are fully embedded in rural communities who have become equal partners in methodological innovation. While this element can be partially encompassed by the participatory research concept, in fact it goes much beyond a particular method of structuring farmer–researcher interaction. It may be best characterized by a broader philosophy of shared learning, inquiry and—perhaps most importantly—mutual respect and friendship. In sites where these interpersonal relations have been strongest and the mind-sets of R&D teams most flexible, methodological innovations have more quickly led to successful outcomes.

Another fundamental piece of the puzzle has been an emphasis on learning-by-doing rather than through pure data capture. Researchers were encouraged from early on in Phase II to "enter the system" rather than simply study it as outsiders. As this approach flies in the face of centuries of empiricism emphasizing the neutral observer, it was perhaps the most difficult challenge faced by AHI researchers. While lessons are therefore still being learned at a rapid pace, the approach was advanced considerably during Phases III and IV in efforts to operationalize the concept of "action research" in the context of INRM. This has led to the development not only of methodological innovations for INRM (action research "outputs"), but to innovations in the methods employed to structure learning itself (action research methods). The latter include tools for planning (e.g. action research protocols), tools for observing "process," monitoring systems (German et al., 2007; Opondo et al., 2005), and approaches for integrating empirical and action research approaches (German and Stroud, 2007).

Action research is exactly what it sounds like—action-oriented research. It focuses explicitly on process, in this case the processes of development and social

TABLE 1.1 Characteristics of African Highlands Initiative benchmark sites

Site attributes	*Benchmark site*				
	Areka, Ethiopia	*Ginchi, Ethiopia*	*Lushoto, Tanzania*	*Kabale, Uganda*	*Kapchorwa, Uganda*
Altitude (meters above sea level)	1800–2600	>2200	1100–1450	1500–2700	1000–2000
Population density (/km^2)	400–600	100–200	200–300	100–300	100–170
Enterprises	Enset, wheat, pea, maize, barley, sorghum, sweet potato, faba bean, horticulture, communal grazing	Barley, pulses, Irish potato, wheat, oilseeds, seasonal rotation from individual cropland to communal grazing	Maize, banana, tea, coffee, horticulture in valley bottoms, high-value trees, zero-grazed livestock	Sorghum, pulses, banana, zero grazed livestock, vegetables, potatoes	Maize, beans, banana, wheat, coffee, barley, Irish potatoes, some cassava
Irrigation	None	None	Seasonal	Seasonal	None
Livestock trends	Low numbers and decreasing; intensive management	High numbers yet decreasing; access to grazing land good	Low numbers and decreasing; zero grazing mostly	Low numbers and decreasing; mostly zero grazed	Low numbers and decreasing
Forest/woodlot access	Medium (tree planting common)	Limited (planting limited; remnant forest is distant)	Medium to high (mostly cultivated; natural forests are protected)	Few trees and decreasing	Limited forests; extensive woodlots
Market integration	Limited; some off-farm employment	Medium	Medium to good (tea, vegetables)	Limited	Medium

change. In the context of agricultural development and natural resource management, this might include testing different approaches to enhancing farmer innovation; mechanisms for linking farmers to markets; strategies for improving governance of landscape processes (such as the movement of water, soil and pests); and approaches to institutional change (for impact-oriented research). By superimposing research or systematic inquiry on development-oriented actions by R&D teams, farmers, and policy makers, new lessons can be learned that may otherwise be lost to observation. These lessons are gained by creating spaces to reflect on processes being implemented at diverse levels—including what was done, how it was done, the outcomes tied to particular approaches, and lessons derived from these experiences. Lessons learning is also strengthened by making observation more systematic, for example by clarifying the area of concern (improved livelihoods, equity and sustainability); the framework of ideas that structure research (for example, key challenges to development, sustainability or equity and related knowledge gaps) (Checkland and Holwell, 1998); the research questions (which often emphasize how to address these challenges); and the methodology (what will be observed and documented, and how). Each of these helps to sift out what is significant from the sum total of what is learned and observed—in other words, to determine which findings really count as knowledge (Checkland, 1991; Checkland and Holwell, 1998).

Action research is different from empirical research in both the questions asked and the methods used. Action research questions are the "how" questions seeking answers to the question, "what works, where and why?" They are questions about development and change. While action research is embedded in an action context (i.e. community-driven watershed management activities), the "research" component helps to promote systematic inquiry about the change process. This can serve two purposes. First, it can encourage systematic reflection at the level where change is taking place (e.g. community, district, institution) on how things are being done so that they can undergo continuous improvement and have a higher chance of success. The second purpose is unique to action research—namely, to derive general principles from the change process that can be of use to other actors (farmers, research and development institutions, policy makers) outside the immediate location. In the context of AHI, for example, we study change processes for the purpose of developing methods and approaches that work in meeting different livelihood or NRM challenges. Without such scrutiny of the method-in-practice, it would be impossible to make reliable claims about the method's usefulness in solving real problems on the ground. This requires both participatory assessments of the methodology and systematic scrutiny at the level of R&D teams.[5] Empirical research, on the other hand, is a more controlled form of research which helps to address the "what" questions. It requires more formal data collection protocols and analysis, but can be equally instrumental in informing decision-making at the local level or among policy makers. Applications of empirical research in watershed management within AHI are explored in subsequent chapters.

Action research starts with participatory action research (PAR). PAR is a process in which the immediate beneficiaries themselves, whether local communities, institutional representatives or policy makers, play the primary role in designing and testing innovations. The objective here is to enhance impact in the context under study—whether community-level change processes, institutional change or policy reforms. However, as applied in AHI, action research does not stop here. AHI has a mandate to generate international public goods in the form of "working methods and approaches," in this case for integrated natural resource management. Therefore, it was essential to move beyond solving site-specific problems to distil lessons of broader relevance for the international community. This requires an additional level of abstraction and analysis that may not be of interest to the immediate beneficiaries.[6] It also requires a particular set of skills to link site-specific circumstances to a broader global community (knowing what challenges and knowledge gaps exist elsewhere); to observe fine details of process (observing how people react to processes when facilitated in certain ways, reading body language, understanding how process relates to outcomes); and to understand how to link the particularities of local-level learning with generalization. While the protagonists (immediate beneficiaries) play a fundamental role in defining research, monitoring progress, adjusting the approach and evaluating impacts, it is generally researchers who play a primary role in managing research quality and bringing a wider body of theory to bear on local innovations. In short, action research encompasses, but is not limited to, participatory action learning for solving localized NRM problems. For a better understanding of how AHI has operationalized the difference between participatory action learning and action research, please see Table 1.2 and Figure 1.2.

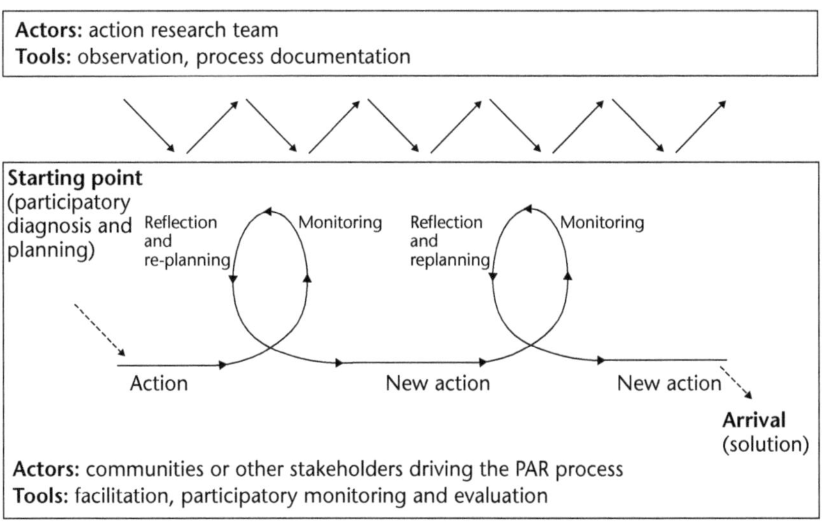

FIGURE 1.2 Illustration of the relationship between action research (upper box) and PAR (lower box) (German et al., 2011)

TABLE 1.2 Distinctions between participatory action research and action research as operationalized within AHI

Learning approach	*Aims and applications*	*Roles in defining the research and learning agenda*	*Characteristics of research design*	*Primary role in designing and managing research*	*Research outputs and applications*
Participatory action research	To guide a change process and strengthen chances of success through systematic reflection and self-learning	Immediate beneficiaries (who integrate lessons into the change process through periodic reflection and re-planning)	Informal; goals and pathways for achieving goals defined at the outset but not rigidly adhered to; "data" capture largely informal	Immediate beneficiaries (whether local communities, institutional representatives or policy makers)	Approaches that "work" relative to the end goals of a development or change process as defined by the immediate beneficiaries
Action research	To help guide the development or change process on/within which research is conducted, or to generate general principles of relevance to managers of change in other locations with similar conditions	End users (immediate beneficiaries or offsite users of results); researchers and facilitators (who may wish to generalize results)	Semi-formal; research questions defined at outset and fixed; methods of data capture may be relatively fixed or opportunistically defined to capture emergent realities	Researchers (specialized skills required to manage research for quality, and to generate lessons and principles relevant to a wider audience)	General principles about development and change processes, including the conditions under which diverse outcomes are reached

Source: German and Stroud (2007).

Before concluding, it is important to mention two questions of common concern by those new to action research. The first relates to validity, and the second to the role of empirical research. Many people are uncomfortable with the inability to structure controlled and replicable processes in action research, given how participation will inevitably lead to the divergence of processes in different sites—independently of how similarly structured the initial steps are. Some well-known action researchers are comfortable with prior clarification of an area of concern, framework of ideas and methodology as means to ensure validity in action research (Checkland and Holwell, 1998). Yet comparison can also play a fundamental role in lessons learning, to understand how diverse approaches and contexts structure outcomes. This comparison can be operationalized in both space and time. One way in which spatial comparison was employed in AHI to learn lessons was cross-site comparison around different AHI thematic thrusts. Learning across cases is also possible within individual benchmark sites, as in the case of negotiation support processes around different types of natural resource problems (excess run-off, spring degradation, free grazing). Yet comparison can also be employed within a single site and action research topic through a systematic temporal comparison of iterative approaches used and their outcomes (as measured by local and/or scientific indicators). While it may not enable broader generalizations to be made, such systematic learning within cases does yield a wealth of observations about processes that do and do not work in particular contexts.

Many researchers also wonder whether empirical research has a role to play in action research, and struggle with the relevance of their training to action research approaches. In watershed management processes, we have found four discrete roles for empirical research, which are highlighted in more detail in Chapter 3:

1. To characterize the social and biophysical dimensions of natural resource management problems, so that interventions may be effectively targeted.
2. To monitor intermediate outcomes of different approaches to solving any particular problem at different phases of an innovation process, particularly in cases where variables are difficult to observe or monitor by local residents but nevertheless can help determine whether the approach is helping to foster community or program level goals (e.g., equity, sustainability).
3. To clarify complex cause-and-effect relationships that are difficult for farmers to observe, such as the effect of different land uses and their spatial arrangement on system hydrology and sub-service water flows.
4. To assess the impacts once the problem is solved, so that something can be said with confidence about the effectiveness of the approach used.

This said, it is important to clearly identify the "critical uncertainties" or program requirements that actually require costly empirical research investments. The tendency is for researchers to want to expand the scope of empirical research within participatory processes as a means to justify their engagement, and generate publications that conform with conventional standards of academic rigor. Ideally, these investments should be chosen carefully, based on gaps in local knowledge and observation capacity (as in the case of sub-service hydrology) *and* need (where the knowledge generated is required to identify an effective solution to a problem). In some cases, empirical research will be needed to fill gaps in local motives or to achieve project objectives. For example, participating farmers will tend to focus solely on their own benefits rather than on how to ensure equitable benefits captured by a broader community. This may require empirical research in social science to identify how opinions on issues differ within any given community, or to monitor how any given approach is structuring the distribution of benefits to different sets of actors. The same may be said about sustainability, given the tendency for farmers to focus on the immediate benefits from any given innovation. Empirical research that exposes the deficiencies of an approach from the perspective of equity or sustainability (e.g., only certain groups are benefiting, the innovation is leading to the depletion of nutrients) can help to raise awareness among the protagonists and encourage new innovations in the approach to address these gaps.

Main achievements

AHI program achievements are of three primary types: methodological innovations for INRM, impacts resulting from the piloting of these innovations, and various knowledge products. Regarding methodological innovations, the program has developed a host of methodological innovations for operationalizing INRM and addressing key natural resource management challenges. A summary of these innovations, and the publications where additional information may be sourced, is provided in Table 1.3.

Regarding the impacts emanating from these methodological innovations, an External Review and Impact Assessment of the program carried out in late 2007 and early 2008 identified the following key outcomes and impacts (La Rovere et al., 2008; Mekuria et al. 2008):

- male and female farmers have increased knowledge of technologies, and greater ability to demand these technologies and seek support from service providers and to freely express themselves with research and extension;
- improvements in crop production and yield owing to improved germplasm, better agronomic practices, better pest and disease management, and increased adoption of conservation practices;
- increased technology adoption owing to efforts to "link" technologies (see Chapter 2 for details);
- benefits associated with collective efforts to address farm-level productivity constraints, including improved access to information, training and credit; improved financial management and business planning; and access to community banks (the last of these unique to Lushoto);

TABLE 1.3 Methodological innovations developed by AHI and selected reference materials

Theme and innovation	*Selected reference materials*
1. Farm level innovations	
Methods for systems intensification and diversification (Areka, W. Kenya)	Amede, T., A. Bekele and C. Opondo (2006) Creating niches for integration of green manures and risk management through growing maize cultivar mixtures in the southern Ethiopian highlands. *AHI Working Papers* No. 14. Amede, T. and R. Kirkby (2006) Guidelines for integration of legumes into the farming systems of the east African highlands. *AHI Working Papers* No. 7. Amede, T., A. Stroud and J. Aune (2006) Advancing human nutrition without degrading land resources through modeling cropping systems in the Ethiopian highlands. *AHI Working Papers* No. 8. Amede, T. and E. Taboge (2006) Optimizing soil fertility gradients in the enset (ensete ventricosum) systems of the Ethiopian highlands: Trade-offs and local innovations. *AHI Working Papers* No. 15.
System-integrated technologies with multiple benefits (Areka, Kapchorwa)	Amede, T. and R. Delve (2006) Improved decision-making for achieving triple benefits of food security, income and environmental services through modeling cropping systems in the Ethiopian highlands. *AHI Working Papers* No. 20.
Linked technologies (Lushoto)	Masuki, K.F.G., J.G. Mowo, T.E. Mbaga, J.K. Tanui, J.M. Wickama and C.J. Lyamchai (2010) Using strategic "entry points" and "linked technologies" for enhanced uptake of improved banana germplasm in the humid highlands of East Africa. *Acta Horticulturae* 879(2): 797–804. Stroud, A. (2003) Linked Technologies for Increasing Adoption and Impact. *AHI Brief* A3.
Use of entry points at farm level (all sites)	Amede, T. (2003) Differential entry points to address complex natural resource constraints in the highlands of eastern Africa. *AHI Brief* A2.
Farmer institutional development for demand-driven service provision, technology innovation and marketing (Kabale)	Stroud, A., E. Obin, R. Kandelwahl, F. Byekwaso, C. Opondo, L. German, J. Tanui, O. Kyampaire, B. Mbwesa, A. Ariho, Africare and Kabale District Farmers' Association (2006) Managing change: Institutional development under NAADS: A field study on farmer institutions working with NAADS. *AHI Working Papers* No. 22.

Continued

TABLE 1.3 *Continued*

Theme and innovation	*Selected reference materials*
Fostering local seed delivery systems	Taye, H., M. Diro and A. W/Yohannes (2006) The effectiveness of decentralized channels for wider dissemination of crop technologies: Lessons from the AHI Areka site, Ethiopia. In: T. Amede, L. German, S. Rao, C. Opondo and A. Stroud (eds.), *Integrated Natural Resource Management in Practice: Enabling Communities to Improve Mountain Livelihoods and Landscapes*, pp. 265–73. Kampala, Uganda: African Highlands Initiative. Wakjira, A., G. Keneni, G. Alemu and G. Woldegiorgis (2006) Supporting alternative seed delivery systems in the AHI–Galessa watershed site, Ethiopia. In: T. Amede, L. German, S. Rao, C. Opondo and A. Stroud (eds.), *Integrated Natural Resource Management in Practice: Enabling Communities to Improve Mountain Livelihoods and Landscapes*, pp. 240–8. Kampala, Uganda: African Highlands Initiative. Woldegiorgis, G., A. Solomon, B. Kassa and E. Gebre (2006) Participatory potato technology development and dissemination in the central highlands of Ethiopia. In: T. Amede, L. German, S. Rao, C. Opondo and A. Stroud (eds.), *Integrated Natural Resource Management in Practice: Enabling Communities to Improve Mountain Livelihoods and Landscapes*, pp. 124–31. Kampala, Uganda: African Highlands Initiative.
Negotiating equitable access to technologies (Areka, Ginchi)	Mazengia, W., A. Tenaye, L. Begashaw, L. German and Y. Rezene (2007) Enhancing equitable technology access for socially and economically constrained farmers: Experience from Gununo Watershed, Ethiopia. *AHI Brief* E4.
Methods for tracking farmer-to-farmer dissemination and impacts (regional/Lushoto)	German, L., J.G. Mowo and M. Kingamkono (2006) A methodology for tracking the "fate" of technological innovations in agriculture. *Agriculture and Human Values* 23: 353–69. German, L., J. Mowo, M. Kingamkono and J. Nuñez (2006) Technology spillover: A methodology for understanding patterns and limits to adoption of farm-level innovations. *AHI Methods Guide* A1.
Catalyzing collective learning and innovation (Kapchorwa, community-based facilitators in NAADS)	Mowo, J., B. Janssen, O. Oenema, L. German, P. Mrema and R. Shemdoe (2006) Soil fertility evaluation and management by smallholder farmer communities in northern Tanzania. *Agriculture, Ecosystems and Environment* 116(1/2): 47–59. Tanui, J. (2005) Revitalizing grassroots knowledge systems: Farmer learning cycles in AGILE. *AHI Brief* D4.

Theme and innovation	*Selected reference materials*
Methodology for designing integrated research	German, L. (2006) Moving beyond component research in mountain regions: Operationalizing systems integration at farm and landscape scale. *Journal of Mountain Science* 3(4): 287–304 and *AHI Working Papers* No. 21. German, L., B. Kidane and K. Mekonnen (2005) Watershed management to counter farming systems decline: Toward a demand-driven, systems-oriented research agenda. *AgREN Network Paper* 45. German, L., A. Stroud, G. Alemu, Y. Gojjam, B. Kidane, B. Bekele, D. Bekele, G. Woldegiorgis, T. Tolera and M. Haile (2006) Creating an integrated research agenda from prioritized watershed issues. *AHI Methods Guide* B4.
2. Watershed management and participatory landscape governance	
Sequenced methods for participatory integrated watershed management: • Participatory watershed diagnosis and characterization • The creation of functional clusters to structure innovations • Participatory watershed planning • Selection of entry points • Participatory monitoring and evaluation	Adimassu, Z., K. Mekonnen and Y. Gojjam (eds.) (2008) *Working with Rural Communities on Integrated Natural Resources Management (INRM)*. Ethiopian Institute of Agricultural Research (EIAR), Addis Ababa. German, L., H. Mansoor, G. Alemu, W. Mazengia, T. Amede and A. Stroud (2006) Participatory integrated watershed management: Evolution of concepts and methods. *AHI Working Papers* No. 11. German, L., H. Mansoor, G. Alemu, W. Mazengia, T. Amede PhD and A. Stroud (2007) Participatory integrated watershed management: Evolution of concepts and methods in an eco-regional program of the eastern African highlands. *Agricultural Systems* 94(2): 189–204. German, L., K. Masuki, Y. Gojjam, J. Odenya and E. Geta (2006) Beyond the farm: A new look at livelihood constraints in the eastern African highlands. *AHI Working Papers* No. 12. German, L. and K. Mekonnen (2006) A socially-optimal approach to participatory watershed diagnosis. *AHI Methods Guide* B2. German, L., A. Stroud, G. Alemu, Y. Gojjam, B. Kidane, B. Bekele, D. Bekele, G. Woldegiorgis, T. Tolera and M. Haile (2006) Creating an integrated research agenda from prioritized watershed issues. *AHI Methods Guide* B4.

Continued

TABLE 1.3 *Continued*

Theme and innovation	*Selected reference materials*
Mobilizing collective action for INRM	Ayele, S., A. Ghizaw, Z. Adimassu, M. Tsegaye, G. Alemu, T. Tolera and L. German (2007) Enhancing collective action in spring "development" and management through negotiation support and by-law reforms. *AHI Brief* E5. Begashaw, L., W. Mazengia and L. German (2007) Mobilizing collective action for vertebrate pest control: The case of porcupine in Areka. *AHI Brief* E3. German, L., W. Mazengia, W. Tirwomwe, S. Ayele, J. Tanui, S. Nyangas, L. Begashaw, H. Taye, Z. Adimassu, M. Tsegaye, S. Charamila, F. Alinyo, A. Mekonnen, K. Aberra, A. Chemangeni, W. Cheptegei, T. Tolera, Z. Jotte and K. Bedane (2011) Enabling equitable collective action and policy change for poverty reduction and improved natural resource management in the eastern African highlands. In: E. Mwangi, H. Markelova and R. Meinzen-Dick (eds.), *Collective Action and Property Rights for Poverty Reduction*. Johns Hopkins and IFPRI, Baltimore and Washington, D.C. German, L., H. Taye, S. Charamila, T. Tolera and J. Tanui (2006) The many meanings of collective action: Lessons on enhancing gender inclusion and equity in watershed management. *AHI Working Papers* No. 17; *CAPRi Working Paper* 52. Tanui, J. (2006) Incorporating a landcare approach into community land management efforts in Africa: A case study of the Mount Kenya region. *AHI Working Papers* No. 19.

Theme and innovation	*Selected reference materials*
Participatory governance of landscape processes: • Negotiation support • Participatory by-law reforms • Solutions to address specific landscape level challenges (spring rehabilitation, controlling free grazing, managing excess run-off, niche-compatible agroforestry, vertebrate pest control, co-management of protected areas)	Adimassu, Z., S. Ayele, A. Ghizaw, M. Tsegaye and L. German (2007) Soil and water conservation through attitude change and negotiation. *AHI Brief* A6. German, L., S. Charamila and T. Tolera (2006) Managing trade-offs in agroforestry: From conflict to collaboration in natural resource management. *AHI Working Papers* No. 10. German, L., S. Charamila and T. Tolera (2005) Negotiation support in watershed management: A case for decision-making beyond the farm level. *AHI Brief* E2. German, L., W. Tirwomwe, J. Tanui, S. Nyangas and A. Chemangei (2007) Searching for solutions: Technology-policy synergies in participatory NRM. *AHI Brief* B6. Mazengia, W., A. Tenaye, L. Begashaw, L. German and Y. Rezene (2007) Enhancing equitable technology access for socially and economically constrained farmers. *AHI Brief* E4. Sanginga P., R. Kamugisha, and A. M. Martin. (2010). Strengthening social capital for adaptive governance of natural resources: A participatory action research for by-law reforms in Uganda. *Society and Natural Resources* 23: 695–710. Sanginga P., A. Abenakyo, R. Kamugisha, A. Martin and R. Muzira 2010. Tracking outcomes of social capital and institutional innovations in natural resources management: Methodological issues and empirical evidence from participatory by-law reform in Uganda. *Society and Natural Resources* 23: 711–25. Sanginga, P.C., R.N. Kamugisha and A.M. Martin (2007) The dynamics of social capital and conflict management in multiple resource regimes: A case of the south-western highlands of Uganda. *Ecology and Society* 12(1): 6. Online at: www.ecologyandsociety.org/vol12/iss1/art6/ Sanginga, P.C., R. Kamugisha, A. Martin, A. Kakuru and A. Stroud (2004) Facilitating participatory processes for policy change in natural resource management: Lessons from the highlands of southwestern Uganda. *Uganda Journal of Agricultural Sciences* 9: 958–70. National Agricultural Research Organization, Kampala. Tanui, J., S. Nyangas, A. Chemangei, F. Alinyo and L. German (2007) Co-management of protected areas is about cultivating relationships. *AHI Brief* B7.
3. District institutional and policy innovations	

Continued

TABLE 1.3 *Continued*

Theme and innovation	*Selected reference materials*
Multi-institutional processes for INRM (district level)	German, L., A. Stroud and E. Obin (2003) A coalition for enabling demand-driven development in Kabale District, Uganda. *AHI Brief* B1. Tanui, J., A. Chemengei, S. Nyangas and W. Cheptegei (2007) Rural development and conservation: The future lies with multi-stakeholder collective action. *AHI Brief* B8.
System for demand-driven technology and information provision	Masuki, K.F.G., J.G. Mowo, R. Sheila, R. Kamugisha, C. Opondo and J. Tanui (2011) Improving smallholder farmers' access to information for enhanced decision making in natural resource management: Experiences from South Western Uganda. In Bationo, A., Waswa, B.S., Okeyo, J. and Maina, F. (eds) *Innovations as Key to the Green Revolution in Africa: Exploring the Scientific Facts* (2): 1145–1160. Opondo, C., L. German, A. Stroud and E. Obin (2006) Lessons from using participatory action research to enhance farmer-led research and extension in southwestern Uganda. *AHI Working Papers* No. 3.
4. Scaling up and institutionalization	
Self-led institutional change	Mowo, J.G., L.N. Nabahungu and L. Dusengemungu (2007) The integrated watershed management approach for livelihoods and natural resource management in Rwanda: Moving beyond AHI pilot sites. *AHI Brief* D5. Opondo, C., P. Sanginga and A. Stroud (2006) Monitoring the outcomes of participatory research in natural resources management: Experiences of the African Highlands Initiative. *AHI Working Papers* No. 2. Opondo, C., A. Stroud, L. German and J. Hagmann (2003) Institutionalizing participation in East African research institutes, Ch. 11, *PLA Notes* 48. London: IIED. Stroud, A. (2003) Self-management of institutional change for improving approaches to integrated NRM. *AHI Brief* B2. Stroud, A. (2006) Transforming institutions to achieve innovation in research and development. *AHI Working Papers* No. 4.
Methods for linking farmers to policy makers	German, L., A. Stroud, C. Opondo and B. Mbwesa (2004) Linking farmers to policy-makers: Experiences from Kabale District, Uganda. *UPWARD Participatory R&D Sourcebook*. Manila: CIP.

Theme and innovation	*Selected reference materials*
5. Improving research–development linkages	
Action research	German, L., W. Mazengia, S. Charamila, H. Taye, S. Nyangas, J. Tanui, S. Ayele and A. Stroud (2007) Action research: An approach for generating methodological innovations for improved impact from agricultural development and natural resource management. *AHI Methods Guide* E1. Opondo, C., L. German, S. Charamila, A. Stroud and R. K. Khandelwal (2005) Process monitoring and documentation for R&D team learning: Concepts and approaches. *AHI Brief* B5.
Use of scientific and local knowledge to ground decision-making	German, L., B. Kidane, R. Shemdoe and M. Sellungato (2005) A methodology for understanding niche incompatibilities in agroforestry. *AHI Brief* C2. German, L., B. Kidane and R. Shemdoe (2006) Social and environmental trade-offs in tree species selection: A Methodology for identifying niche incompatibilities in agroforestry. *Environment, Development and Sustainability* 8: 535–52; *AHI Working Paper* 9. Wickama, J. and J.G. Mowo (2001) Indigenous nutrient resources in Tanzania. *Managing African Soils* 21, IIED.
Planning for integrated research and development interactions	German, L. and A. Stroud (2004) Integrating learning approaches for agricultural R&D. *AHI Brief* B4. German, L. and A. Stroud (2007) A framework for the integration of diverse learning approaches: Operationalizing agricultural research and development (R&D) linkages in eastern Africa. *World Development* 35(5): 792–814 and *AHI Working Papers* No. 23.

- improved livelihoods resulting from significant increases in agricultural income, and related improvements in housing, nutrition, and ability to pay school fees;
- improvements in water quality and quantity owing to the by-laws protecting water sources, restricted cropping near springs, and the cultivation of water-conserving vegetation;
- improved livelihoods owing to water conservation, resulting from reduced burdens on women, reduction in conflicts over water, the increased availability of irrigation water, and reduction in waterborne diseases and related medical expenditures;
- significant increase in the prevalence of collective action to solve NRM issues and cooperate on matters of common concern, and improved negotiation of resource conflicts;
- greater harmony at community level when dealing with the management of water sources/springs, boundary trees, and soil conservation issues;
- increased tendency to participate collectively in addressing NRM issues and comply with by-laws;
- increased confidence among farmers in their ability to solve NRM problems;
- significant improvements in access to information (i.e., on input and output prices, technology, financial services);
- more positive attitudes among farmers toward research;
- increased awareness and appreciation of watershed management in particular, and INRM in general, among many high level officials, leaders of institutions, and policy makers.

Impacts associated with wider dissemination of lessons learned and methodologies are impossible to assess, but the report notes that "the process of disseminating AHI outputs, successes, and methods is fairly effective at the international level"—in large part owing to the publication series, website, and regional trainings carried out for ASARECA member countries to disseminate select methods developed by the program. The assessment team also notes the program's role as a "think tank" for developing tools and methods, and for institutionalizing INRM at the regional level. The study concludes that INRM works so effectively owing to the interaction between AHI's bio-physical and socio-economic components, and to AHI's community-driven approach. "The capacity to put INRM to work is a rare achievement within CGIAR centers" (Mekuria et al. 2008: 17).

Regarding AHI knowledge products, the program launched a series of AHI Briefs in 2003 and followed this up with a set of Working Papers, Methods Guides and Proceedings in 2006. These may be found at: http://worldagroforestry.org/projects/african-highlands/archives.html. In addition to contributing to this series, site and regional team members have collectively contributed to working papers of other organizations such as ODI's AgREN, CAPRi Working Papers, Managing Africa's Soils and IIED's PLA Notes. They have also published papers in academic

TABLE 1.4 Number of contributions made to different types of publications by different contributing partners

Publication type	*Number of contributions from site team members*	*Number of contributions from regional team members*	*Number of contributions from other partners*
Peer reviewed journal articles	54	31	31
Books and book chapters	97	20	4
Working papers	75	51	25
Methods guides	25	8	5
Briefs	31	39	2
Conference proceedings	134	37	32
Program reports	19	15	0
Totals	424	200	99

journals such as Acta Horticulturae; Agriculture, Ecosystems and Environment; Agriculture and Human Values; Agricultural Systems; Development in Practice; Environment, Development and Sustainability; Human Ecology; Journal of Mountain Science; Society & Natural Resources; Uganda Journal of Agricultural Sciences; and World Development, among others. They have collectively produced 195 publications, including 26 peer reviewed journal articles, two books, 23 book chapters, 39 working papers, seven methods guides, 33 briefs, 49 papers in workshop proceedings, and 16 program reports[7] (Annex IV). The project also produced a number of knowledge products oriented toward farmers. As may be seen by the authorship, the vast majority of these publications were developed collaboratively by site team members, regional research team members, and other partners. The numbers of contributions made by different contributing partners are summarized in Table 1.4.

Conclusions

This chapter provides a brief overview of the integrated natural resource management concept, as defined in both the literature and in the research from which this volume emanates. A few key concepts are presented and defined to clarify the conceptual foundations of INRM and the chapters that follow. The chapter also provided an introduction to the African Highlands Initiative, the eco-regional program operating in the eastern African highlands under which the methodological innovations presented in this volume were developed and piloted. Following an introduction to the program's mandate and evolution, key strategies for putting INRM into practice and deriving lessons from experience were presented—along with a summary of key program achievements to date.

When reflecting back on the investments in time, energy, and funding to AHI over the years, a key question that emerges is "What is the value added of such a program, relative to other research programs in the region and beyond?" AHI is unique in the region in its efforts to identify gaps in professional practice that undermine the effectiveness of agricultural research and development, to identify innovative approaches—through action research—that can help to fill these gaps, and to document and share lessons learned for the global community. It therefore sits squarely on the interface of research and development, bringing systematic learning to bear on contemporary development challenges. Another important aspect of added value lies in navigating complexity. On the one hand, this refers to the complexity of the farming systems and mountain landscapes in which AHI works, for which multiple enterprises and aims are the norm and where tightly coupled interactions among system components and adjacent land users means that changes in any one component or farm may induce any number of (often unanticipated) spin-off effects. On the other hand, AHI has had to grapple with the complexity of approaches for contributing to rural livelihood improvements while also fostering more sustainable and equitable natural resource management. By drawing on the skills of biophysical scientists from diverse disciplines and social scientists, and on the vast array of skills that development practitioners bring to the table when working with rural communities, AHI has worked to integrate and strategically sequence technological, social, governance, and market innovations for "win–win" outcomes. The need to navigate complexity also encompasses the methodological challenges associated with facilitating innovations within benchmark sites while attempting to distil lessons and "international public goods" for a wider audience.

As "AHI's comparative advantage lies in its implementation of 'INRM that works'," a program review proposes that

> evolution of the program is towards a center of excellence on INRM, that … international organizations, NGOs and national institutes can refer to in order to initiate and implement successful INRM approaches to that work. Failures or partial successes of others to implement INRM effectively offer elements for reflection on AHI and open for AHI a niche for which there is increasing demand.
>
> *Mekuria et al. 2008: 23*

This volume seeks to support the knowledge management component of this vision by documenting methods, experiences, and lessons learned to date in efforts to implement INRM in practice. While the book emphasizes AHI successes, it is important to recognize that many of these came on the back of early failures to operationalize integrated and demand-driven research, to move beyond extractive research and "enter the system" in an action research and social learning mode, and to move beyond the status quo in research and development practice. And as highlighted in the "Missing links" section at the end of each chapter, many challenges remain—some of these despite concerted efforts to bring change. It is hoped that these will form the basis of productive efforts by AHI and other NRM researchers and practitioners to learn new lessons on "what works, where, and why" and to fill these knowledge gaps.

The chapters that follow summarize the methodological innovations carried out in efforts to operationalize INRM at diverse levels. Chapter 2 summarizes experiences operationalizing INRM and systems intensification at farm level. Chapters 3 and 4 provide a detailed description of AHI efforts to foster participatory landscape-level innovation. The first of these chapters takes an in-depth look into methods for operationalizing participatory integrated watershed management, while Chapter 4 focuses in on efforts to address landscape level NRM problems through a participatory governance lens. Chapter 5 then explores district institutional and policy innovations in support of local level NRM efforts. The book concludes with a chapter summarizing the program's early experiences with scaling out pilot innovations and institutionalizing participatory, integrated approaches within national research and extension organizations.

Notes

1 These include community-based natural resource management (Fabricius and Koch, 2004), participatory watershed management (Rhoades, 2000; Shah, 1998), adaptive collaborative management (Colfer, 2005), and integrated natural resource management (Sayer and Campbell, 2003a).
2 See also: www.icarda.cgiar.org/INRMsite/index.htm
3 Social-constructivist approaches foster collective understandings through deliberative process (open discussion and debate among actors), peer learning, and other means of engaging diverse sets of interests and knowledge (Fosnot, 1996).
4 With the notable exception of the Swiss Development Cooperation and the *Ministerie van Buitenlandse Zaken*, who have historically funded the AHI program as a whole.
5 The latter has been done through process documentation, a research tool that formalizes data collection on any facilitated change process and ensures changes are made to adapt the tools to the challenges faced during implementation. PM&E and process documentation are defined, with examples provided, in Chapter 3.
6 In AHI, we have found rather that the liaison function of drawing explicit linkages between site-level experiences and the interests and concerns of a broader global community tends to empower local actors (farmers, development partners) to care for what they do and to want to share their experiences with others.
7 This list is far from complete, with a number of annual and phase reports and this volume not captured in these statistics. We also experienced difficulty tracking down early AHI team members and publications derived from their experiences with AHI.

References

Amede, T., T. Belachew, and E. Geta (2001) Reversing the degradation of arable land in the Ethiopian highlands. *Managing Africa's Soils* 23. London: IIED.

Anderson, D. (1984) Depression, dust bowl, demography, and drought: The colonial state and soil conservation in East Africa during the 1930s. *African Affairs* 83(332): 321–343.

Bekele, M. (2003) Forest property rights, the role of the state and institutional exigency: The Ethiopian experience. Ph.D. Thesis, Swedish University of Agricultural Sciences, Uppsala. 220 pp.

Blomley, T. and H. Ramadhani (2006) Going to scale with Participatory Forest Management: Early Lessons from Tanzania. *International Forestry Review* 8(1): 93–100

Blomley, T., H. Ramadhani, Y. Mkwizu, and A. Böhringer (2010) Hidden harvest: Unlocking the economic potential of community-based forest management in Tanzania. In: L. German, A. Karsenty and A.-M. Tiani (eds.), *Governing Africa's Forests in a Globalized World*. London: Earthscan.

Borrini-Feyerabend, G., T. Banuri, T. Farvar, K. Miller, and A. Phillips (2002) Indigenous and local communities and protected areas: Rethinking the relationship. *Parks* 12: 5–15.

Brown, K. (2003) Integrating conservation and development: A case of institutional misfit. *Front Ecol Environ* 1(9): 479–487.

Burgess, N.D., J. Fjeldså, and R. Botterweg (1998) Faunal importance of the Eastern Arc Mountains of Kenya and Tanzania. *Journal of East African Natural History* 87(1): 37–58.

Byerlee, D. and R. Tripp (1988) Strengthening linkages in agricultural research through a framing systems perspective: The role of social scientists. *Experimental Agriculture* 24:137–151.

Campbell, B. (2001) What is INRM? In: Adamo and Hagmann (eds.), *Integrated Management for Sustainable Agriculture, Forestry and Fisheries. Proceedings of a Workshop of the INRM Task Force, Cali, Colombia, August 28–31, 2001.*

Chambers D.E., Wedel K.R. and Rodwell M.K. (1992) Qualitative research methods and the constructivist paradigm in the evaluation of social programs and policies. In: Chambers D.E. (ed.) *Evaluating Social Programs.* Boston: Allyn & Bacon, 288–332.

Checkland P. (1991) From Framework through Experience to Learning: the essential nature of Action Research. In: H.-E. Nissen (ed.), *Information Systems research: Contemporary Approaches and Emergent Traditions.* Amsterdam: Elsevier.

Checkland, P. and Holwell, S. (1998) Action Research: Its Nature and Validity. *Systemic Practice and Action Research*, Volume 11, Issue 1, February : 9–21.

Colchester, M. (2004) Conservation policy and indigenous peoples. *Environmental Science & Policy* 7(3): 145–153.

Colfer, C.J.P. (2005) *The Complex Forest: Communities, Uncertainty, and Adaptive Collaborative Management.* Washington, D.C. and Bogor, Indonesia: Resources for the Future and Center for International Forestry Research.

Consultative Group on International Agricultural Research (CGIAR) (2001) Integrated Management for Sustainable Agriculture, Forestry and Fisheries. *Proceedings of a Workshop of the INRM Task Force, Cali, Colombia, August 28–31, 2001.*

Consultative Group on International Agricultural Research (CGIAR) (2002) Putting INRM into Action. *Proceedings of a workshop of the INRM Task Force, Aleppo, Syria, September 16–19, 2002.*

Fabricius, C. and E. Koch (eds.) (2004) *Rights, Resources and Rural Development: Community-Based Natural Resource Management in Southern Africa.* London: Earthscan.

Farrington, J. and N. Martin (1987) Farmer Participatory Research: A review of concepts and practices. *Agricultural Administration Network Paper* 19. London: ODI.

Fosnot, C.T. (1996) *Constructivism: Theory, Perspectives and Practice.* New York: Teachers College Press.

Gebremedhin, B., Pender, J. and Tesfay, G. (2002) Collective action for grazing land management in mixed crop-livestock systems in the highlands of northern Ethiopia.

Socio-Economics and Policy Research Working Paper No. 42. Nairobi, Kenya: International Livestock Research Institute.

German, L. (2006) Moving beyond component research in mountain regions: operationalizing systems integration at farm and landscape scale. *Journal of Mountain Science* 3(4): 287–304.

German, L., W. Mazengia, H. Taye, M. Tsegaye, S. Charamila, and J. Wickama (2009) Minimizing the livelihood trade-offs of natural resource management in the eastern African highlands: Policy implications of a project in 'Creative Governance'. *Human Ecology* 38(1): 31–47.

German, L., J. Ramisch, and R. Verma (eds.) (2010) *Beyond the Biophysical: Knowledge, Culture and Politics in Agriculture and Natural Resource Management.* Dordrecht: Springer.

German, L.A., A.M. Tiani, T. Mutimukuru Maravanyika, A. Daoudi, E. Chuma, N. Beaulieu, H. Lo, C. Jum, N. Nemarundwe, E. Ontita, G. Yitamben and V. Orindi (2011) *The Application of Participatory Action Research to Climate Change Adaptation: A Reference Guide.* Ottawa: IDRC.

German, L., W. Mazengia, S. Charamila, H. Taye, S. Nyangas, J. Tanui, S. Ayele, and A. Stroud (2007) Action Research: An Approach for Generating Methodological Innovations for Improved Impact from Agricultural Development and Natural Resource Management. *AHI Methods Guide* E1.

German, L. and A. Stroud (2007) A framework for the integration of diverse learning approaches: Operationalizing agricultural research and development (R&D) linkages in eastern Africa. *World Development* 35(5):792-814.

German, L. and H. Taye (2008) A framework for evaluating effectiveness and inclusiveness of collective action in watershed management. *Journal of International Development* 20: 99–116.

German, L., H. Taye, S. Charamila, T. Tolera, and J. Tanui (2006) The many meanings of collective action: Lessons on enhancing gender inclusion and equity in watershed management. *AHI Working Papers* No. 17.

Hagmann, J. (1999) *Learning Together for Change: Facilitating Innovation in Natural Resource Management through Learning Process Approaches in Rural Livelihoods in Zimbabwe.* Weikersheim: Margraf Verlag.

Hagmann, J., E. Chuma, K. Murwira, M. Connolly, and P.P. Ficarelli (2003) Success factors in integrated natural resource management R&D: Lessons from practice. In: B.M. Campbell and J.A. Sayer (eds.), *Integrated Natural Resource Management: Linking Productivity, the Environment and Development.* Wallingford, UK: CABI Publishing, pp. 37–64.

Haverkort, B., J. van der Kamp, and A. Waters-Bayer (1991) *Joining Farmers' Experiments: Experiences in Participatory Technology Development.* London: IT Publications.

Hobley, M. (1996) *Participatory Forestry: The Process of Change in India and Nepal. Rural Development Forestry Study Guide.* London: Overseas Development Institute.

Hughes, R. and F. Flintan (2001) Integrating conservation and development experience: A review and bibliography of ICDP literature. *Biodiversity and Livelihoods Issues* 3. London: International Institute for Environment and Development.

La Rovere, R., M. Mekuria, and J. Szonyi (2008) *External Review and Impact Assessment of the African Highlands Initiative (AHI): Program Evaluation Report.* Mexico City: CIMMYT.

Lubell, M., M. Schneider, J.T. Scholz, and M. Mete (2002) Watershed partnerships and the emergence of collective action institutions. *American Journal of Political Science* 46(1): 148–163.

Marshall, G. (1998) *A Dictionary of Sociology*. New York: Oxford University Press.

Meinzen-Dick, R., A. Knox, B. Swallow and F. Place (2002) Introduction. In: R. Meinzen-Dick, A. Knox, F. Place, and B. Swallow (eds.), *Innovation in Natural Resource*

Management: The Role of Property Rights and Collective Action in Developing Countries. Baltimore, MD: Johns Hopkins University Press, pp. 1–11.

Mekuria, M., R. La Rovere, J. and Szonyi (2008) *External Review and Impact Assessment of the African Highlands Initiative (AHI).* Program Evaluation report. February 2008 Mexico City: CIMMYT.

Munk Ravnborg, H. and J.A. Ashby (1996) Organising for local-level watershed management: Lessons from Rio Cabuyal watershed, Colombia. *AgREN Network Paper* 65: 1–14.

Nsibambi, A. (ed.) (1998) *Decentralisation and Civil Society in Uganda: The Quest for Good Governance.* Kampala: Fountain Publishers.

North, D.C. (1990) *Institutions, Institutional change and economic performance.* Cambridge: Cambridge University Press.

Oates, J.F. (1999) *Myth and Reality in the Rain Forest.* Berkeley: University of California Press.

Omiti, J.M., K.A. Parton, J.A. Sinden, and S. K. Ehui (1999) Monitoring changes in land-use practices following agrarian de-collectivisation in Ethiopia. *Agriculture, Ecosystems and Environment* 72: 111–118.

Opondo, C., L. German, S. Charamila, A. Stroud, and R. K. Khandelwal (2005) Process monitoring and documentation for R&D team learning: Concepts and approaches. *AHI Brief* B5.

Ostrom, V. (1994) *The Meaning of American Federalism: Constituting a Self Governing Society.* San Francisco, CA: Institute for Contemporary Studies.

Oyono, P.R., J.C. Ribot and A.M. Larson (2006) Green and Black Gold in Rural Cameroon: Natural Resources for Local Governance, Justice and Sustainability. Environmental Governance in Africa Working Papers No. 22. Washington, D.C. and Bogor: WRI and CIFOR.

Pattanayak, S.K. (2004) Valuing watershed services: Concepts and empirics from Southeast Asia. *Agriculture, Ecosystems & Environment* 104(1): 171–184.

Pender, J. and Scherr, S.J. (2002). Organizational development and natural resource management: Evidence from central Honduras. In: *Innovation in natural resource management: The role of property rights and collective action in developing countries,* (ed.) Meinzen-Dick, R., A. Knox, F. Place, and B. Swallow. Baltimore, MD: Johns Hopkins University Press.

Rhoades, R. (2000) The participatory multipurpose watershed project: Nature's salvation or Schumacher's nightmare? In: R. Lal (ed.), *Integrated Watershed Management in the Global Ecosystem.* London: CRC Press, pp. 327–343.

Rocheleau, D. and D. Edmunds (1997) Women, men and trees: Gender, power and property in forest and agrarian landscapes. *World Development* 25(8): 1351–1371.

Sanginga P, R. Kamugisha, and A. M. Martin. (2010) Strengthening social capital for adaptive governance of natural resources: A participatory action research for by-law reforms in Uganda. *Society and Natural Resources* 23: 695–710.

Sanginga, P.C., R.N. Kamugisha, and A.M. Martin (2007) The dynamics of social capital and conflict management in multiple resource regimes: A case of the south-western highlands of Uganda. *Ecology and Society* 12(1): 6. Available online at: www.ecologyandsociety.org/vol12/iss1/art6

Sayer, J.A. (2001) Integrated Natural Resource Management: A progress report. In: Adamo and Hagmann (eds.), *Integrated Management for Sustainable Agriculture, Forestry and Fisheries. Proceedings of a Workshop of the INRM Task Force, Cali, Colombia, August 28–31, 2001.*

Sayer, J.A. and B.M. Campbell (2003a) *Integrated Natural Resource Management: Linking Productivity, the Environment and Development.* Wallingford, UK: CABI.

Sayer, J.A. and B.M. Campbell (2003b) Research to integrate productivity enhancement, environmental protection, and human development. In: J.A. Sayer and B.M. Campbell (eds.), *Integrated Natural Resource Management: Linking Productivity, the Environment and Development.* Wallingford, UK: CABI.

Schroeder, R.A. (1993) Shady practice: Gender and the political ecology of resource stabilization in Gambian garden/orchards. *Economic Geography* 69(4): 349–65.

Scott, C.A. and Silva-Ochoa, P. (2001) Collective action for water harvesting irrigation in the Lerma-Chapala Basin, Mexico. *Water Policy* 3: 555–72.

Shah, A. (1998) Watershed development programmes in India: Emerging issues for environment-development perspectives. *Economic and Political Weekly* (June, 1998): A66–A79.

Soil Conservation Research Programme (SCRP) (1996) Data base report (1982–1993), Series II: Gununo Research Unit. Berne: University of Berne.

Stroud, A. (2001) Proposal—2001: One-year extension of Phase 2. A proposal to SDC. Kampala, Uganda: AHI.

Stroud, A. (2002) Broadening horizons: Institutional, policy and technical innovations for improving NRM and agricultural productivity in the east and Central African highlands. Phase 3 Report. Kampala, Uganda: AHI.

Sultana, P., P.M. Thompson, and M. Ahmed (2002) Methods of consensus building for collective action: Community based aquatic habitat and floodplain fisheries management in Bangladesh and the Mekong delta. Paper presented at the workshop Methods for Studying Collective Action, Nyeri, Kenya, February 25–March 1, 2002.

Swallow, B. M., D. P. Garrity, and M. van Noordwijk (2001) The effects of scales, flows and filters on property rights and collective action in watershed management. *CAPRi Working Paper* No. 16. Washington DC: IFPRI.

Tanner, C.L. (1995) Class, caste and gender in collective action: Agricultural labour unions in two Indian villages. *The Journal of Peasant Studies* 22(4): 672–698.

Terborgh, J. (1999) *Requiem for Nature.* Washington, D.C.: Island Press and Shearwater Books.

Throup, D. (1988) *Economic and Social Origins of Mau Mau, 1945–53.* London: J. Currey.

Voss, J. (2001) Integrated management for sustainable agriculture, forestry and fisheries. In: Adamo and Hagmann (eds.), *Integrated Management for Sustainable Agriculture, Forestry and Fisheries. Proceedings of a Workshop of the INRM Task Force, Cali, Colombia, August 28–31, 2001.*

Walters-Bayer, A. (1989) Participatory technology development in ecologically-oriented agriculture: Some approaches and tools. *Agricultural Administration Network Paper* 7. London: ODI.

Wells, M. (1992) Biodiversity, affluence and poverty: Mismatched costs and benefits and efforts to remedy them. *Ambio* 21: 237–243.

Western, D. (1994) Ecosystem conservation and rural development: The case of Amboseli. In: D. Western, R.M. Wright, and S.C. Strum (eds.), *Natural Connections: Perspectives in Communitybased Conservation.* Washington, D.C.: Island Press, pp. 15–52.

Zhou, G., N. Minakawa, A.K. Githeko, and G. Yan (2004) Association between climate variability and malaria epidemics in the East African highlands. *PNAS* 101(8): 2375–2380.

2

PARTICIPATORY FARM-LEVEL INNOVATION

Tilahun Amede, Charles Lyamchai, Girma Hailu, Bekele Kassa, Leulseged Begashaw, Juma Wickama, Adugna Wakjira, and Gebremedhin Woldegiorgis

Context and rationale

Improved natural resource management in the densely settled highlands of eastern Africa must begin at the farm level, as this is the basic decision unit and the locus of key inputs (land, labor, capital) which have a defining role in agricultural production systems. It is also here where concrete livelihood improvements can be made most directly, thus enhancing the likelihood that farmers will invest in natural resource management (at farm and landscape scale) with slower returns, or wider landscape governance initiatives (covered in Chapter 4) with more diffuse or uncertain returns.

Throughout the centuries, farmers have on their own initiative devised, developed, adopted, and adapted ingenious strategies and technologies for ensuring food security and economic welfare for their households (O'Neil, 1995). Farmer innovations such as the plough and the domestication of plants and animals that revolutionized agriculture date back more than 10,000 years. Farmer innovation is a process through which individuals or groups discover or develop new and better ways of managing available resources to suit their particular conditions. The resulting innovations or outcomes of the innovation process may be technical or socio-institutional (Waters-Bayer and Bayer, 2009). Innovation also occurs within a wider socio-economic and institutional context which conditions the extent and directionality of innovation processes, including the interaction among individuals or groups, policies and norms, and institutional and societal cultures. Innovative farmers are those who have tried or are trying out new, often value-adding agricultural or NRM practices, using their own knowledge and wisdom while also appropriating outsiders' knowledge (Assefa and Fenta, 2006). For instance, improving crop varieties through careful selection of seed, harvesting rainwater from roads, and implementation soil and water conservation measures, among others, are often carried out in the absence of any outside facilitation or support. Traditional irrigation systems (e.g., of the Chagga and Sonjo in Tanzania and Qantas in Iran), local knowledge on weather forecasting, biological control technologies, production of new pesticide concoctions, use of different plants and roots for soil fertility improvement, and local cures for different animal and human ailments are some of the well-documented farmer innovations (Lyamchai et al., 2006). These innovations have clearly played a significant role in the improvement of rural livelihoods and will continue to do so in the future.

Yet while local knowledge is essential to household food security, income generation and risk management vis-à-vis market fluctuations and an uncertain climate, formal research can play an instrumental role in supporting farmer innovation. This is either because certain techniques are beyond farmers' reach, as in the case of complex crop and livestock breeding technologies, or because researchers have access to a wider range of information and technologies that can assist farmers in capturing new opportunities or coping with a fast pace of change. As stated by O'Neil (1995: 1):

> Farmers are the ultimate integrators of the information they receive to increase production, stabilize yields, use pesticides, etc. It is the farmer that "lives the problem," gains the benefits and suffers the consequences. Therefore, a combination of farmers' and scientific knowledge will increase the rate of success and identify new areas of effort that neither group alone would have discovered.

Characteristics of crop–livestock systems in the eastern African highlands

The highlands of eastern Africa, with an average altitude of 1,500 metres above sea level, occupy 23 percent of the total land area but support 80 percent of the population (Alumira and Owiti, 2000). Mixed crop–livestock systems predominate, where 70 percent of the total human population and approximately 80 percent of the cattle and small ruminants can be found (Thornton et al., 2002). These systems are predominantly small-scale and highly diversified, combining annual and perennial crops with livestock. The area receives relatively high annual rainfall (>1,000 mm), the soils are generally more productive than the adjacent lowlands and in some countries irrigation water is also prevalent. Given this endowment, the highlands have been a major locus of human settlement historically, as well as a source of food and nutritional security within the wider subregion. These areas also produce important export crops such as tea, coffee, khat, and other horticultural crops that contribute to hard currency earnings. Livestock is a major component of these systems and makes significant contributions to food production, income, and social security. The importance of livestock is most apparent in countries such as Ethiopia and

Sudan, where crop production depends heavily on animal traction and nutrient recycling between crop and livestock components and livestock plays a critical role in expanding the area under cultivation and enhancing labor productivity. It also improves environmental processes such as the turnover of nutrients and soil carbon (van Keulen and Schiere, 2004).

In these crop–livestock systems, livestock and crops are produced within the same farm unit, whereby the by-products of one enterprise serve as a valuable resource for the other (Tarawali et al., 2004). Accordingly, the crop and the livestock components are strongly integrated through livestock feed, nutrient cycling (via manure), draught power, and input–output markets. The crop–livestock systems in the region are not homogeneous but vary from place to place depending on rainfall, soil fertility, population density, socio-economic characteristics, and access to capital and markets, among others. They are also at different levels of intensification, integration, and productivity. The production systems vary from the perennial banana–coffee gardens in the mid highlands of Uganda to the maize–beans systems of western Kenya and the more temperate barley-based systems of Ethiopia. These mixed farming systems have also evolved over time in response to changes in relative access to land, labor, and capital (van Keulen and Schiere, 2004).

These low-input systems are heavily reliant on the recycling of internal and organic nutrient resources and rainfall. Annual food production and availability in the region varies widely according to the seasonal climate, with the number of food-insecure people increasing significantly in seasons characterized by an uneven distribution and/or shortage of rainfall. Meanwhile, in good years not only food production but also national economies recover rapidly.

Yet while these highland areas are relatively rich in natural resources, livelihoods in the region are negatively affected by the following system constraints, which together with climate change limit people's coping and adaptive capacity:

- High and growing population density, leading to small landholdings, high degrees of fragmentation of landholdings and land tenancy regimes limiting systems intensification.
- Declining crop and livestock yields, owing in large part to declines in soil fertility, limited access to improved seeds and breeds, and increasing incidence of disease.
- Limited access to reliable markets, thus discouraging farmers from intensifying their systems and investing in their land.
- Limited capacity to develop water resources for multiple uses, including irrigation.
- The erosion of local genetic diversity and management systems, leading to declining ability of crop and livestock enterprises to resist climatic and disease-related shocks.
- Widespread poverty, with limited investment in yield- and value-enhancing technologies and practices.

- Limited and/or top-down extension services that do not reflect socio-economic realities, and limited institutional and policy support for enhancing access to inputs, credit and markets.

The high human and livestock population densities, land shortage, and steep slopes jointly contribute to resource degradation through overgrazing, nutrient mining, soil erosion, and water depletion. The outcome of this is that in some countries such as Ethiopia, a quarter of the crop–livestock systems are seriously eroded—of which approximately 15 percent are so seriously affected that it will be difficult to make them economically productive in the near future (Amede, 2003). Also in Ethiopia, three out of the five principal farm-level problems listed by farmers to be critically affecting farm productivity were the loss of seed and fertilizer from excess run-off, soil erosion, and increasing cost of fertilizers owing to soil fertility decline (Amede et al., 2006). Above all, nutrient depletion is a much more serious concern to food security in sub-Saharan Africa than in any other part of the world (Smaling, 1993).

Degradation of natural resources in the region is therefore partly a consequence of rural poverty, resulting from the interplay between rising population density, shrinking landholdings and livestock numbers, unreliable markets, and weak institutional and policy support in responding to emerging challenges. Livelihood strategies and assets management at farm level are also changing or need to change as families repeatedly face food deficits, livestock deaths, and degradation of the resource base.

Intensification of crop–livestock systems

Intensification is one option for fulfilling the growing demand for food, feed, and energy in the region. Intensification is a process to increase production levels, both in terms of amount, quality and fulfillment of local priorities and preferences (see also Morrison, 1994). In simple terms, it has been defined as an increase in average inputs of labor or capital on smallholdings, either on cultivated or uncultivated land, for the purpose of increasing the value of output per land area (Tiffen, 2003). Reardon et al. (1999) distinguish between capital-led and labor-led intensification. While labor-led intensification involves excessive dependence on labor as a key input to production, capital-led intensification refers to intensification based on substantial use of non-labor inputs such as chemical fertilizers and herbicides that enhance the productivity of land resources. Intensification has also been facilitated or induced by exogenous factors including population pressure, land shortage, and increasing labor availability. It can also be induced by external factors including increasing demand for livestock products and improved access to markets. In some cases, farmers are forced to shift from extensive cereal-based to intensive crop–livestock systems. The recent shift in policy towards market-led agriculture in Ethiopia, for instance, has influenced the way farmers are managing their resources. Introduction of small-scale irrigation into the cereal-based systems has induced farmers to diversify their cropping systems by growing high-value vegetables, fruits, coffee, and other crops. It has also led to the intensification of production systems through increased investment in high-yielding seeds and livestock breeds, chemical fertilizers, pesticides, and soil conservation practices. In these cases, both diversification and intensification are outcomes of farmers' responses to market opportunities while simultaneously striving to satisfy household food demands. Similarly, in places where dairy enterprises are becoming a major source of household income, such as in central Kenya, farmers have integrated fast-growing forage grasses (e.g., Napier grass) and fodder shrubs (e.g., Calliandra) in an effort to intensify the dairy component of their maize–beans–dairy systems.

As the intensification of agricultural production commonly requires external inputs, small-scale farmers in Africa tend to rely on government support in the form of loans and subsidies to finance external inputs.[1] Access to credit is one of the three pre-conditions identified by Reardon et al. (1999) for sustainable agricultural intensification in Africa, along with the availability of productive labor and high returns to investment resulting from accessible input and output markets. For the majority of households intensifying their systems based on scarce endogenous resources, the process is much slower. Kelly et al. (1996) suggest that in order to obtain sustainable intensification of agriculture in Africa, agricultural policies must also consider:

- improved access to quantity and quality seed/breeds;
- improved strategies in restoring soil fertility;
- functional land tenure policy; and
- increasing rural cash income and investments to improve food security and input access.

From the social welfare-based perspective of AHI, agricultural intensification should be a means to improved livelihoods and household income of rural communities without degrading the natural resource base (water, nutrients, vegetation), irrespective of its manifestations.

This chapter summarizes AHI experiences in supporting farm-level innovation for improved livelihoods and more sustainable management of natural resources underpinning agricultural production. While its focus is on the farm level and individual households as decision units, it also describes group initiatives that supported, in one way or another, farm-level innovations. The chapter presents the methods used and the main approaches developed to support farmer-led system intensification and diversification as pathways to achieve farmers' own livelihood and resource management goals. It displays strategies used to minimize resource degradation, and highlights the impact they have had on people's livelihoods and production systems. By providing detailed steps in each thematic section, it aims to facilitate the dissemination and adoption of the various approaches used by AHI and its partners within the wider eco-region.

Understanding systems and clients

The crop–livestock systems within this region differ in their agro-ecological and socio-economic features and in the wider policy, economic, and institutional contexts in which they are embedded. Efforts to understand the farming systems and farmers' needs and priorities are therefore required as a first step to any innovation process.

Approach development

Approach 1—Understanding systems and clients through participatory approaches

The first approach, used by all site teams, was to conduct a participatory appraisal of system problems and possible solutions as a means of setting the participatory research process in motion. By attempting to understand a system from the perspectives of farmers themselves, the diagnostic process in effect integrates the deep knowledge farmers bring with them on the constraints they face and the pathways most likely to unlock the potential for change.

Following a preliminary reconnaissance survey for the identification of research sites, a general meeting was called by local government officials, involving farmers from the area and researchers. This meeting consisted of the following steps:

1. An introduction to the purpose of the meeting and a clarification of expectations.
2. The division of farmers into subgroups based on different system components (crops, livestock, soil, and socio-economics). Using participatory rural appraisal (PRA) techniques, farmers identified constraints they faced—often identifying more than 20 per group, both researchable and non-researchable.
3. Return to plenary, where farmers ranked and prioritized the identified constraints.
4. Key informants were then asked to group farmers into different wealth and gender groups. The important constraints identified by the general community, along with potential solutions proposed, were presented to these subgroups to solicit their reactions. Practical solutions were discussed with these groups and final decisions were made on what potential solutions they would like to pursue.
5. Proposals were then developed collaboratively by groups of researchers and farmers, who were organized into small working groups according

to sub-themes derived from the priorities set by farmers. These farmer groups critically discussed the various proposals put forward by farmers and researchers, and farmers were consulted on methodological issues and possible treatments to be employed in the design of on-farm trials. During proposal formulation, explicit attention was given to wealth and gender considerations.

The major system constraints identified in step 4 included soil erosion, soil fertility maintenance, livestock feed, lack of credit, and limited access to improved varieties of various crops (Amede et al., 2006). The aforementioned steps show how farmers were involved in problem identification, priority setting, the identification of solutions, and technology evaluation. Subsequent steps of engagement consisted of hands-on training, implementation of on-farm trials, evaluation, and replanning—with regular contact with technical assistants (hired or seconded by the project, and present on a daily basis) and researchers to discuss on-farm challenges and to evaluate performance. These steps will be discussed in greater detail in the sections that follow.

Approach 2—Understanding systems through optimization models

Models can also play a useful role in understanding systems and enhancing the capacity of farm managers and development actors to make strategic decisions. In AHI, models were employed to predict the short-term risks and long-term impacts of agricultural interventions. They also provided a mechanism for evaluating the vulnerability of livelihoods and assets to external factors, targeting potential technologies to clients, and extrapolating and synthesizing knowledge for wider use.

One modeling approach was designed to enhance household food security in the context of limited resources such as land, water, nutrients, and labor (Amede et al., 2004). The aim of the modeling process was to select the best crop combinations to produce the required amount and quality of food for rural households through the reallocation of cropland to expand the area under crops with high content of nutrients in deficit, considering resource availability and local preferences. Stated in another way, the model helps communities and households to maximize the returns from investments of fertilizer, labor, land, water, manure, and other resources by helping them to optimize investments to enhance system productivity and household nutrition.

Use of a participatory, multiple-criteria decision model was considered to be particularly suited to assessing appropriate resource allocation strategies, considering local resources, socio-economic preferences, and market options.[2] Participatory steps in the modeling process helped to define key production objectives of farmers and to define potential pathways to achieving these—in recognition that there are many alternative ways to maximize farm productivity. By considering the production objectives and socio-economic preferences of farmers together with the available resources, the modeling process aids in exploring what is possible and—with the help of farmers—feasible for achieving established objectives.

When employing the model to optimize land resources for food and nutritional security, the modeling process consisted of the following steps (see also Box 2.1):

1. Identify representative households to participate in modeling, based on locally-defined social categories. A community meeting was organized to identify households keen to gain a deeper understanding of their resources and design cropping strategies for improved income, food security, and environmental protection. Based on these objectives, community members were asked to classify themselves into different social groups based on locally defined criteria. These could relate to farming system characteristics, the location and features of landholdings or resource endowments. The established groupings were then facilitated to identify their major production constraints, to discuss the causes of recurrent food insecurity and to suggest potential and practical solutions based on local resource endowments.
2. Quantify household and farm resources. Household resource inventories were carried out during the growing season. Researchers recorded farm size, distribution of land for crop production and grazing, type of crops grown, amount of land allocated for each crop, frequency of cropping, grain yield, and crop residues for each participating household. Household data were also obtained from women related to household demographics (including household members, resident guests), amount and type of food consumed per day and other relevant data. Community leaders played an important role in cross-checking household information.
3. Quantify the amount and distribution of resources. Once yields of the various crops and livestock were established, and annual household production levels estimated, additional household resources such as labor, nutrients, cash, and other assets were assessed.
4. Compare household resource endowments with established or calculated norms to identify levels of vulnerability and nutrients in deficit or excess. In this case, community averages, resource holdings of a representative household or internationally established norms (e.g., for household nutrition, the recommended daily dietary nutritional allowance of the World Health Organization (WHO)) were employed to establish optimal levels of resources (e.g., amount of nutrition required per person per day). This was used to identify households vulnerable to malnutrition, famine, or drought; to identify factors involved (e.g., labor shortage); and to identify nutrients that are in excess or deficit in the system.
5. Use optimization models to suggest an improved resource allocation strategy. Optimization and trade-off models were then used to optimize scarce resources, with provisions for placing constraints on certain parameters

based on household objectives or constraints. It was used, for example, to discuss the cropping and resource (e.g., labor, nutrient) allocation implications for maximizing different household production objectives such as human nutrition or cash income.

6. Validate model outputs with communities. Commonly, model outcomes represent optimal solutions, but do not represent realistic solutions unless they are validated by end-users and modified to match socio-economic realities. This process was used to update farmer preferences, such as a desire to maintain some enterprises and get rid of others—which were then newly incorporated as model parameters. In addition to discussions with individual households on the implications of different land and resource allocation decisions, this step should include an intensive and iterative process of community visioning. This is useful for deepening researchers' understanding of farmers' decision frames; for raising awareness among farmers on mechanisms to improved income and food security, reduce resource degradation, and reduce vulnerability to famine; and for articulating the agronomic management implications of desired future conditions.
7. Establish potential trade-offs of different farming system innovations. Trade-off analysis could be done either as an integral component of optimization models or as *ex ante* analysis once optimum choices for the intended objective function are made. In the mixed crop–livestock systems of the Ethiopian highlands, farmers considered both the crop and livestock sub-systems when deciding to integrate new interventions such as crop varieties into their farms. Hence, there is generally a need to establish how changes in one or more components affect other system components (Box 2.1). In other words, there is a need to quantify how possible changes in one enterprise affect the performance of other enterprises, with all other factors remaining constant.

This modeling process was found to be a powerful tool not only for helping farmers better understand and manage their farming systems, but also for enhancing collaboration between farming households and researchers by deepening mutual understanding.

BOX 2.1 OPTIMIZATION OF ENSET-BASED SYSTEMS FOR ENHANCED FOOD SECURITY IN ETHIOPIA

The site: The Areka site, in southern Ethiopia, is characterized by a multiple cropping system, with heavy reliance on perennials such as enset and coffee but a high level of diversification also achieved with sweet potato, taro, maize, wheat, and many other crops. The population pressure is high (>400 people/km^2), with average land holdings of less than 0.5 ha (about 816.8 m^2/person). The most apparent problems include small landholdings, limited livelihood options beyond farming, limited flow of information and investments, and very low income from farm operations.

The system: Currently, more than 50 percent of the land is allocated to root/corn crops, in particular sweet potato, Irish potato, enset, and taro, with land allocations in the order of 25.8, 16.2, 10.1, and 2.75 percent of household landholdings, respectively. Most of these crops are grown in the homestead or in fields just outside homestead areas (where nutrients are concentrated). Another 45 percent of the land area, on average, is allocated to cereals—predominantly maize. The total land allocated to legumes and vegetable crops is less than 5 percent. The current production system was found to be deficient in its ability to satisfy human nutritional needs for almost all nutrients. The daily energy supply of resource-poor households was only 75 percent of that recommended by the WHO. Extremely high deficits were found for vitamin A, vitamin C, zinc, and calcium, at 1.78, 12, 26.5, and 34 percent of the required levels, respectively. The trend was similar even for relatively resource-rich farmers, for whom all nutritional indices other than energy were deficient.

Results: To enhance household nutrition, the model recommended a significant shift from cereals and root crops to an enset–bean dominant system. The shift was significantly high, from about 10 to 36 percent and from 0.1 to 40 percent for enset and the common bean, respectively. However, during a feedback meeting, farmers revealed that there is a need for modification on the outcomes of the model for them to adopt the recommendations as a risk minimization measure. Their main concern was about retaining high proportions of sweet potato, considered by them to be a crop essential to household food security. This is because sweet potato can be planted throughout the year and is available when other crops are not yet ready for harvest or have failed (e.g., due to rainfall shortage). Following further iterations in which farmers' objectives were considered alongside nutritional ones, an agreed goal was reached to reduce the targeted cropland allocation for enset from 40 to 18 percent while increasing the target for sweet potato to 20 percent.

Lessons learned

The following lessons were learned from AHI efforts to understand farming systems and clients in Phase II:

- Efforts to understand farmers' socio-economic realities from the start of any innovation process strengthen researcher–farmer communication and, ultimately, increase the chance of technology adoption and impact.
- Feedback mechanisms between researchers and farmers and timely responses from research are critical for identifying solutions likely to be effective in addressing the challenges faced by agricultural systems and clients.

Building rapport and farmer confidence to innovate

One of the first steps in any farmer innovation process supported by outside actors and institutions is to build rapport between farmers and service providers (research, extension, NGOs) and to boost farmers' confidence in their ability to solve their own problems. Although research in natural resource management needs to take a holistic view as well as acknowledge the complexities and diversity of farming systems, research with farmers should also address critical problems that they have identified and prioritized (Amede et al., 2001). While a range of options were available to address the wide array of constraints identified by farmers in the diagnostic phase, it was important that those first tested would advance these aims, so as to encourage farmer engagement in more complex endeavors later on. This section therefore focuses on the use of "entry points"—defined here as an initial action that is strategically applied to enhance the likelihood of success of early innovations, and thus to build rapport between actors jointly engaged in an innovation process.

The choice of entry points has been proven to have a significant effect on whether farmers will be keen to invest in a partnership with researchers and extension agents for the purpose of experimentation, and whether farmers will continue to innovate in solving their problems without the support of external actors (Amede et al., 2001; Amede et al., 2006). Entry points can be an intervention in the form of an attractive technology or incentive. Entry points are essential to build trust between the community and outside actors, arouse their interest and keep their spirits high as the innovation process evolves—despite ever-complex challenges that may be tackled.

Approach development

Entry points utilized by AHI were commonly crop varieties, which could be identified on the basis of key constraints identified by farmers (e.g. market requirements), tested and disseminated rapidly. In some sites, varieties of high-value vegetable crops were used as entry points while in others fast growing forage grasses, such as napier, were used—based on farmer preferences. Researchers involved in AHI used "entry points" as a strategy to quickly get engaged with the farmers by providing some "best bet" technical solutions to priority problems (Wickama and Mowo, 2001).

Key steps in the process included the following:

1. Identification of the constraints faced by different wealth and gender groups, along with potential solutions, as discussed above.
2. Identification of criteria for selecting entry points. These often included:
    a) of high priority, addressing felt needs of intended beneficiaries;
    b) capable of bringing quick benefits (often economic in nature); and

c) low risk (e.g., involving limited cost or having been tested and validated in similar agro-ecological zones previously).

3. Generate a basket of options for farmers by matching specific technologies (sourced from research stations, extension agencies, sites with similar agro-ecologies, innovative farmers within the benchmark site) to farmers' felt needs and goals.
4. Facilitate the formation of "interest groups" for technology testing. Researchers facilitated the formation of thematically-based 'interest groups' based on technology preferences, to work as a group in carrying out a number of experiments on a particular theme.
5. Participatory testing of a wide range of technologies by farmer interest groups. Based on group plans, group members implement experiments assigned to them. Group members periodically visit each others' experiments, monitor performance, share information (e.g., on yield, observed characteristics or performance of technological options) and disseminate popular technologies to other group members.
6. Facilitate cross-group sharing of popular technologies likely to be of interest to the wider community.

In some cases, as a result of proper selection and implementation of entry points, AHI and its partners managed to reach more than 75 percent of farmers with income-enhancing technologies (Box 2.2). Entry points actually

BOX 2.2 EXAMPLES OF ENTRY POINTS USED IN ADDRESSING MORE COMPLEX SYSTEM CONSTRAINTS IN SOUTHERN ETHIOPIA (AMEDE ET AL., 2006)

Case 1—Sweet potato, a major staple crop planted year-round as a monocrop or intercropped with maize, is frequently damaged by the sweet potato butterfly. Controlling the pest is one strategy for increasing household food security. By planting sticky vines of desmodium around sweet potato fields, farmers reduced the incidence of the pest. They have also used desmodium as a protein source for dairy cows (together with carbohydrate-rich elephant grass). This technology became popular among farmers.

Case 2—Tephrosia and Canavalia are effective legume cover crops (LCCs) to restore soil fertility. Farmers started to integrate these LCCs as short-term fallows. Tephrosia was adopted in part because of farmer interest in its reputation for controlling mole rats, a pest affecting many crops. Farmers in Areka used to invest at least four hours to dig out and kill just one or two mole rats. Thus, it was an effective entry point by addressing an issue of high concern with short response time, while also contributing to the high-priority but medium-term aim of restoring soil fertility.

adopted by farmers were also found to vary according to their social status and agro-ecologies, evidencing the need for effective targeting and participatory selection of entry points to be tested by different households. A detailed analysis of selection preferences in the Areka benchmark site, for example, showed that resource-rich farmers with fertile plots and many livestock (and ample manure) preferred high yielding crop varieties, while resource-poor farmers with degraded land and limited access to manure preferred interventions contributing to soil fertility improvement as entry points (Amede et al., 2006). Examples of successful entry points with win–win effects from AHI sites are presented in Box 2.3.

BOX 2.3 TOMATO VARIETIES MEETING MARKET REQUIREMENTS—A SUCCESSFUL ENTRY POINT IN LUSHOTO BMS

During the PRA with farmers in Kwalei village, Lushoto in 1998, low crop productivity was reported to be the major problem in the village. On the other hand, vegetable crops were identified as the best options for generating much-needed cash income throughout the year because they can be produced three times a year. As a result, the majority of farmers, especially the youth, prioritized tomato and cabbage as top priorities for production and marketing innovations. Small-headed cabbage and firm tomato varieties that can withstand transportation and with a long shelf life were said to fetch a better price in the market than those cultivated locally. Tengeru Horticultural Research Institute supplied the required varieties of tomato (Tengeru 97 and Tanya) and cabbage (Glory F1) for testing. Farmers were then taught improved agronomic practices, from nursery management to transplanting, spacing, integrated soil fertility management, weeding, disease and pest control, harvesting, packaging, and marketing. After several seasons of bumper harvests and good marketing, more than 50 percent of farmers in Kwalei were found to be eagerly producing the introduced varieties—which had spread through family members and friends to distant areas of the district (German et al., 2006b). When consulted about the benefits, different households claimed to have used the income to pay school fees for their children, improve their houses, purchase more land, save up for marriage, and/or adopt improved land management practices. Farmers are now responding to market demands through grading, improvements in the quantity and quality of produce, and timely delivery. They are also in contact with traders in Dar es Salaam and Arusha via telephone, to keep an eye on current market prices. In this way they are making more informed marketing decisions (see Plate 1).

Lessons learned

The following lessons were learned from AHI's experiments with entry points:

- Interventions that bring immediate and visible benefits to farmers and their families are essential within any INRM initiative, as they build farmers' confidence in their ability to solve their own problems and help to build trust and rapport between farmers and support services. This is particularly true when the entry point addresses multiple concerns simultaneously (at least one of which brings quick returns).
- Characteristics of good entry points include their high priority for farmers, their ability to address concrete problems of local concern, their ability to generate quick benefits—particularly income, their simplicity for and accessibility to a wide range of households (so that unequal benefits do not compromise the enthusiasm of large portions of the community), their high chance of success (as early successes go a long way to enhance enthusiasm and trust) and the ease with which they can be managed and multiplied.
- To maximize impact, entry points need to be matched to household preferences and constraints, as well as to local agro-ecological and marketing conditions. For example for teff, the staple crop in Ethiopia, women's major selection criterion was color (white grain fetches more money than red, and is preferred for cooking the local bread *enjera*), while men considered yield and resistance to lodging as the most important criteria. Meanwhile, in Tanzania farmers preferred high yielding and firm tomato varieties with long shelf life and that can withstand transportation because better markets are more than 300km away.
- Owing to the simplicity and low-cost, low-risk nature of entry point technologies, they can often be effective in reaching less advantaged social groups.
- To maximize the contribution of entry points to addressing more complex system or NRM challenges, it is important to consider entry points that can enhance the subsequent adoption of other NRM technologies.
- Where benefits, especially monetary benefits, accrued from entry point technologies, farmers are often more willing to engage in more complex and integrated technological innovations.
- Initially unaware of the potential benefits of a lasting partnership with researchers, farmers may initially come to the innovation process with expectations of quick rewards such as fertilizers and seeds. With a lack of experience working with research and understanding its value, they may be unprepared to take risks associated with adopting complex technologies and practices. Finding means to respond to their immediate demands without creating dependency while working on more complex innovations with slower returns can go a long way in fostering interest in the latter and in moving towards more sustainable farming practices.

Supporting farmer innovation to address farm-level constraints

There is general agreement in the agricultural research and development community in the region that low agricultural productivity and resource degradation in Africa are not owing to the absence of technologies, but to the limited adoption, adaptation, and dissemination capacity of farmers and the ineffectiveness of methodologies employed to support these processes. Farmer experimentation and innovation are recognized as essential in efforts to improve productivity and reverse natural resource degradation. This innovation is not only technological in orientation, but may also encompass networking and communication, the strengthening of local institutions, planning and monitoring, or accessing resources and marketing—anything that may be considered "new ways of doing business" (Assefa and Fenta, 2006).

While diverse approaches were employed in supporting farmer experimentation in AHI, some elements were common to all. For example, different actors tended to make different types of decisions. Decisions on the location of trials, choice of crops, and harvesting time are usually made by farmers. Experimental design and implementation were carried out jointly, but researchers had a strong input into the basic research design to ensure adequate replication and controls. Researchers also participated in identifying parameters to be tested and in carrying out the analysis, but farmers were involved in managing experiments and evaluating technologies. Officers of local offices of agricultural ministries were also involved in the decision-making process, given their familiarity with wider areas over which technologies could be applicable. Farmers generally tried to address their specific problems by testing a wide range of technological options selected in response to the problems they face. In certain cases, researchers also assisted client farmers in resource mobilization, leadership, building organization skills, group management, and conflict resolution.

Approach development

AHI experimented with at least three different approaches to supporting farmer innovation: local testing and adaptation of the farmer field school (FFS) approach, approaches for inducing innovation based on local knowledge, and approaches for linking complementary technologies to achieve synergies between livelihood and natural resource improvements.

Approach 1—Farmer field schools

The FFS approach is an innovative, participatory, and interactive learning approach developed by FAO in the 1980s to address pest and disease problems of rice farmers in Southeast Asia (Pontius et al., 2002). It builds farmers' capacity to understand their systems, identify system constraints, and to test and adopt technologies and practices matched to those constraints. With more than 1,000 FFS active in Kenya, the FFS approach is not new to the eastern Africa region. The FFS approach employs non-formal adult education methods, particularly experiential learning techniques. Farmers are selected based on their interest and willingness to follow the proposed methodology or action plan and their commitment to invest and allocate their time in the program. Typically, a group of 20 to 25 neighboring farmers meets regularly, commonly once a week, on one of the farmers' fields during the entire experimental cycle. The school is not meant to introduce farmers to new technologies developed outside their environment, but to provide them with tools and methods that will enable them to analyze their own production practices and identify possible solutions.

The AHI team and its partners used FFS in selected cases where there was a need for intensive interaction between researchers and farmers. It was applied particularly when a farm constraint demanded a comprehensive package of knowledge, practices, and technologies, such as controlling crop diseases. It was also tested for its relevance to addressing complex natural resource management challenges and community organizing. Communities were empowered to establish FFS to organize, test, adopt, and disseminate improved technologies and practices. They were also facilitated to sustain the learning process by building the capacity of colleague farmers and communities to enable them to respond to emerging local challenges. The approach was used to build local capacity and interest in sustaining farmer experimentation on their own, even in the absence of external material and technical support.

The following steps were involved in adapting the FFS approach to achieve the program aims:

1. Facilitate a dialogue among farmers and with outside agencies (research, extension) to enhance local awareness and refresh people's memory about key system constraints identified in the diagnostic phase (e.g., through PRAs) and their implications for food security, income, and natural resources.
2. Plan and facilitate discussions together with local institutions on how to solve these system constraints using local solutions, skills and collective action.
3. Identify farmers with similar farm-level problems or constraints and a shared interest to find practical solutions, and assist them in organizing themselves into thematic groups (generally consisting of a commodity and related NRM innovations) to identify and test endogenous and exogenous innovations.
4. Organize and conduct a formal, classroom-based training program to help farmers to analyze the biological and socio-economic causes of the problem. This included the development of a detailed theoretical and practical curriculum to enable farmers to understand the causes and develop skills to solve the given constraint. In some cases, "classroom" learning was

supported by laboratory experiments, field days, and other practical methods.
5. Identify a basket of technological options based on local economic and social criteria and introduce them to farmers through cross-site visits, exposure to on-station trials, or other means.
6. Conduct formal trainings to equip farmers with the skills needed to enable them to successfully compare options, including formal experimental methods.
7. Organize farmers and support their efforts to test, adapt, and adopt the interventions in their own fields, assisting them to capture data on key parameters such as yield, income, and labor and to compare the performance of different enterprises and management options.

Box 2.4 describes a case where FFS were used to solve farm-level constraints.

BOX 2.4 USE OF FARMER FIELD SCHOOLS IN PROMOTING POTATO-RELATED INNOVATIONS IN THE ETHIOPIAN HIGHLANDS

At 3,000m above sea level, the Ginchi BMS has a temperate climate with barley and Irish potato the major crops. There are two cropping seasons, the first from February to April (the short rainy season) and the second from June to October (the main season). Farmers often fail to grow potatoes in the main growing season owing to late blight infestation. A technology development and dissemination activity was undertaken by using the Farmers Field School (FFS) approach to develop potato technologies suitable to local conditions. The purpose was to assist farmers in developing healthy potato farms, which are more productive, profitable, and sustainable. Using this approach, experiments including varietal evaluation and fungicide-by-variety interactions were conducted. In order to differentiate the natural variability of potato clones in response to major potato diseases such as late blight, two blocks were protected with fungicides while the other two blocks were left without fungicide application. The FFS approach was found to be effective in stimulating farmer participation by considering their goals in the targeting and design of innovations. Outcomes included the following:

- A very popular potato variety was identified.
- The FFS approach helped the farmers to better understand complex environmental interactions in the process of identifying disease-tolerant potato varieties.
- The FFS approach enhanced the efficiency and effectiveness of the extension system.
- Many of the farmers involved in the FFS were encouraged to continue research on their own.

Approach 2—Farmer experimentation using local knowledge as a starting point

Efforts to support farmer innovation to address farm-level constraints must begin with local innovation processes, a critical starting point for building partnerships of mutual respect between different actors in an innovation system. It starts with looking at what farmers are already trying to do to solve problems or grasp opportunities they have identified (Waters-bayer and Bayer, 2009).

Acknowledging this reality, the AHI research team in collaboration with NARIs and international research organizations employed several participatory techniques for integrating local knowledge into agricultural experimentation. The aims were to: 1) develop strategies to address complex NRM issues; 2) foster a change from a commodity orientation to a more holistic systems and participatory approach in the research system; and 3) develop and improve technologies and approaches that could be used by policy makers, development actors, and farmers to address identified NRM challenges. Farmers were in the forefront throughout the processes of technology development, technology dissemination, and impact assessment—a process that included the following basic steps:

1. Participatory identification of problems and opportunities from the standpoint of farmers and researchers;
2. Characterization of various local innovations employed by different farmers, how widespread they are and their potential benefits;
3. Scientific validation of local knowledge or innovations to better understand their features and benefits, and to explore how to link them to scientific knowledge in addressing system constraints;
4. Feedback of findings and discussion with the holders of local knowledge and other community members;
5. Demonstration and experimentation of ways to link local innovations with exogenous technologies (an optional step, employed only where possible synergies are identified); and
6. Promotion of best performing innovations as integrated packages to the wider community through training and awareness creation.

Box 2.5 presents experiences from Areka, southern Ethiopia where farmers employed local knowledge to control mole rats, a vertebrate pest causing yield reductions of up to 60 percent in root crops. Conventional methods of controlling the pest (e.g., use of poisonous substances) had proven to be ineffective in addressing the problem. In addition to the expense, the mole rats quickly learn to dodge them once they detect they are poisonous. Fortunately, a combination of local experiences and conventional techniques proved to work, and were both less costly and more environmentally friendly. At the time of writing, more than 50 percent of the farmers in the village where experiments were carried out were using the technology.

BOX 2.5 CASE STUDY ON MOLE RAT CONTROL IN AREKA, ETHIOPIA

Mole rat is the most troublesome wild pest affecting home garden crops in southern Ethiopia owing to its effects in exacerbating food insecurity. Conventional control methods such as fumigation and baits are costly for the resource-poor farmers. The AHI team collaborated with farmers to identify effective control measures. The few individuals with knowledge of how to control the pest were identified. These individuals used to make money by hunting mole rats without sharing their knowledge with others. After a facilitated dialogue between the knowledge-bearers and other farmers, these individuals agreed to share their methods, which involve the use of local attractant herbs and traps. The trap is composed of a metal hook tied with sisal string on a bended stick (see Plate 2). The bait—banana, sweet potato, or local spices—is placed behind the metal hook in the burrow of the mole rat. In order to reach the bait, the mole rat has to bite and cut the string. When the string is cut, the metal hook is swiftly pulled out of the hole by the bent stick. It is this sudden action that causes the hook to pierce and kill the mole rat.

While this proved to work initially, the mole rats were eventually able to distinguish bait that had been contaminated by human hands. To rectify this problem and enhance the effectiveness of the trap, farmers started treating their hands with the soil dug by mole rats to reduce the human "smell." In doing so, some farmers were able to control mole rats in their homesteads and farms.

AHI scientists are cognizant of the moral dilemmas and ecological challenges associated with vertebrate pest control practices, and have made an effort to ensure complex spin-offs on local ecosystems are identified and managed in the process of putting local livelihoods needs first.

Approach 3—Promoting linked technologies

One key challenge in supporting farmer innovation is to identify and integrate technologies addressing one or more identified problems without negatively affecting other system components. For instance, during the colonial era, promotion of conservation technologies was led by conservation programmes while technologies for improving agricultural production were facilitated by agricultural research institutes. Separating conservation and production, and piecemeal promotion of technologies and management practices (using a "commodity" or "single factor" approach) did not bring real benefits to farmers; in fact it failed to create the desired impact. In response to this challenge, AHI developed and tested approaches for facilitating farmer experimentation with an explicit effort to link conservation with production-enhancing technologies.

The term "linked technologies" was coined to define technologies that when applied simultaneously at plot or farm level render multiple benefits by enhancing adoptability of discrete technologies or fostering synergies that would not exist had technologies been applied in isolation. For instance, given the steep slopes, intensive cropping, and high rainfall intensity in most of the AHI sites, soil fertility decline was very apparent. The research teams employed several participatory techniques in order to develop the capacity of farmers and researchers in integrated soil fertility management; foster partnerships among stakeholders to avail best-bet technologies; and foster a change from commodity-oriented to a more holistic and participatory approach placing farmers at the forefront of technology development and evaluation (see, for example, Stroud, 1993). The guiding hypotheses were the following (Stroud, 2003):

- Technologies with win–win benefits (e.g., increased income, improved soil fertility) will build farmers' confidence to test more complex NRM technologies, and strengthen the demand side in the technology innovation process.
- Problem-solving technologies with multiple benefits will bring more food and cash income to farmers of different resource endowments by solving multiple problems simultaneously, with solutions attractive enough to "sell" to others.

The methodology for developing linked technologies starts with the methods described in the above section "Understanding systems and clients"—namely, participatory rural appraisal techniques at village level to identify constraints faced by different social groups, followed by participatory testing of a wide range of technologies by thematically based farmer groups—starting with identified entry points. The next steps were specific to the linked technology approach, as follows:

1. Once solutions are found to the most pressing issues (addressed through the testing of entry points), researchers facilitate access to more complex technologies. These technologies often relate in one way or another (e.g., through nutrient or capital flows, or labor savings) with technologies that have already been tested. The latter could have already been adopted, or could face some constraint that a new technology can assist in alleviating. The new innovations can also be unrelated to the entry point, and build on farmers' enthusiasm to innovate rather than on proven technologies as a means to propel interest in more complex innovations. Importantly, however, the linked technologies bring immediate benefits while also fostering farmer investments in more complex NRM technologies with slower returns. For instance, while soil and water conservation was a key intervention to minimize erosion, increase water infiltration and increase input efficiency at farm and landscape scales, farmers had a difficult time engaging in such labor-intensive practices without immediate financial returns to their

labor. By combining the testing of conservation bunds with forage grasses, organic nutrient management, and multipurpose trees as linked innovations, farmers were able to generate immediate benefits such as livestock feed, improved yields (crops, milk, manure), and fuel wood from investments in soil stabilization and fertility improvements (see also Box 2.4).

2. Gradually farmers intensify and specialize in a system such as horticulture that renders the much-needed economic as well as social benefits and sustains or expands NRM investments. As the economic returns from NRM investments begin to materialize from the high-value crop and/or livestock enterprises, farmers are often propelled to invest more or expand the area over which the innovation is carried out. Thus, complementary innovations not yet tested in the first step such as integrated pest management (IPM) or other high-value crops with complementary growing cycles (for intercropping or relay cropping) can be brought into the innovation system.

The role of research was to facilitate access to technologies, train lead farmers, support farmer experimentation, and guide and monitor what different farmer groups did to integrate the various technological options—and their perceived impacts. The gradual, iterative process of planning, testing, evaluation, and replanning in a system that becomes ever-more diversified and integrated, is portrayed in Figure 2.1.

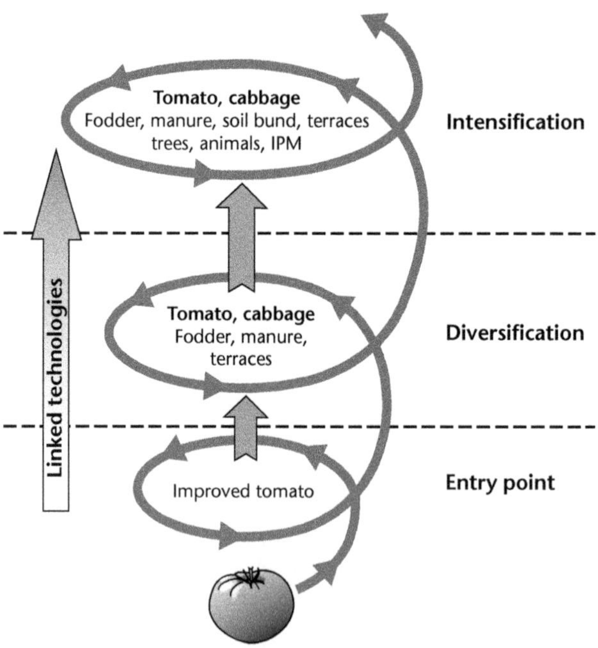

FIGURE 2.1 Simplified model of farm level entry point and linked technologies

Integrating technologies is a function of time, space, demand, and appropriateness of the interventions under the given circumstances. It should be done through targeting clients and system niches and by providing problem-solving interventions addressing the most important household priorities. This will improve the confidence of both farmers and researchers and the rapport between them as they seek to address more complex system constraints that may need more than one technological and institutional intervention.

It should be noted that the process of developing evermore complex linkages between technologies in wider system-wide innovations is not a one-off process but rather a time-consuming, stepwise engagement whereby farmers integrate options to supplement earlier investments for increasing returns from their farms and investments. Figure 2.2 illustrates where the stepwise approach is used to foster integrated soil fertility management in the Ethiopian highlands. These innovations can then be integrated into watershed-level innovations at a later stage (see Chapter 3 for details).

The approach should give emphasis to building the capacity of the communities and R&D teams to implement a systems approach and address the needs of diverse social groups. Farmers play a key role in linking technologies. Box 2.6 presents another success case, linking soil conservation with fodder production in Ethiopia to conserve soil while enhancing livestock production.

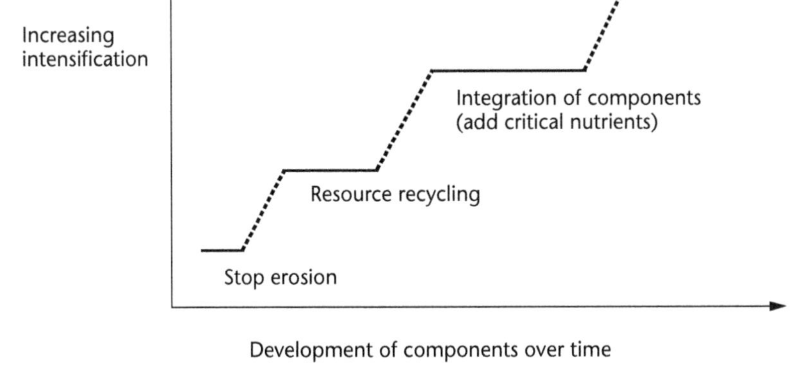

FIGURE 2.2 Stepwise integration of various technologies and approaches to improve natural resources management in the Ethiopian highlands

BOX 2.6 LINKED TECHNOLOGIES FOR LIVESTOCK PRODUCTION AND SOIL CONSERVATION IN AREKA BMS

Farmers in Areka rated soil erosion as one of their major production constraints. Government agencies such as the Bureau of Agriculture and the Wolaita Agricultural Development Unit had made various attempts to promote soil and water conservation structures in the area. With farmers perceiving these initiatives to be externally imposed, they met with limited success. Moreover, with small farm sizes, farmers were unwilling to allocate strips of land for the construction of conservation bunds.

AHI and its partners organized consecutive community meetings to create awareness and to seek solutions jointly. Soil bunds were selected as a practical solution for minimizing erosion and reducing loss of seed and fertilizer from excess run-off. Farmer Research Groups (FRG), established to test interventions, were used as a platform for farmer organization and collective action. By-laws were first developed by farmers to establish the working principles and arrangements for organizing collective action in soil bund construction. Based on periodic meetings to evaluate progress, modifications were made based on farmers' recommendations. This included expanding the technical spacing recommendations between two adjacent bunds to allow sufficient space for the "U-turn" of an oxen-pulled plough. The land allocated for conservation bunds was used to grow food and fodder crops. In addition to its role as a soil stabilizer, Napier grass attracted farmers' attention as a quality feed. This was further expanded by distributing cuttings to more communities using the FRGs; but also through encouraging farmer-to-farmer seed dissemination across villages.

Lessons learned

AHI's experience in supporting farmer innovation confirmed the need for scientists to facilitate a dialogue based on mutual respect and learning by accepting and respecting farmers' knowledge. Scientists have important roles to play by bringing in information, methods and analyses that complement what farmers already know and can do themselves. Key lessons learned on efforts to foster farmer experimentation and innovation include the following:

- A host of considerations and decision criteria enter into any innovation process, many of which are specific to the local setting or cultural preferences, posing a challenge to diffusion of innovations. The same constraint is not necessarily resolved in the same way in different locations, even within the same agro-ecological region. This implies the need to replicate farmer innovation as an approach to problem solving, not the solutions generated through these approaches.

- A critical element to developing effective partnerships between farmers, researchers, and development actors in supporting farmer innovation is overcoming the widespread tendency to underestimate farmers' knowledge and innovation capacity and treating them as equal partners. Learning to listen to farmers and take their feedback on board whenever they report challenges faced in testing and adopting interventions is essential if researchers are to play an important role in farmer-led innovation processes.
- Research and extension practices that build on farmers' knowledge, engage farmers' creativity, and allow for their active involvement in outreach activities are capable of producing results that far exceed and outlast those possible through more conventional approaches.
- The "linked technology" approach enabled farmers, development agencies, and research organizations to address poverty and natural resources degradation in a holistic manner.
- Market opportunities are an important impetus for technology adoption and systems intensification, and efforts to identify and meet market demands within a wider innovation effort can go a long way in catalyzing change.
- Farmer–researcher partnerships for farm-level innovation require flexibility when defining the role of research. Some interventions do not require formal experimentation, as the returns are quickly visible. In some cases, the researcher's role became one of conceptualizing a system so as to introduce new ideas rather than the design and implementation of experiments, of monitoring with the aim of understanding farmers' innovations and evaluations, and of providing support to dissemination and scaling-up processes.
- Interventions with win–win benefits are effective in bringing about immediate impact at household and community scales.
- Mechanisms to involve innovative farmers as local champions of an innovation process can be an effective means of stimulating local innovation, providing technical backstopping to other farmers and facilitating dissemination—a topic to which we now turn.

Disseminating proven technologies and approaches

Conventional approaches for technology dissemination are usually top-down and commodity-oriented, with the mode of technology dissemination assumed to be "linear"—namely, from research to extension to farmers. Critical factors affecting adoption such as socio-cultural, policy and institutional conditions were not considered in this approach. Furthermore, most of the technologies were generated on station through researcher designed and managed trials. Direct feedback from farmers as well as several formal adoption studies have clearly shown that technologies developed using this conventional approach were often not appropriate to local circumstances, thus leading to low adoption

(Amede et al., 2001). This was largely owing to the limited involvement of farmers in the development and dissemination of technologies, as well as to weak institutional support to facilitate the adoption capacity of target groups. Horizontal and geographical spread of technologies is limited even when facilitated by public institutions and NGOs. The challenge for AHI was therefore to identify socio-economic and biophysical incentives to facilitate the scaling up of innovations proven to work in select locations. Using more "bottom-up" approaches, AHI has seen a gradual increase in farmers' interest and adoption of different technologies, resulting in higher incomes and food security for households (Mowo et al., 2002).

Technology dissemination beyond partner households and villages has been hindered by blanket recommendations and poor packaging. Contrasting production systems and socio-economic circumstances demands a diversity of technological innovations and approaches. The diversity of household production objectives, for example, with some households concentrating on cash crops and others focused more on achieving food self-sufficiency, requires careful targeting of technological interventions. Resource-poor farmers, especially those distant from markets, face difficult decisions over the allocation of scarce resources (e.g., land, labor, nutrients, and water). Decisions on the allocation of resources are often associated with immediate financial gains and food security, with limited assessment or appreciation of the impact of management decisions on long-term effects or other system components (e.g., soils). There is therefore a need to explore mechanisms for matching technologies to specific recommendation domains, as defined by agro-ecological conditions, cropping systems, cultural values, system niches, or socio-economic variables.

Besides technologies being poorly adapted to different agro-ecologies and socio-economic circumstances, some technologies and approaches demand collective decisions and policy support to be adopted—further limiting their spread when these factors are not taken into consideration. Farmer-to-farmer dissemination of technologies through existing social networks—be they defined by area of residence, friendship, kinship, marriage, religion, or other factors—has been found to be one successful approach (Adamo, 2001), though the reach is limited. AHI has developed approaches that have significantly increased the spread and adoption of technologies within benchmark sites where spontaneous farmer-to-farmer sharing was limited or socially biased (German et al., 2006b). Research played a critical role in helping farmers to organize themselves, access and multiply preferred technologies, and sustainably utilize these interventions and promote them within the locality and beyond.

Approach development

Various technologies and practices may demand different dissemination approaches. This is because some technologies are easy to disseminate while others are more knowledge-intensive and difficult to scale up unless accompanied by intensive mentoring, guidance and institutional support. AHI used three different approaches for scaling up: farmer research groups, externally mediated diffusion in which dissemination is governed through locally formulated by-laws, and a formal approach based on research to identify social and biophysical barriers to adoption.

Approach 1—Farmer research groups

In addition to being employed as a platform for farmer experimentation, FRGs were used as a means to scale up impacts through technology dissemination and the testing of additional innovations that might work synergistically with technologies to enhance impact (e.g., micro-credit). As mentioned above, the process of establishing FRGs for technology dissemination included the organization of farmers into thematically based groups for testing technologies on behalf of the wider community, formal training, the identification and implementation of experiments and evaluation of results. With this approach, the transition between experimentation and scaling up is a relatively seamless one—with steps in the latter a natural transition from the former. This means that the use of FRGs as a platform for scaling up is informed as much by the natural progression of farmers' interests and experiences as it is by a set of discrete steps. However, in most cases it consisted of the following components:

1. Members of different thematic groups presenting their findings to the wider community (including FRGs working on other themes) at different stages of experimentation.
2. Research teams and FRG members developing a scaling-up strategy and jointly organizing field days, farm visits, posters, and demonstration trials.
3. Farmers beyond pilot communities or groups seeking support from research to expand the FRG methodology, requiring both continued support to farmer experimentation and a proper strategy for ensuring continuity and sustainable delivery of technological options.
4. FRG members starting their own community seed multiplication initiative as a business venture, often on their own initiative.
5. FRG experimentation with other non-technological innovations (e.g., credit provision, marketing) to alleviate the constraints to adoption and thereby enhance technology adoption and dissemination either directly or indirectly (see Box 2.7 for the case of savings and credit associations in Tanzania).
6. Research and development teams facilitating linkages between successful FRGs and local authorities to disseminate proven innovations beyond pilot sites.

It is interesting to note that the dissemination phase exposed a number of weaknesses in the FRG methodology as a whole, highlighting some of the challenges that need to be managed for the successful implementation of FRG-based experimentation and dissemination efforts. With the majority of researchers initially lacking experience in applying principles and concepts of participatory methodologies, they were ill-prepared to assist moving farmers away from individualistic attitudes and to support the evolution of cohesive farmer groups. In some instances, this led to limited interest in fostering collective benefits, with FRG members seeing technologies as their own property rather than something they have been given a mandate to test on behalf of the wider community. This became apparent at the time of scaling up. In another instance where participatory approaches were tested in a site where government programmes were providing cash and inputs free of charge to farmers implementing soil and water conservation, some of the technologies used as entry points were accepted by farmers owing to these benefits rather than the technologies themselves. This hindered subsequent efforts to scale up technologies requiring significant financial or labor inputs, as the underlying motives for uptake were weak. Yet these experiences were more the exception to the rule, with most farmer groups realizing the benefits of working with researchers during participatory technology testing and dissemination.

BOX 2.7 COMMUNITY DRIVEN MICRO-CREDIT SYSTEMS: BUILDING THE FINANCIAL CAPITAL OF SMALLHOLDER FARMERS IN LUSHOTO DISTRICT, TANZANIA THROUGH FRGS

In 1998, farmers of Kwalei village identified low livestock productivity and land degradation as major challenges. They also identified limited financial capital as one of the barriers to adopting promising technologies, and therefore requested financial assistance to enable them to test and adopt the improved technologies. Limited availability of capital had impaired the adoption of technologies owing to the ever-increasing cost of farm inputs. In response, farmers were sensitized on establishing their own savings and credit cooperative society (SACCOS). Although the farmers were skeptical about the success of such an initiative owing to negative past experiences with cooperatives, they formed a SACCOS in 2000 after undergoing formal trainings, exchange visits to successful SACCOS, and carrying out group negotiations. They were able to officially register their society in 2002 under the name Kwalei SACCOS through support from the district cooperative department. Over the next five years, membership grew from a village association with 36 members to a membership of 182 involving farmers from six neighboring villages, with a credit-worthiness of 120,000,000 Tanzanian shillings (US$ 100,000).

Based on local records, farmers have borrowed money from the SACCOS to purchase agricultural inputs as well as to address other pressing family matters

(Table 2.1). The majority of borrowers are women, choosing to invest their money in establishing businesses and to cover family emergencies. Many of those going into businesses have begun marketing agricultural produce in distant markets and bringing back merchandise such as clothes and farm inputs. Fifteen percent of the loans has been used to purchase farm inputs such as fertilizer, improved seeds, and pesticides—which are normally expensive to the average farmer. About 5 percent of the loans has been used to construct soil and water conservation structures and establish tree nurseries; the majority of those investing in nurseries doing so to produce seedlings for sale. At the time of writing, Kwalei SACCOS had loaned 2,000,000 Tsh (US$ 1,668) to two nursery groups.

TABLE 2.1 Local credit arrangements in Lushoto

Purpose of borrowing	*Number of borrowers*			
	Men	*Women*	*Groups*	*Total loans*
Payment of school fees	3	1	2	6
Soil and water conservation	10	1	1	12
Establishing tree nursery	–	–	2	2
Purchase of farm inputs:				
• Vegetables	20	10	–	30
• Food crops	9	1	–	10
• Perennial crops, e.g. coffee, tea, banana	1	1	–	2
Purchase of land	10	–	–	10
Building improved houses	10	–	–	10
Establishment of business	30	70	–	100
Emergency (e.g., sickness, death, school fees)	31	75	–	106
Grand Total	11	145	4	288

Approach 2—Externally mediated diffusion

AHI's experience in several benchmark sites suggests that the spread of knowledge-intensive technologies is not as fast and simple as crop varieties and forage, even within a village. Moreover, scaling up "fast-moving" technologies does not mean these innovations will reach different social groups or locations without external support and facilitation by local institutions, extension departments, or research institutions. AHI and its partners therefore experimented with different approaches for mediating technology dissemination and farmer-to-farmer "spillover" to ensure equitable access to technologies being tested by different FRGs. The question of how to equitably share knowledge and technology among male and female farmers, and to reach farmers with different resource endowments, was discussed at community meetings involving community leaders, local authorities, and early adopters. This was done in part to ensure seeds continue to spill over from one FRG to the next, to counter the tendency for FRG members to take project seed as their own property and stop there. The participatory formulation of local by-laws and their subsequent endorsement by local administrative authorities was also considered as a means to guide or govern the technology dissemination process according to guidelines agreed upon by all. In short, this approach may be characterized by a very direct mediation from outside the community designed to enhance equity.

The following steps were used to facilitate externally mediated technology diffusion:

1. Mobilize and sensitize community members on key crop and livestock issues identified through participatory diagnostic procedures such as PRAs, to explore the extent to which those who wish to access technologies are able to do so and to identify barriers to technology access and uptake.
2. In cases where inequitable access to technologies is observed (either by gender or any other factor), hold community meetings to generate by-laws to govern modes of technology dissemination (Box 2.8).
3. Establish technology testing sites with beneficiary farmers.
4. Form marketing committees and higher level organizations from village to sub-county levels to facilitate market linkages, including establishment of collection centers.
5. Establish village libraries for publicity (reading materials, pamphlets regarding NRM and other agricultural activities).
6. Improve the capacity of farmers to multiply seeds, including phyto-sanitary measures during production (as required with Irish potato) up to post-harvest handling.
7. Promote the products using dramas, role plays, shows, demonstrations, and other tools.
8. Encourage farmers to keep records and use them for monitoring and tracking progress.

Key stakeholders such as local leaders and extension agents were involved in planning and implementation at all levels.

BOX 2.8 CASE STUDY ON THE USE OF BY-LAWS FOR EQUITABLE TECHNOLOGY DIFFUSION IN AREKA, ETHIOPIA

In a participatory diagnostic process in Areka BMS, gender-disaggregated focus group discussions highlighted very inequitable patterns of technology access and extension delivery by gender and wealth—with a tendency to focus on wealthier male farmers with larger landholdings. Efforts were made to better "govern" extension services and the spread of technologies by negotiating collective choice rules and endorsing these as formal by-laws. The proposed by-laws included the following:

- FRG leaders must select farmers who will receive seeds to multiply fairly from among women and men, and from poor, medium and better-off farmers each year. At least one-third and two-third of the farmers should be women and poor farmers, respectively.
- FRG leaders should coordinate, facilitate, and follow up with seed multiplication and dissemination by identifying who is multiplying which varieties, in what amount, and where.
- A farmer who multiplies seed has to return the same amount of seed she or he took for multiplication to a farmer selected by FRG leaders. If a farmer loses a portion of the harvest owing to natural factors, a similar proportion of the seed taken has to be transferred to a farmer selected by FRG leaders. A farmer who lost the improved seed owing to natural hazards will be free from the sanctions for non compliance. Reasons for loss should be justified and verified by FRG leaders.
- Farmers that take improved seed for multiplication should apply all the necessary improved agronomic practices and should not lose or consume the seed, unless due to a situation beyond his/her control.
- A farmer who multiplies improved seeds should ask the FRG leaders or PA leaders before selling the seed in the market whether there are farmers in the watershed who want to buy the seed.
- Local and external institutions are governed by this by-law and must work with the local administration and FRGs when selecting farmers for technology dissemination.
- When a farmer or an institution goes against this bylaw, the PA social court should see it as disrespect to the PA regulations and should pass judgment accordingly.
- PA leaders must facilitate the implementation of this by-law, charge non-compliances and implement the judgments passed by social courts.

Approach 3—Targeting systems and clients for dissemination of technologies

While technological innovations are vital in solving farm-level constraints to food security and sustainable NRM, their adoption and utilization by local communities can be limited unless interventions effectively target clients and key system constraints, and contribute to overall household objectives. Although farmers are keen to learn about technologies through farmer field schools and on-farm testing, not all farmers are involved in piloting technologies and other farmers will need time to test and adapt them to their own, often sub-optimal, conditions. Yet managing such a process with each and every household is costly, and tools for predicting adoptability under different conditions can help to reduce such costs.

Predicting the likelihood of adoption of different technologies and formulating relevant recommendations are difficult owing to the variable nature of biological and social-economic systems and the trade-offs that characterize production and resource/input management decisions. Thus, the generation of decision support tools based on detailed analyses of farming systems may provide a complementary tool to more participatory techniques in identifying technologies and their potential socio-economic and biophysical niches. With the ability to consider multiple variables simultaneously, the tools described below can enable more accurate targeting of innovations and clients to foster multiple household objectives simultaneously (e.g. increased productivity of crop and livestock systems, income, and food security). Key steps in the approach include the following:

1. Characterization of systems and clients. As systems with different characteristics will differ in their capacity to intensify and the pathways through which this occurs, and different drivers of change will influence enterprise choices and their management, the targeting of interventions should start with a characterization of the system. This includes both socio-economic and biophysical perspectives. The former includes household resource endowments by wealth, gendered perspectives on constraints and priorities, household involvement in institutions of collective action, and access to technologies and innovations. The latter includes a characterization of the production system, access to water and nutrient resources, soil fertility, and other relevant biophysical parameters.
2. Identification of socio-economic and biophysical factors affecting adoption (Box 2.9). Farmers employ multiple criteria when deciding whether a technology in question is appropriate for their circumstances, and whether it can be productively integrated into their farming practices. This step involves identifying variables affecting the adoption of a particular technology by households with variable economic and demographic characteristics, resource endowments, and system constraints. These include the extent to which the technology is aligned or compatible with household preferences and cultural values, its actual or anticipated performance in different farm and landscape niches, the immediacy of benefits derived from its adoption, complementarities or conflicts with other system sub-components and users, and the potential of the intervention to address multiple challenges simultaneously. These factors are determined based on past experience by some or all households with the same or similar technologies.

3. Prioritization of major socio-economic and biophysical determinants of adoption using pair-wise ranking, community validation, and case study analysis.
4. Development of decision guides to assist development agencies, extension personnel, and farmers to target systems and clients for a specific intervention.

BOX 2.9 SOCIO-ECONOMIC CRITERIA TO INTEGRATE LEGUMES INTO FARMING SYSTEMS OF THE HIGHLANDS OF EASTERN AFRICA (AMEDE AND KIRKBY, 2004)

Through the above process, the following factors were identified by farmers as influencing the adoption of legumes in smallholder systems in the region:

- Good performance, in terms of biological productivity, under given agro-ecological conditions. The most favorable candidate is one with relatively high yield of both grain and biomass under variable agro-ecological conditions, namely precipitation, temperature, soil fertility, and variable management conditions.
- Positive effect of legume incorporation on grain yield of the subsequent crop. If the effect on subsequent crops is negligible, adoption will be limited.
- Minimal competition with food crops for land and water. Because of land scarcity, farmers may not be willing to grow legume cover crops as a monocrop. Therefore, those legumes that do not strongly compete with the companion food crop for water, nutrients and light when grown in combination are best options.
- Contribution to minimizing soil erosion. LCCs with firm root systems capable of protecting the soil against erosion (determined based on the strength of the plant during uprooting) are favored by farmers with plots on steep slopes.
- Rapid decomposition. The rate of decomposition when incorporated into the soil (determined by the strength of the stalk and/or the leaf to be broken by hand) is considered as an important indicator to predict whether the organic resource applied is in a position to release nutrients for the subsequent crop in a short period of time or not.
- Mulching capacity. Mulching capacity, determined by farmers as the moisture content of the soil under the canopy of each LCC species, is an indication of the water use efficiency of the respective legume and its compatibility in multiple cropping systems.

- Drought resistance when exposed to dry spells. Crops less susceptible to drought will yield returns to labor invested under variable climatic conditions, and therefore be favored by farmers.
- Compatibility with other staple and cash crops. Whether the LCC is found to compete with food legumes for space and resources, and its effects on land productivity, were critical in this regard.
- Value as a feed. Palatability for livestock and ability to produce high quality feed for the dry season are important considerations for farmers, owing especially to the high calf mortality during the dry season.
- Early soil cover. LCCs with fast mulching characteristics not only conserve water through reduced evapotranspiration, but make the soil easy to work with—thereby reducing the labor burden for farmers. It also reduces the kinetic effects of heavy rain on the soil and soil erosion.

In addition to these biophysical factors, a number of socio-economic indicators affecting adoption were also identified. These included farm size, marketability, toxicity of the pod to children and animals, risk (e.g., from the introduction of new pests), and farm ownership and management (e.g., whether the land was managed by the landowner or sharecroppers who would have less interest in investing in long-term productivity).

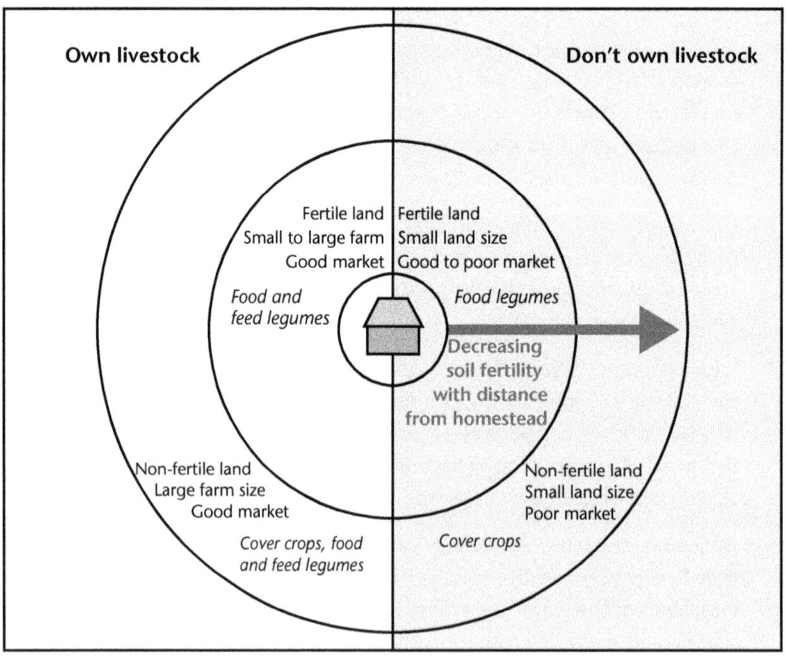

FIGURE 2.3 Integration of food and feed legumes and legume cover crops into small-scale farms as a function of resource endowments and market conditions

After comparing these factors in a pair-wise analysis, it was possible to identify and rank the five major indicators that would influence a farmers' decision on whether or not to adopt LCCs. These variables and the way they were employed in local decision processes were then employed to construct decision guides. An example of a resulting decision guide that integrates farmer resource endowments and market conditions as priority indicators is illustrated in Figure 2.3. For more information, see AHI Brief A5 (Soil Fertility Decision Guide Formulation), available at: http://worldagroforestry.org/projects/african-highlands/archives.html#briefs.

Lessons learned

The following lessons were learned in our efforts to test different strategies for local technology dissemination:

- Building the skills and capacity of "elite" farmers may help to kick-start the technology innovation process; however, technologies developed by these farmers are not guaranteed to reach the broader community. This may be owing to limitations in farmer-to-farmer sharing or, in cases where the early innovators are wealthier farmers, to different resource endowments.
- Contrary to common perception, communities are not homogeneous entities in which benefits to some households will automatically "trickle down" to all. Explicit strategies are often needed to ensure resources brought from the outside and intended for the collective benefit are well governed based on principles of equity.
- Where select individuals step forward to test technologies on behalf of the group, the acquired technologies will often be considered to be their personal property unless the individual responsibilities to the group (e.g., subsequent sharing of information or seed) are clarified in advance.
- Building farmers' confidence, trust and collegial spirit will go a long way in building strong groups and enhancing farmer-to-farmer sharing of technologies. The initial trust between farmers and research and development teams was an important factor contributing to building strong local institutions.
- Building farmers' capacity to access loans and services and linking them to district- and national-level financial institutions will significantly contribute to agricultural productivity, rural livelihoods, and ability to invest in natural resource management.
- Targeting potential clients and system niches can help to facilitate technology dissemination and adoption by providing a cost-effective tool for predicting adoptability of agricultural innovations. There is a need to develop these tools together with the potential users through participatory

and data-based approaches. Testing and validating decision tools in diverse settings can help to expand the tool's reach beyond pilot sites.

- Different technologies and practices may demand different dissemination approaches as some (e.g., crop varieties) are easy to disseminate, while other technologies (e.g., conservation agriculture) are knowledge-intensive and difficult to scale up unless the process is strongly facilitated by intensive mentoring, guidance and external or internal institutional support.
- Finally, and most importantly, the establishment of strong local institutions for technology dissemination requires that the demand for such institutions comes from the grassroots, whether the community at large or historically disadvantaged groups therein.

Tracking technology spread and impacts

In addition to proactively engaging in technology testing and dissemination strategies, it is often useful to understand the actual fate of technologies following such formal interventions. This can help to identify adoption bottlenecks, whether social, economic, or biophysical. It can also highlight the spontaneous ways in which farmers adapt technologies or their management to enhance their compatibility with local farming systems or increase the benefits derived from them, so as to ensure these innovations are popularized. Finally, it can help to identify areas where complementary innovations are needed, for example to minimize negative social or biophysical impacts resulting from technology adoption.

This section describes a methodology for tracking the spontaneous "spillover," or farmer-to-farmer sharing, of introduced technologies. Conventional adoption studies emphasize identification of factors influencing adoption and evaluation of impact in terms of the numbers of adopters and the area over which the technology is applied. The proposed methodology operates under an expanded set of objectives and research questions. *Identification of pros, cons, and adoption barriers* for different technologies can assist the targeting of improvements on the technology or its mode of delivery. *Identification of the characteristics of adopting households and farming systems* enables our understanding of who benefits from introduced technologies and can improve technology targeting for diverse social groups. *Characterization of social networks through which technology flows* in the absence of outside intervention can enable us to tap into existing social networks or to target strategies to overcome social biases inherent in these (i.e., gender bias within patrilineal or male-dominated societies). *Identification of social and biophysical innovations* made by farmers can help in our understanding of how technologies may be modified to better fit the farming system, and integrated into scaling out efforts. Finally, *identification of positive and negative social and agro-ecological impacts* can shed light on how to maximize positive while minimizing negative spin-offs of technological innovation (German et al., 2006a, 2006b).

Approach development

This expanded scope is achieved through a number of variations in conventional adoption studies, which tend to follow four basic steps: 1) researcher identification of variables likely to influence adoption; 2) structured household questionnaires focusing on key variables; 3) statistical analysis to correlate key variables with technology adoption; and 4) researcher interpretation of observed patterns. The modified methodology includes these same steps, but systematically builds local perceptions into the approach. Focus group discussions with different social groups (adopting and non-adopting farmers, or by gender and wealth) during Step 1 of the methodology aid in identifying basic patterns of adoption and technology sharing, as observed by farmers. Newly identified variables from these focus group discussions are then integrated into the standard household surveys, to enable quantification of relevant variables. Focus group discussions are also utilized during Step 4 of the methodology to integrate farmers' interpretation of observed patterns into the analysis. While researchers may believe an observed pattern may be explained in one way, farmers will often have their own explanation that differs considerably from researchers' interpretations. Each of these steps ensures that the methodology is sensitive to patterns of adoption and social interaction specific to the local context.

Household survey methods used in Step 2 also differ in two important respects. Sampling of interviewees can be done through the standard random sampling approach or through a form of "snowball sampling" in which social networks are traced from the original "project farmers" (L0) to "level one adopter" (L1) (farmers adopting from project farmers) to "level two adopters" (L2), and so on as presented in Figure 2.4. While random sampling may be better for rigorous econometric analysis of adoption variables, snowball sampling is best for understanding social networks through which technologies spread in the absence of outside interventions and how adoption levels and technologies themselves change through successive levels of "spillover." The latter also provides a picture of local adoption dynamics and pathways. The household survey methods employed here also differ by the integration of more in-depth qualitative interviews in a selected number of households. This aids in understanding social and biophysical innovations, livelihood and environmental impact, and the steps associated with technology adoption—generally, information requiring qualitative inquiry.

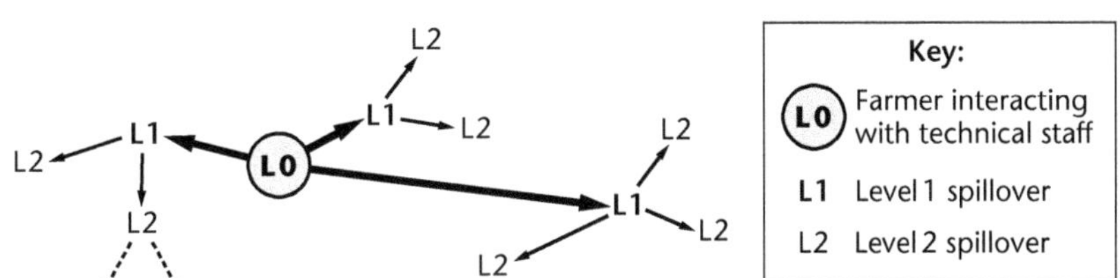

FIGURE 2.4 Levels of technology "spillover" relative to project interventions

As a whole, this methodology helps us to move from a view of technological innovation as a one-off step (introducing new technologies) to a process that proceeds from problem definition to technology targeting, testing, monitoring, troubleshooting, and dissemination or discontinuation. This is of fundamental importance in ensuring that patterns and lessons are not lost, and to minimize the risks introduced through technological innovation—such as negative agro-ecological impacts or socio-economic gap-widening resulting from biases towards wealthier or male farmers.

The methodology is applied for at least two consecutive growing seasons after technology dissemination, so that patterns of farmer-to-farmer sharing may be identified. The steps in the modified methodology may be summarized as follows:

1. Reaching a common conceptual understanding and agreeing on technologies to be tracked.
2. Focus group discussions to identify basic adoption patterns.
3. Identification of networks through which technologies flowed from source farmers ("Tracking surveys").
4. On-farm interviews with new adopters.
5. Data analysis to identify patterns of technology spillover.
6. Focus group discussions to interpret emerging findings.

For a detailed description of the methodology, including research instruments and sample findings, please see German et al. (2006b).

Following application of the methodology, new technologies and dissemination approaches are targeted to overcome identified problems. These problems might include social, economic or technical barriers to technology adoption, or negative social and agro-ecological impacts of adoption. Table 2.2 presents some of the agro-ecological impacts that have been identified through application of the methodology. This table illustrates the substantial spin-offs, both positive and negative, that often accompany technological innovation. These impacts are generally obscured under conventional adoption studies, but have a profound impact on the technology's success and system sustainability.

Examples of social and economic barriers to technology adoption are summarized in Box 2.10. Different types of barriers lend themselves to different types of solutions. Negative agro-ecological impacts can be addressed by testing complementary technologies that help to minimize negative effects of innovation, or by further research (breeding or on-farm experimentation) to further improve upon the technology itself. Economic barriers to technology adoption may require coupling technology dissemination activities with credit systems, facilitating negotiations among early and late innovators prior to technology testing, and dissemination to establish rules for technology dissemination that will ensure technology access by low-income farmers. Gendered barriers to technology access can also be addressed through

TABLE 2.2 Positive and negative[a] agro-ecological impacts associated with technologies introduced in Lushoto, Tanzania

Type of impact	*Banana germplasm and management*	*Soil and water conservation*	*Tomato germplasm and management*
Impact on other system components	**Favorable effects on other crops when intercropped**	**Positive effect on banana (soil fertility and moisture) and livestock (fodder production)**	**Increased fallowing of hillside plots as more time is allocated to cash crop cultivation in valley bottoms**
Input requirements	*Increased demand for scarce inputs at farm level given high organic matter inputs during establishment*	No outside inputs identified	*High demand for pesticide and inorganic fertilizers given crop demands and extended periods of cultivation*
Land, labor and nutrient allocations	*Recommended spacing takes up land; increased labor investments during planting and mulching*	*Organic nutrients and labor diverted from other activities during terrace establishment*	*Substantial diversions of land, labor and nutrients from coffee and maize*
Pests and disease	None observed	**Reduction in maize stem borer**	*Increase in pests and wilting disease owing to decreased crop rotation*
Soil	**Mulching increases soil fertility and water holding capacity; reduces erosion**	**Positive** or *negative*, depending on levels of organic amendments	**Increased water holding capacity and fertility from manure usage**
Weeds	**Sharply reduced through mulching**	*Increase in weeds near Napier grass*	*Increased along with soil fertility*

Note: [a] Positive impacts, as viewed by farmers, are in bold font and negative impacts in italics.

negotiation of rules for equitable technology access, as was done in Areka benchmark site.

In addition to its application as a retrospective impact study, this methodology can be applied within an iterative process of technology targeting, dissemination and monitoring. In this case, adoption barriers or negative effects of new technologies are periodically captured and addressed through further technological or methodological innovations. The methodology would need to be simplified for regular use, focusing on the most salient observations of farmers and perhaps minimizing the level of formal data collection.

BOX 2.10 PATTERNS OF TECHNOLOGY SHARING IN LUSHOTO, TANZANIA (GERMAN ET AL., 2006)

Gendered patterns of exchanges for Lushoto (north-eastern Tanzania) and Vihiga (western Kenya) are highlighted in Table 2.3. While an initial attempt was made by project personnel to enhance gender equity by working with equal numbers of men and women, inherent social dynamics caused male farmers to capture more of the benefits over time. Furthermore, since the percentage of source farmers who are female declines with successive levels of spillover owing to gender biases at lower levels of spillover, these differences are even more striking than they seem. In Lushoto, for example, only 22 percent of all farmers at level 1 were female, with much lower numbers of women (13.2 percent) obtaining technologies from source farmers who are male. For cash crops, exchanges with women were found to be negligible in Lushoto site, indicating that this gender bias in the spontaneous sharing of technologies could have far-reaching implications for women's ability to capture cash income.

TABLE 2.3 Gendered patterns of technology sharing in Lushoto and Western Kenya

Site	*Source farmer*	*Level L1 Adopters (%)*		*Level L2 Adopters (%)*	
		Female	*Male*	*Female*	*Male*
Lushoto	Female	50	50.0	60.6	39.4
	Male	13.2	86.8	25.1	74.9
W. Kenya	Female	66.3	33.7	55.6	44.4
	Male	34.5	65.5	0.0	100.0

TABLE 2.4 Exchange of different types of technologies among farmers in Lushoto

Technology	*Exchange characteristics*
Banana germplasm and management	88% given free of charge; the remaining 12% was sold
Soil conservation measures	75% given free of charge; the remainder through in-kind exchange
Tomato germplasm and management	57% was given for free; the remaining 43% was sold
Soil fertility management	67% was given for free; the remainder was exchanged

Data on types of exchanges in Lushoto site (Table 2.4) further reveal that most exchanges occurred at no cost to adopting farmers. This represents a positive trend with regard to maximizing access by resource-poor farmers. However, while knowledge-intensive natural resource management technologies are never characterized by cash exchanges, 12 to 43 percent of exchanges of cash crop technologies are. This suggests that financial barriers to technology access may exist for those technologies that can make the most immediate livelihood impact.

Lessons learned

Lessons learned from efforts to develop methods to track technology spread and impacts are several:

- Technological innovation often involves substantial spin-offs, both positive and negative; failure to identify and address these can reduce demand for the technology or introduce a set of problems that are propagated along with the technology. Identifying them provides an opportunity for corrective measures to be designed and tested and for "linking" multiple technologies for improved impact.
- "Sharing biases" within rural communities can propagate inequitable access to technologies, irrespective of efforts by extension agencies to work with equal numbers of male and female farmers. Systematically tracking sharing patterns can help to identify such biases and to design and test strategies to overcome them for improved adoption and equity.
- The new approach systematically integrates farmers' perceptions and experience on the introduced technology into the formal methodology, broadening the scope of what is learned and integrating farmer recommendations into research and dissemination strategies designed to overcome identified problems.
- The new approach provides an opportunity for adapting the introduced technology to address its negative effects and better fit the targeted farming system, by identifying local innovations introduced during the technology's spontaneous spread or proactively identifying adoption niches and negative impacts.

Missing links

While substantial progress has been made in identifying effective approaches for enabling livelihood improvements while also countering the degradation of resources at farm level, a few key areas of methodological innovation remain to be explored. These include the following:

1. There is a need to consider how to tap into a wider set of opportunities and drivers in designing interventions, so as to tap into potential motivating factors (e.g., market outlets) and to move from a reactive to a proactive approach in supporting farmer innovation. The choices of interventions and innovations were often based exclusively on local preferences, without considering wider market opportunities and policy drivers and the existing institutional capacity to scale up interventions beyond pilot sites. By placing attention on wider market opportunities and policies shaping farmer behavior, it may be possible to tap into wider motivating factors and thus support more widespread adoption. By embedding innovation processes within existing institutions, it would be possible to embed the innovations—and the innovation process itself—within organizational structures capable of sustaining the innovation process within and beyond benchmark sites.
2. Systems optimization through the use of models requires an analytical simplification of the system that may depart from real life decision processes and management principles. The more detailed the analysis of system features, community needs and preferences, market opportunities, and drivers of change, the more likely that optimization models and participatory optimization processes will be effective. More effort is needed to develop and test cost-effective methods that simultaneously enhance system understanding by farmers and researchers while targeting "best bet" facilitation processes for system change (including policy reforms).
3. One critical gap was in the development of approaches for building on the knowledge and skills of innovative farmers (a source of learning and innovation) to bring change across a wider area, and thus achieve watershed-wide farm productivity gains. Most "early innovators" are either isolated from others, lacking the mechanisms or motives to support innovation at a wider scale, or have unique characteristics that enable them to take risks and try out new innovations—thus limiting the extent to which proven innovations will be automatically accessible to others. The development and testing of methods for linking such early innovators with the needs and capacities of a wider set of actors at village and landscape scales is needed, including mechanisms to incentivize efforts expended for the collective good rather than for personal gain.
4. There is a need to explore how to move beyond "linked technologies" to "linked innovations." The success of efforts to couple technological innovations with credit facilities and with social and governance innovations, and the tendency for farmers themselves to employ social innovations when adopting new technologies (German et al., 2006a), illustrate the promise of linking social and technical innovations. More effort is needed to bring social scientists and marketing specialists into efforts to support farmer innovation, so as to identify and test social

and marketing innovations that work in synergy with technologies in enhancing impact at farm level.

5. More can be done to explore opportunities for enhancing impact by going further "downstream" along the farmer-to-farmer dissemination pathway. The technology spillover methodologies identified a number of factors constraining adoption and positive impacts, both social and biophysical, which could be the subject of further experimentation and innovation. This is likely to be a very fertile area of technological and social innovation, and thus impact, as it is informed by actual adoption bottlenecks. While the program engaged in a few such innovations, for example to address germplasm constraints to the spread of banana in Lushoto BMS, much work remains to be done in this regard.

Conclusions

The crop–livestock farming systems in the highlands of eastern Africa are characterized by low-input farming, heavily reliant on the recycling of internal resources. Resource degradation is aggravated by high human and livestock population densities, which lead to overgrazing, nutrient mining, erosion, and water depletion. Intensification and diversification of these systems is one important pathway for improving rural livelihoods. This chapter sought to share AHI experiences in farm-level intensification and diversification through approaches for characterizing systems and clients (and thus potential adoption niches), supporting farmer experimentation, reaching larger numbers of farmers in benchmark sites through technology dissemination, and ex-post tracking of the spontaneous farmer-to-farmer spread of innovations. Ultimately, a combination of strategies is needed at different stages of an innovation process to effectively support farmers to generate greater returns from a limited resource base. Farmer field schools, farmer-managed experimentation, and farmer-led dissemination enhance farmers' capacity to make informed choices and test them through an experimental learning approach, whereas researcher-led development of decision support tools and documentation of farmer-to-farmer dissemination provide a means to identify strategic interventions to enhance impact for further testing with farming communities.

Farmers' choices of livelihood strategies substantially influence crop and livestock decisions and welfare and resource outcomes. Based on our experience to date, adoption of technological innovations often depends on a few key factors, including: 1) the type of technical and material support farmers receive from extension and research; 2) the level of familiarity of farmers with the suggested interventions; 3) the demands placed on limited resources by innovations; 4) the associated benefits that are derived, both financial and other; and 5) the time required to derive these benefits. Bottom-up processes for engaging communities are essential in integrating these and other considerations into the innovation process, and in motivating farmers to individually and collectively address production constraints and capture new livelihood opportunities.

There are two possibilities for achieving wider impact from the innovations presented in this chapter. The first is to scale up the *actual technological or social innovations* that were successfully employed to intensify or diversify local farming systems. The second is to scale up the *approaches and tools* used to generate these innovations or to target specific niches for further uptake among development agencies. The latter approach is the preferred approach for accounting for the diversity in local resources, preferences, and conditions. Each approach will be treated in greater depth in Chapter 6.

Notes

1 As observed in Areka, these subsidies also entail risks if they are not continued, given the tendency for farmers to invest less in organic nutrient management when using chemical fertilizers. If farmers have limited ability to continue purchasing these inputs, this means they will have less fertile soils to fall back on.
2 The model had three basic modeling components: 1) an objective function, which minimizes or maximizes a function of the set of activity levels; 2) a description of the activities within the system, with coefficients representing their productive responses; and 3) a set of constraints that define the operational conditions and the limits of the model and its activities.

References

Adamo, A. (2001) Participatory agricultural research processes in Ethiopia using farmers' social networks as entry points. *CIAT Occasional Papers Series* No. 33. Cali, Colombia: CIAT.

Alumira, J. and A.O. Awiti (2000) *The African Highlands Eco-region: Situation Analysis.* Kampala, Uganda: AHI.

Amede, T. (ed.) (2003) *Natural Resource Degradation and Environmental Concerns in the Amhara Regional State, Ethiopia: Impact on Food Security*. Bahir Dar, Ethiopia: Ethiopian Society of Soil Sciences.

Amede, T., T. Belachew, and E. Geta (2001) Reversing the degradation of arable land in Ethiopian highlands. *Managing African Soils* No. 23. London: IIED.

Amede, T. and R. Kirkby (2004) Guidelines for integration of legume cover crops into the farming systems of the east African highlands. In: A. Bationo (ed.), *Managing Nutrient Cycles to Sustain Soil Fertility in Sub-Saharan Africa.* Nairobi, Kenya: Academy Science Publishers, pp. 43–64.

Amede, T., A. Stroud, and J. Aune (2004) Advancing human nutrition without degrading land resources through modeling cropping systems in the Ethiopian Highlands. *Food and Nutrition Bulletin* 25(4): 344–353.

Amede, T., R. Kirkby, and A. Stroud (2006) Intensification pathways from farmer strategies to sustainable livelihoods: AHIs' experience. *Currents* 40/41: 30–37. Uppsala, Sweden: SLU.

Assefa, A. and T. Fenta (2006) Harnessing local and outsiders' knowledge: Experiences of multi-stakeholder partnership to promote farmer innovation in Ethiopia. *Prolinnova Working Paper* No. 12.

German, L., J.G. Mowo, and M. Kingamkono (2006a) A methodology for tracking the "fate2 of technological innovations in agriculture. *Agriculture and Human Values* 23: 353–369.

German, L., J. Mowo, M. Kingamkono and J. Nuñez (2006b) Technology spillover: A methodology for understanding patterns and limits to adoption of farm-level innovations. *AHI Methods Guide* A1. Kampala, Uganda: AHI.

Kelly, V., B. Diagana, T. Reardon, M. Gaye, and E. Crawford (1996) Cash crop and food grain productivity in Senegal: Historical view, new survey evidence and policy implications. *MSU International Development Paper* No. 20.

Lyamchai, C.J., M.N. Kingamkono, J.G. Mowo, and L. German (2006) Farmers innovations in natural resource management: Lessons and challenges from Lushoto, Tanzania. In: T. Amede, L. German, S. Rao, C. Opondo and A. Stroud (eds.), *Integrated Natural Resource Management in Practice: Enabling Communities to Improve Mountain Livelihoods and Landscapes*. Kampala, Uganda: AHI.

Morrison, K.D. (1994) The intensification of production: Archeological approaches. *Journal of Archaeological Methods and Theory* 1(2): 111–159.

Mowo, J.G., S.T. Mwihomeke, J.B. Mzoo, and T.H. Msangi (2002) Managing natural resources in the West Usambara Mountains: A glimmer of hope in the Horizon. Mountains High Summit Conference for Africa, Nairobi, Kenya, May 6–10, 2002. Available at: http://www.mtnforum.org/sites/default/files/pub/213.pdf (accessed May 29, 2009).

O'Neil, R.J. (1995) Farmer Innovation and Biological Control. *Midwest Biological control Newsletter* 2(3): 1–2.

Pontius, J.C., R. Dilts and A. Bartlett (2002) From farmer field school to community IPM: Ten years of IPM training in Asia. RAP/2002/15, FAO Regional Office for Asia and the Pacific, Bangkok.

Reardon, T., C. Barrett, V. Kelly, and K. Svadog (1999) Policy reforms and sustainable agricultural intensification in Africa. *Development Policy Review* 17(4): 375–86.

Smaling, E.M.A. (1993) Soil nutrient depletion in Sub-Saharan Africa. In: H. Van Reuler and W.H. Prins (eds.), *The Role of Plant Nutrients for Sustainable Food Production in Sub-Saharan Africa*. Leidschendam: Dutch Association of Fertilizer Producers. Leidschendam, pp. 53–67.

Stroud, A. (1993) *Conducting On-farm Experimentation—A Guide Book*. Cali, Colombia: CIAT.

Stroud, A. (2003) Linked technologies for increasing adoption and impact. *AHI Brief* A3. Kampala: AHI.

Tarawali, S., J.D.H. Keatings, J.M. Powel, P. Hiernaux, O. Lyasse, and N. Sanginga (2004) Integrated natural resources in West African crop–livestock systems. In: T.O. Williams, S.S. Tarawali, P. Hiernaux and S. Fernandez-Rivera (eds.), *Proceeding of the International Conference on Sustainable Crop–livestock Production for Improved Livelihoods and Natural Resource Management in West Africa*. Ibadan, Nigeria: IITA, pp. 349–370.

Tiffen, M. (2003) Transition in sub-Saharan Africa. *World Development* 31(8): 1343–1366.

Thornton, P.K, R.L. Kruska, N. Henninger, P.M. Kristanson, R.S. Reed, F. Ateno, A. Odero, and T. Ndegwa (2002) *Mapping Poverty and Livestock in the Developing World*. Nairobi, Kenya: International Livestock Research Institute.

Van Keulen, H. and J.B. Schiere (2004) Crop–livestock systems: Old wine in new bottles. In: *New Directions for a Diverse Planet: Proceedings of the 4th International Crop Science Congress.* Brisbane, Australia, September 26–October 1, 2004. Available at: www.cropscience.org.au (accessed May 29, 2009). Waters-Bayer, A. and W. Bayer (2009) Enhancing local innovation to improve water productivity in crop–livestock systems. *The Rangeland Journal* 31: 231–35.

Wickama, J. M. and J.G. Mowo (2001) Using Local resources to improve soil fertility in Tanzania. *Managing African Soils* No. 21. London: IIED.

3

PARTICIPATORY INTEGRATED WATERSHED MANAGEMENT

Laura German, Waga Mazengia, Simon Nyangas, Joel Meliyo, Zenebe Adimassu, Berhanu Bekele, and Wilberforce Tirwomwe

Context and rationale

Most NRM interventions in the eastern Africa region tend to focus on farm-level innovations and facilitate change through individualized decision processes. This has left many NRM problems unresolved, including natural resource conflicts, negative transboundary interactions among neighboring farms and villages, absence of collective action (CA) in addressing common concerns, and the degradation of common property resources. AHI sought to address these challenges through methodological innovation at landscape scale. This work was conducted under the conceptual umbrella of participatory integrated watershed management (PIWM). Conceptual evolution of this approach has gone hand in hand with methodological innovations and research findings at site level. An introduction to the conceptual grounding of the approach as interpreted within AHI will clarify reasons for the specific methodological innovations which follow.

Interest in the watershed management approach has increased in recent years in response to water deficits in urban and lower catchment areas (Constantz, 2000; van Horen, 2001), and as a framework for enhancing livelihoods through more efficient and sustainable use of water and other natural resources in rain-fed areas and upper catchments (De and Singh, 1999; Shah, 1998; Turton and Farrington, 1998). In recognition of the causal linkage between NRM and poverty reduction and between water and other natural resources (CGIAR, 2002), watershed approaches are gaining in popularity in a host of countries in Africa, Asia and Latin America. The government of India has chosen to invest in rural development through the provision of public finance to community-based watershed management (Shah, 1998; Turton and Farrington, 1998). Several eastern African governments are considering similar approaches.

Despite this upsurge in interest in watershed management, the large range of projects and approaches falling under this umbrella has led to confusion in goals, lack of consistency in approaches, and limited success in putting the concept into practice (Bellamy et al., 1998; Rhoades, 2000; Shah, 1998). Current practice in the eastern Africa region is biased toward soil and water management for agriculture despite a wide range of NRM concerns among local actors. Approaches for operationalizing watershed management in ways responsive to local NRM concerns and attentive to trade-offs among system components and user groups are therefore sorely needed.

Time–space interactions between plots and common-pool resources, lateral flows of materials (water, nutrients, pests), and interdependence between users in terms of resource access and management, require decision-making and intervention strategies beyond the farm level (Johnson et al., 2001; Knox et al., 2001; Ravnborg and Ashby, 1996). The latter requires effective mechanisms to ensure *participation* of diverse interest groups and stakeholders, as well as *integrated* decision-making that acknowledges system linkages (among water, soils, crops, trees, and livestock) and multiple spin-offs from any given intervention. "Participation" and "integration" are two concepts that have helped to ground the conceptual evolution and methodological innovation of watershed management in AHI.

Participatory watershed management may be defined as a process whereby users define problems and priorities, set criteria for sustainable management, evaluate possible solutions, implement programs, and monitor and evaluate impacts (Johnson et al., 2001). Participation implies that broad-based livelihood concerns will guide the watershed management agenda, where water and soil are likely to be only two of many important components. Watershed development is known to work best when there is a perceived deficiency in a vital resource, when integrated with other means of enhancing livelihoods, and when benefits of NRM are localized (Bellamy et al., 1998; Datta and Virgo, 1998; Turton and Farrington, 1998). AHI therefore decided to ground methodological innovations at landscape scale in a systematic assessment of local priorities beyond the farm level, and in mechanisms to unblock pathways from *motivation* (local concern) to *action* (solutions) in addressing felt needs. Enabling such processes has meant crafting and testing methodological innovations for ensuring effective representation in decision-making at watershed level; fostering collective contributions to common NRM problems; supporting the negotiation of solutions among groups with divergent interests to minimize the social and environmental costs of current and alternative land uses; equitably monitoring benefits capture; and reformulating by-laws to align the behavior of individuals with collective decisions. While such social and institutional dimensions are part and parcel of participatory integrated watershed management, many of these dimensions are captured in Chapter 4 ("Participatory Landscape Governance") owing to the depth at which these issues were explored and the scope of lessons learned.

As with the participation concept, *integrated* watershed management may be understood in a number of ways (German, 2006). As presented in Chapter 1, "component integration" emphasizes the interrelatedness of components and acknowledges the impacts of changes within any given component on other parts of the system. Within the agricultural research paradigm, "system components" are understood to roughly correspond to the boundaries of biophysical disciplines: crops, livestock, trees, and soil. While these components capture much of the "structure" of single plots or farms, they are inadequate for capturing structures and processes at landscape level. While water is present at farm level as a resource for agricultural production, its non-productive function (water for domestic rather than agricultural use) only becomes visible at landscape level. It is at this level where the sum total of management practices on individual plots and farms becomes apparent in terms of the effects on the quality and quantity of water in springs and waterways. Yet the social function of water remains invisible within agricultural research and development institutions, whose institutional mandates are restricted to agricultural production. At landscape level, public and common property resources such as forests, waterways, and communal grazing areas become visible, requiring one to think about more collective decision-making processes. In short, component integration implies moving beyond component-specific objectives (i.e., maximizing the yield of edible plant products) to broader systems goals whereby the relationship between components—as opposed to the individual components alone—becomes a foundation of professional practice. This might include optimizing returns to diverse system components (tree, crop, and livestock) or increasing the yield of any of these components without depleting system nutrients or water. Similarly, the sectoral and constructivist integration concepts featured in watershed-level work in the synergies fostered between social, biophysical and policy innovations on the one hand, and in efforts to systematically identify and integrate diverse interest groups in the innovation process on the other.

The concepts of participation and integration were instrumental to methodological innovation in AHI, and form a conceptual thread that is intricately woven throughout the thinking and methodological interventions presented in the text that follows. Key methodological innovations to be covered in this chapter include methods for landscape-level diagnosis (watershed delineation and characterization, participatory diagnosis and prioritization), planning (at "community" and R&D team levels), participatory management of change, and approaches for putting empirical research at the service of farmers and policy makers to support decision-making. Key knowledge gaps and remaining challenges in methodological innovation for participatory landscape-level innovation are also highlighted.

Watershed delineation

Watershed delineation is the process of defining, identifying, and marking biophysical boundaries to be used for subsequent interventions. Watershed delineation in each site was needed to inscribe both the collection of baseline data on social and biophysical characteristics of the watershed and the eventual innovations to be tested. It was also needed to define stakeholders and to enable future impact assessment of interventions to follow.

While the standard approach is to delineate watersheds on the basis of strict hydrological boundaries, many landscape-level NRM problems involve spatial and temporal processes that have no bearing on hydrology or hydrological boundaries per se. Therefore, the watershed concept employed within AHI has been a flexible one, with a provisional boundary set to guide baseline studies and participatory diagnosis but subsequent flexibility in boundary delineation based on the spatial characteristics of specific challenges to be addressed and the social dynamics therein. Context also matters in the way in which specific landscape-level NRM problems, such as free grazing, are manifested in the different sites. Therefore, ways in which watersheds were defined vary across AHI benchmark sites. This section describes and discusses the methods and approaches used for watershed delineation, and the relative strengths and weaknesses of each.

Approach development

Approach 1—Hydrological delineation

The approach to delineation employed in Gununo watershed followed most closely the standard approach using strict hydrological boundaries. The output, in the form of a digital elevation model, is presented in Figure 3.1. The potential benefit of the approach is its effectiveness in encompassing the biophysical processes involved in effective soil and water management. Employing hydrological units for watershed delineation can enable soil conservation structures and drainage ways throughout the catchment area to be interconnected, thus minimizing the potential negative effects of isolated conservation structures on neighboring cropland (through their effect in shifting drainage patterns). Furthermore, by taking the catchment as the implementation unit, if all households were to conserve their fields, structures lower on the landscape would be protected from excess run-off from upslope practices. It also facilitates the identification of areas to be targeted for soil and water conservation for optimum returns (in terms of both quality and quantity) to water resources affected by these interventions. Finally, the aggregate effect of structures on water resource recharge can be enhanced. However, this approach also had its weaknesses. Watersheds are not meaningful units for mobilizing collective action, for example. Furthermore, the process of "dissecting" social units (villages, kin, leadership domains) falling within and outside watershed boundaries, can cause resentment among those who were excluded and undermine collective action in addressing common watershed problems.

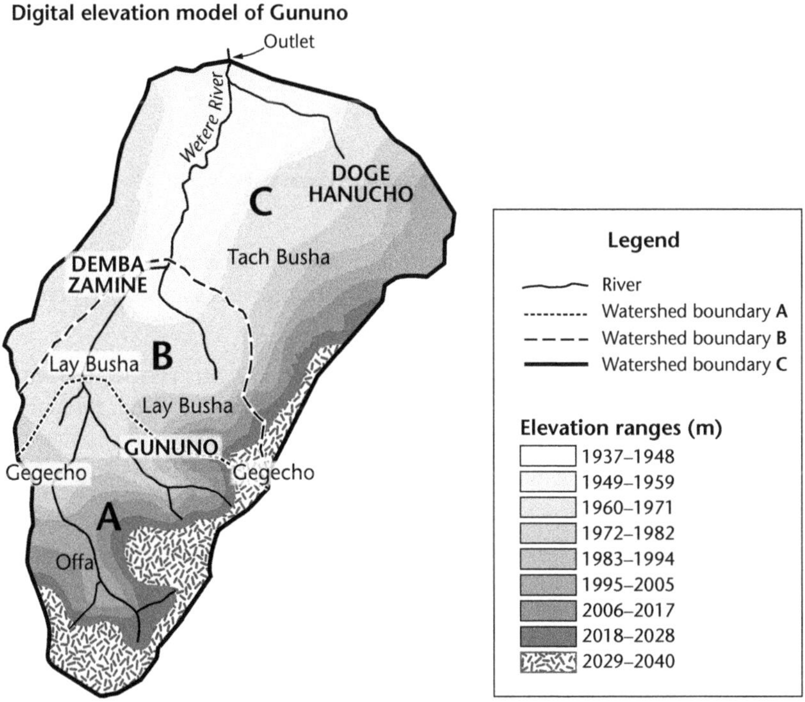

FIGURE 3.1 Digital elevation model illustrating hydrological boundaries and features of Gununo watershed, Ethiopia

Examples of watershed issues that conform and do not conform to hydrological boundaries are provided in Box 3.1.

BOX 3.1 WATERSHED ISSUES THAT DO AND DO NOT CONFORM TO HYDROLOGICAL BOUNDARIES

"Watershed" issues conforming to hydrological boundaries:

1. Soil erosion and excess run-off. Flows of soil and water across the landscape follow topographical variations within hydrological units, and require collective action within hydrological units to manage upslope–downslope interactions and achieve "aggregate" benefits from enhanced infiltration.
2. Spring degradation. Several causal processes leading to spring degradation conform to hydrological units, including siltation and the effects of tree species selection and forest cover on water quantity and quality, and

require collective solutions to manage interlinked landscape processes. The latter include the influence of land use on springs, and consequences of spring protection on lower slopes.

"Watershed" issues that do not conform to hydrological boundaries:

1. Crop and livestock pests. Vertebrate and invertebrate pests wander freely irrespective of hydrological boundaries, with crop pests such as porcupine roaming up to 14 km in a single night.
2. Conflicts over resources in protected areas. While isolation of protected area resources from certain land uses has a direct influence on hydrology through its effects on land use, related conflicts and co-management efforts conform to the spatial dimensions of the protected areas themselves, not watersheds.
3. Free grazing. Conflicts resulting (or opportunities lost) from free grazing have spatial dimensions related to the distribution of grasslands and the administrative units from which grazing households emanate.

Approach 2—Administrative delineation

The second approach, employed by the Lushoto site team, utilized political–administrative boundaries to demarcate the target area. Although the project used the term "watershed" to refer to the area, a hydrological approach to demarcation was not adopted because of the difficulties that would be experienced in community mobilization. Political–administrative boundaries of individual villages were instead considered, with the entire (micro-)watershed encompassing six villages (Figure 3.2).

The use of administrative boundaries in watershed delineation had the important advantage of facilitating the mobilization of watershed residents around issues of collective concern. While the presence of areas within the watershed but falling outside hydrological boundaries may complicate efforts to coordinate soil and water management at landscape level (Box 3.2), this was found to be of minor concern. Other watershed problems having a landscape dimension but not conforming to hydrological boundaries (i.e., free grazing, trees incompatible with crops on farm boundaries, people–park interactions, and pest control) will be less negatively affected by taking administrative boundaries as the basis for "watershed" delineation, provided flexibility is used when determining how many administrative units to involve in addressing the issue. Controlling pests and free grazing at landscape level, for example, requires collective action over a larger area than solving boundary conflicts between adjacent landowners and flexibility is therefore required not only in how boundaries are defined but in the spatial scale over which watershed innovations are organized.

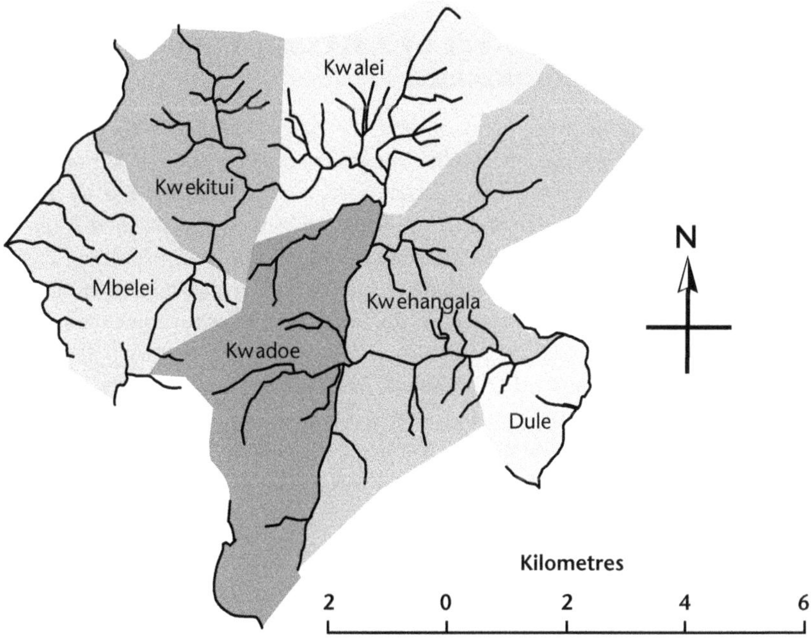

FIGURE 3.2 Baga watershed demarcated using (village) administrative boundaries

Approach 3—"Hybrid" delineation: Hydrological and village boundaries

The major criterion used for the third delineation approach was the hydrological boundary. However, a flexible approach to boundaries was taken to include villages that were dissected by the hydrological boundary, so as to include parts of villages falling outside hydrological boundaries in the delineated watershed. The advantage of this approach is that it accommodates both biophysical and administrative boundaries, which are important for soil and water management, community mobilization, and addressing landscape-level problems whose spatial dimensions extend beyond the hydrological boundaries of the watershed. The disadvantage of this approach is that delineation of the target area tends to be subjective, lacking strict criteria to include or exclude different areas. Ultimately, delineation becomes an art rather than a science, which must be flexibly adapted to emerging challenges and the spatial scale over which these are manifest both socially and physically (Box 3.3).

Approaches used for watershed delineation in AHI benchmark sites are summarized in Table 3.1.

BOX 3.2 THE IMPORTANCE OF A FLEXIBLE CONCEPT OF WATERSHED BOUNDARIES

Limited coverage of areas lying outside the hydrological boundary of the watershed can hinder implementation for either social or biophysical reasons. Figure 3.3 below illustrates two adjacent watersheds (A and C). B is an area hydrologically part of watershed C, but included as part of watershed A during watershed delineation. This is because the support of local institutions and local government residing in area B are crucial for the effective implementation of watershed innovations in area A. At the same time, construction of soil erosion and run-off controlling measures in area B will not be effective unless the upstream part of watershed C is treated. Depending on topography, run-off generated from the upper part of watershed C could also destroy soil conservation structures in area B unless also treated. There may therefore be a need to expand certain watershed interventions among landowners in watershed C so as to improve the effectiveness of soil conservation interventions in area B.

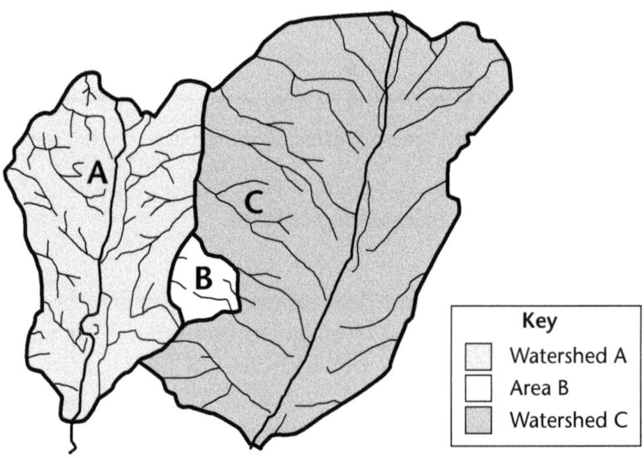

FIGURE 3.3 Relationships between adjacent watershed units and the need for a flexible interpretation of watershed boundaries

TABLE 3.1 Delineation approaches used by different benchmark sites

Approach used to delineate watershed boundaries	*Benchmark Site*				
	Lushoto	*Ginchi*	*Areka*	*Kapchorwa*	*Kabale*
Hydrological			✓		
Administrative	✓			✓	✓
Hybrid		✓			

BOX 3.3 MANAGING HYDROLOGICAL AND ADMINISTRATIVE BOUNDARIES: BIOPHYSICAL AND SOCIAL "BALANCING ACTS"

Case 1—Biophysical balancing acts

Watershed boundaries are generally set according to hydrological processes and units, with a focus on soil and water conservation. When taking a participatory approach to watershed management, however, other types of biophysical issues emerge that are not readily inscribed by hydrological boundaries. Negative effects of free grazing such as crop destruction and constraints to agricultural intensification are examples. Multiple tenure systems that overlap in time and space define the movement of livestock in Galessa watershed. While all land is publicly owned in Ethiopia, in practice all cropland is allocated to individual households. During the cropping season, a household's livestock may only graze in "private" outfields owned by them and a small group of adjacent households which lie fallow (restricted access grazing). During the dry season, however, after all crops are harvested, outfields are managed as open-access resources and any given household can graze their livestock anywhere inside or outside of the watershed. No rules govern livestock movement on one's own fields, and freely roaming livestock come from villages inside and outside the watershed. Efforts to intensify outfields through integrated interventions (soil conservation structures, high-value multi-purpose trees, or perennial crops) must involve decisions by all outfield users to restrict livestock movement, which would otherwise destroy technological innovations. Outfield intensification and free grazing are therefore "watershed" or "landscape" issues whose boundaries and solutions extend far beyond the arbitrary confines of the watershed.

Case 2—Social balancing acts

Another form of balancing act relating to watershed boundaries involves human motivations to participate or "opt out" of any innovation. Two cases help to illustrate this dynamic. In Ginchi, spring development and value addition through a shift from ware to seed potato were very popular interventions among watershed residents. The watershed falls inside larger administrative units (Peasant Associations—PAs) which were not fully covered by these activities. When watershed meetings were called, PA leaders residing outside the watershed held mandatory meetings on the same days to "sabotage" watershed activities as a form of protest for their non-inclusion. The team therefore had to expand membership in some activities to adjacent villages as a means of managing the social challenges faced.

The second case comes from Kapchorwa, where four villages were involved in watershed management activities. Free grazing came up as an issue of concern to them. However, managing this problem required curtailing access to communal areas for their own livestock as well as for non-participating villages. Non-participant households were in effect asked to participate in an activity with detrimental effects to their livelihood without otherwise benefiting from other watershed activities bringing concrete benefits, and therefore had no incentive to engage in collective action. The team was therefore challenged to come up with innovative ways to include them in a broader set of activities, so as to foster a stronger collaborative spirit among watershed villages. A two-pronged strategy was adopted: to seek technological alternatives (i.e., alternative feed) to minimize the costs of cutailing free grazing, and to invite these households into other watershed management activities bringing more concrete benefits.

Lessons learned

The following lessons may be distilled from the application of different approaches and their consequences for subsequent stages of implementation:

- Delineation together with local leaders enabled both parties to take cognizance of the landscape dimensions of NRM problems and the magnitude of degradation experienced in watershed villages, and heightened local ownership in the activities to come.
- It is difficult to strictly follow hydrological boundaries in delineating watersheds. Delineation may be carried out on the basis of social dynamics, administrative boundaries, hydrological boundaries, boundaries of landscape-level NRM problems that do not conform to hydrological boundaries, or a combination of factors. When employing combined criteria, it is possible for delineation to accommodate both biophysical and social processes, thus facilitating implementation.
- It is important to let the context—in terms of the specific dimensions of landscape-level NRM problems found within each particular site or niche—determine how flexibility in boundary definition will be defined. This implies keeping a flexible definition of boundaries during planning and implementation stages to ensure that the spatial dimensions of identified problems are considered in the intervention area.
- Regardless of whether the "watershed" is delineated according to administrative or biophysical criteria (or both), the boundaries of any given intervention should be kept flexible to accommodate social or biophysical influences from outside the pilot area and to enable them to be adapted to the spatial configurations of issues subsequently identified during planning and implementation.

Watershed characterization

During watershed characterization, biophysical and socio-economic baseline data is collected prior to intervention to enable R&D teams and communities to measure progress during implementation, and to identify socio-economic and environmental "hotspots" and opportunities for intervention. Collection of baseline data is crucial for organizations specializing in methodological innovation such as AHI, as it facilitates subsequent assessment of impacts from diverse innovations.

Approach development

Socio-economic aspects of watershed characterization

Household surveys using pre-tested questionnaires were carried out with a representative number of households in AHI watersheds to gather basic information on the five capital assets (human, social, natural, physical, and financial capital), and on household livelihood portfolios and related constraints. Households were selected using purposive sampling techniques based on household wealth status, as determined through standard participatory wealth ranking methods (Rietbergen-McCracken and Narayan, 1998). For a summary of data collected, see Box 3.4.

BOX 3.4 DATA COLLECTED IN HOUSEHOLD SURVEYS IN AHI BENCHMARK SITES

Human capital

- Household demographics (family size, gender, age, education, labor force, disability)
- Awareness of soil erosion and other NRM challenges

Social capital

- Conflict resolution mechanisms
- Perceived importance, levels of enforcement and effectiveness of different by-laws in solving identified watershed problems; awareness of by-law formulation processes
- Significance of and access to resources through kin relations
- Membership and role within local institutions, and benefits derived from the same

- Importance of different local institutions, and degree to which they help meet livelihood objectives
- Barriers to, and willingness to invest in, diverse collective action institutions and activities
- Coping strategies (sources of assistance) and assistance given to others during financial crises

Natural capital

- Distance to seasonal and year-round potable water sources
- Source of potable water, and observed changes in springs/rivers over time
- Landholdings (size, number of plots, soil quality, landscape location, conserved land) and perceived tenure security
- Access to irrigation water
- Livestock holdings (number, type, and breed)
- Trees and woodlots (species and area), and changes in tree diversity/cover over time
- Common property resource access (grazing land, forest products)
- Changes in farming system, yields, and productivity over time
- Energy access/use

Financial capital

- Income from different sources (crops, livestock, trees, and off-farm)
- Household investments in order of importance
- Changes in income sources and investments over time
- Loans received and sources

Physical capital

- Levels of adoption of different introduced technologies
- Housing, sanitation
- Tools, equipment, livestock structures
- Transportation and communication

Livelihood portfolios and constraints

- Major constraints to improved livelihoods and agricultural production
- Most important crops/on-farm activities in household livelihood portfolios
- Crop/livestock pests and diseases most affecting livelihoods

Wealth- and gender-disaggregated analysis facilitated the identification of enterprises and constraints common to different groups. In most sites, R&D teams also identified and characterized local institutions that either currently influence NRM or might play a role in NRM in the future. Local institutions were given an important consideration during characterization of the watershed because they were assumed to be important for community mobilization and technology dissemination. The characterization also involved an identification of diverse types of institutions, from formal groups to local norms and by-laws, traditional beliefs influencing NRM and influential leaders. The description of each institution included its function, its influence on community well-being, how respected it is by different social groups (by gender, wealth, and age), and its possible role in NRM. Identified institutions, classified according to their functions, are summarized in Table 3.2 (see also Mowo et al., 2006 for details).

TABLE 3.2 Local institutions in Areka, Ginchi and Lushoto benchmark sites

Type of institution	*Lushoto site*	*Ginchi site*	*Areka site*
Faith-based	Church groups, Mosque groups, traditional healers	*Mahiber* and *Senbete* (Orthodox followers), *Jabir / Jarssuma*	*Mahiber* and *Senbete* (Orthodox followers)
Financial	Rotational credit associations	*Edir, Ek'ub*	*Edir, Ek'ub*, Meskel Banking
Agricultural	Rotational livestock associations	*Ribi*, Sharecropping, Contracting	*Hara, Kota*, Sharecropping, Contracting
Collective action for heavy tasks	*Ngemo*	*Debo*	*Debo, Zaye*
Other		*Quallu*	

TABLE 3.3 Local institutions most linked to livelihood goals by wealth category (German et al., 2008)

Type of Institution	*Areka (% respondents listing institution)*			*Ginchi (% respondents listing institution)*		
	Low	*Medium*	*High*	*Low*	*Medium*	*High*
Labor sharing (*debo/zaye*)	27.3	0.0	0.0	19.4	33.3	28.6
Livestock sharing (*hara/ribi*)	27.3	0.0	0.0	12.9	6.7	14.3
Revolving fund (*equb*)	9.1	0.0	0.0	9.7	6.7	7.1
Contracting	0.0	0.0	0.0	3.2	0.0	7.1
Sharecropping	72.2	50.0	6.5	48.4	40.0	42.9
Idir	27.3	5.6	9.7	0.0	0.0	0.0
Meskel banking	0.0	0.0	0.0	–	–	–
Kota	27.3	5.6	6.5	–	–	–

To understand which institutions were most valued for their social or economic functions, and therefore most likely to be effective in mobilizing collective action, interviewees were asked during household surveys which institutions "are most valued" or "contribute most to livelihood goals." From our experience, answers to the two questions were very different so they provided complementary information. Outputs for the second question are summarized in Table 3.3 for two Ethiopian sites. Results show the fundamental importance of shareholding to livelihoods, particularly for low-income households.

Biophysical aspects of watershed characterization

The biophysical characterization involved land resources assessment, including soil, water, vegetation, and types of crops grown. Local soil classes were identified using local knowledge and indicators across sites. To complement the local soil classification system, the FAO soil classification system (FAO-UNESCO, 1987) was used for one village in Lushoto and results extrapolated to other villages with similar soils. Water resources were characterized according to location and degradation status through the use of global positioning systems (GPS), ethno-historical accounts and physical observation. Participatory mapping techniques were employed to identify key land uses and the location of environmental "hot spots" or highly degraded areas. Aerial photos, topographical maps and satellite images were used to develop preliminary land-use maps and/or digital elevation models (DEMs). Outputs of these techniques included geo-referenced watershed maps (Figures 3.1 and 3.2, shown earlier), land-use types and their spatial extent, water resources location and status, slope classes

TABLE 3.4 Watershed characterization and baseline methods used in different sites

Scope and methods of watershed characterization	*Benchmark Site*[a]		
	Lushoto	*Ginchi*	*Areka*
Social:			
• Semi-structured questionnaires	✓	✓	✓
• Participatory evaluation of local institutions	✓	✓	✓
Biophysical:			
• GPS readings	✓	✓	✓
• Aerial photos and satellite images	✓	✓	✓
• Digital Elevation Model		✓	✓
• Water resource characterization	✓		
• Soil classification using FAO system	✓		
• Assessment of land use types	✓	(✓)[b]	(✓)
• Participatory mapping	✓	✓	

Notes:

a These characterization methods were not conducted in Kabale or Kapchorwa owing to the diversity in approaches being tested, donor funding and related commitments, and the stronger development orientation of partners.

b Parentheses are used where the method was applied but with less detail (i.e., percentage coverage of each land use type was not assessed).

(where DEMs were generated), and the location of highly degraded areas. Table 3.4 summarizes the methods used for watershed characterization in different sites. While water source characterization was not carried out during the watershed characterization phase in Ginchi and Areka, it was later included following the participatory diagnosis of watershed problems given the high priority of water quantity and quality to watershed residents in all sites. Soil classification was only carried out in Lushoto given the presence of a PhD student hosted by AHI and the expense associated with doing so in other sites.

It is important to consider how data are to be utilized once collected, so that watershed characterization does not remain a purely academic exercise. One such use is to gather baseline data for subsequent impact assessments. In this case, data should explicitly focus on variables or parameters expected to change—whether biophysical (productivity, biodiversity, hydrology), social (prevalence of conflict and cooperation), institutional (attitudes and practices of researchers and extension agents) or economic (household income and investments). Another use is for the effective targeting of interventions. This targeting may also cut across diverse areas of impact. Economic data may help to target interventions to address the production strategies of different households. Table 3.5, for example, provides an indication of the crop preferences of households from different wealth categories across the Baga watershed. This

TABLE 3.5 Average income from selected crops by wealth category in Baga watershed, Tanzania (Tsh)

Wealth category	*Tea*	*Tomato*	*Cabbage*	*Sweet pepper*	*Beans*	*Irish potato*	*Maize*	*Banana*
High	176,784	260,000	72,000	110,000	83,856	20,267	68,033	59,240
Medium	46,099	187,248	53,400	144,340	59,410	111,870	53,096	86,936
Low	22,684	113,643	95,180	38,907	66,124	112,840	28,163	86,085

TABLE 3.6 Standard deviation (SD) in household income from selected crops in Baga watershed

Wealth category	*Tea*	*Tomato*	*Cabbage*	*Sweet pepper*	*Beans*	*Irish potato*	*Maize*	*Banana*
High	218,392	364,623	144,810	124,499	50,650	22,521	30,024	54,543
Medium	50,433	105,790	53,712	145,051	17,515	102,893	9,462	69,851
Low	50,724	90,951	125,677	39,354	40,353	104,617	30,181	81,968

table suggests that support for the production and marketing of tomato, pepper, and potato are likely to have implications for a broad cross-section of the population. However, to improve the status of lower income groups, a focus on cabbage and banana (a crop with lower investment costs) may be warranted.

Looking solely at income averages may, however, be deceptive in assessing whether all villages and the poorest households will benefit from technologies that are highly dependent on access to prime cropland, such as valley bottoms. In the case of the highest value crops (tea, tomato, cabbage, pepper), standard deviations are significant—illustrating the high variability in income sources among households within any wealth category (Table 3.6). Targeting interventions to different households may therefore require understanding not only the most important income earners in the aggregate, but also key income earners for the poorest households.

Priority areas of intervention may also be derived from data on environmental "hot spots" or areas of extreme degradation. A participatory mapping exercise combined with detailed field observations helped to identify priority areas for intervention at watershed and village levels (see Plates 3 and 4).

Lessons learned

The diversity of approaches utilized and the extent to which collected information was utilized in subsequent stages enabled lessons to be learned on the characterization process, including the extent to which AHI approaches have added value to standard methods and procedures used in watershed characterization. These include the following:

- The need to balance costs and benefits of watershed characterization. While the integration of diverse methods has the potential to generate important data on the integrated nature of problems and their solutions, and to facilitate proper targeting of technological, social, and policy interventions, comprehensive characterization work requires time and resources and may generate fatigue within watershed communities. Therefore, characterization work must be justified by program requirements (e.g., baseline data for subsequent impact assessments), additionality (e.g., inability to solicit the same information through participatory techniques) and balanced by the need to effectively capture farmer enthusiasm at early stages of any watershed management initiative.
- The importance of an iterative approach to watershed characterization. Collection of voluminous data on the watershed prior to participatory diagnosis of problems of concern to local residents may represent an inefficient use of resources. A basic understanding of watershed boundaries and features is often sufficient at this phase, provided this is followed up with a more in-depth characterization of problems prioritized by watershed residents for intervention.
- The importance of considering social variables in watershed characterization. Research on variables such as local institutions, traditional beliefs and norms in NRM, and how residents rank local leadership (traditional, political, religious, and opinion leaders) may provide important information on the best means to mobilize the community for different types of activities. Including questions such as willingness to participate in collective action for different watershed activities, perceived land tenure security for different ecological niches, perceptions on the status of common property resources (e.g., rangelands, forests), and forms of social capital most essential to the livelihoods of different groups also provide important insights into watershed problems and solutions.
- The value of farmer participation in social and biophysical characterization, which can enhance understanding by the research team of important problems and opportunities to be captured within intervention strategies.

Participatory watershed diagnosis and planning

When agricultural research organizations have taken an interest in watershed management, the approaches used often place undue emphasis on soil and water conservation without integrating livelihood concerns and other priority landscape-level NRM challenges (e.g., crop destruction from free grazing, competition of fast-growing trees with springs and crops, or water resource degradation). Other NRM investments seek to maximize returns from specific components (trees, crops, livestock, or water) rather than from integrated interventions designed to bring multiple returns and synergies, and disseminate technologies in isolation from complementary social and policy interventions. Furthermore, research organizations tend to plan in isolation from local government, community-based organizations (CBOs), and NGOs. While some development agencies have evolved much more integrated approaches to NRM, common deficiencies remain in ensuring that diverse local 'voices' are effectively captured during planning processes.

Approaches used in AHI have attempted to overcome these limitations in a number of ways. First, collective and negotiated decision-making became part and parcel of watershed planning. Disaggregated watershed diagnosis and prioritization strategies were also tested in some sites as a means to identify approaches effective in capturing diverse or divergent perspectives. While there is still much to learn, we also strove to develop more integrated approaches to planning to address a wider range of issues through collective action and identify opportunities for fostering synergies between different system components (trees, crops, water, soil, livestock) and strategies (social, technological, policy, and marketing). Some planning strategies were also unique in fostering partnerships among complementary institutions—and in bridging institutional gaps between research and development agencies, different sectors (i.e., agriculture and water), and among agencies with livelihood and conservation mandates.

Approach development

Four different strategies for participatory watershed diagnosis and prioritization were tested in AHI benchmark sites. These are described in detail below along with their relative strengths and weaknesses.

Approach 1—Demand-driven approach to diagnosis and stakeholder engagement

This approach focused on enabling community members residing in a watershed area to articulate their concerns and to demand broader stakeholder engagement in support of subsequent actions. The approach consisted of the following steps:

1. Identify emerging leaders concerned about landscape or "watershed"-level NRM problems.
2. Carry out village-level meetings in all watershed villages to identify problems affecting farmers and their livelihoods in the watershed and prioritize the most urgent issues to be addressed.
3. Task villages with the formation of Village Watershed Committees (VWC) (see Plate 5).
4. Task villages with the selection of members from VWCs to serve on higher-level Parish Watershed Committees.
5. Task the Parish Watershed Committees to call a meeting with all VWCs; Local Councilors from village, parish and sub-county levels; the Local Council Chairperson from sub-county level; local opinion leaders; and

staff from relevant line ministries to map the watershed and assist the community in articulating demand for support from relevant actors.

6. Carry out a field visit with technical staff from district line ministries and Watershed Committees to the areas most affected by urgent NRM problems (i.e., excess run-off, landslides).
7. Hold meetings with Village Watershed Committees and technical staff to develop provisional work plans.
8. Conduct technical assessments of the areas most affected in each village with Village Watershed Committees to map the watershed and identify hotspots associated with key NRM problems.
9. Hold meetings at watershed level involving all stakeholders (including all watershed residents) to give feedback on the draft work plans and technical recommendations on ways to address priority problems, and harmonize the two work plans.

This approach enabled the community to own and fully participate in the process of planning and implementation, while also consolidating the commitment of other development agencies to support communities in collectively addressing their priority concerns.

Approach 2—Watershed entry through local leadership and local NRM structures

The second approach entailed working through established leadership structures and existing local NRM institutions with a history of involvement with development agencies to inculcate responsibility on their behalf for mobilizing communities for improved NRM. The steps in this approach included the following:

1. Hold district-level meeting with representatives of targeted sub-counties (in this case, Sub-County Chiefs, Secretaries for Production, Farmer Fora Chairpersons, National Agricultural Advisory Services (NAADS) Coordinators, and concerned farmers from villages in each sub-county) to build commitment, empower them with facilitation skills and generate a general strategy for supporting participatory NRM in their sub-counties.
2. Hold meetings at sub-county level to consolidate NRM institutions and initiatives in the sub-county through:
   a) Election of members of sub-county NRM committees by sub-county representatives participating in the above meeting, Local Council representatives from village level, and farmers with an active commitment to NRM.[1]
   b) Selection of priority areas for project intervention based on villages experiencing severe degradation or demonstrating the most commitment to NRM.

   c) Establishment of a schedule for monthly review and planning meetings by these newly constituted committees to evaluate progress on NRM strategies.
   d) Appointment of core teams from the sub-county committees to spearhead sensitization and the formation of other committees in each target village.
3. Task the core teams to mobilize village meetings for the purpose of sensitizing all village members on NRM and encouraging them to elect members of village-level NRM committees. During these meetings, each village identifies the major NRM challenges that have caused widespread misunderstanding or conflict at village level, and prioritizes the most pressing challenges that the project can help them address collectively.
4. Hold a series of meetings in each village to orient newly constituted village committees on their roles and responsibilities in NRM. This is done through joint reflection on what is required from them (their envisaged roles) to support their respective villages in mobilizing collective action to address previously identified priorities. Identified responsibilities may include awareness creation, mobilizing local residents to formulate NRM by-laws, the selection of demonstration sites within identified environmental hotspots, and conducting training needs assessments. These are then integrated into Natural Resource Management Planning Committee (NRMPC) work plans in support of village-level collective action.[2]
5. Hold joint meetings between village NRM committees and local government structures to enable a participatory process of by-law formulation to address identified watershed problems (Box 3.5), and to aid in compliance and enforcement of agreed responsibilities.

This approach builds the capacity of local government in supporting communities and ably fulfilling their responsibilities toward their constituents. It also builds the capacity of local institutions in articulating and addressing local concerns.

BOX 3.5 REFORMULATED AND HARMONIZED BY-LAWS IN RUBAYA SUB-COUNTY[a]

Soil and water conservation:

- Everyone shall dig water trenches (soil erosion structures) especially on hill-sides in their own land prior to any cultivation. Anyone who violates the above by-law will be liable to a fine, which will be decided by the sub-county (LC3) council, in collaboration with representatives of policy task forces (PTFs).
- Napier/Elephant grass and other grasses (and/or trees) shall be planted in landscapes where water trenches are not feasible, such as in very rocky or rugged terrain.

- Every farmer should consult neighboring landowners prior to breaking down the terrace or contour bund along the common land demarcations or borders.
- No one shall cultivate his/her land without digging water trenches and planting trees and grasses, to conserve soil and water in their own land.
- Prior to cultivating, everyone should excavate trenches and construct steps and "A" frames.

Those who violate these by-laws shall be fined Sh. 5,000; or else they will be forwarded to the LC3 council authorities for punishment.

Grazing:

- No one shall graze in the valley bottoms, irrespective of whether or not the land is one's own.
- Everyone shall graze in his/her own land, and if not, seek permission to graze in others' land. Any abandoned land—including hill tops—should be utilized for growing agroforestry trees.
- No one is allowed to come from another country and graze in Uganda. [Ref: Rwanda].

Those who violate these by-laws will be fined Sh. 10,000.

Water:

- Everyone who draws water from a communal water source or well shall cooperate with others in its cleaning or maintenance.
- Anyone utilizing land near a communal well, road, foot path or water trench, should reserve a stretch of 1–2 meters of uncultivated land between their land and the said communal structures.
- No one is allowed to graze or cultivate near water sources/wells, or wash clothes from them.

Those who violate this by-law will be fined Sh. 5,000.

Other:

- Burning of grasses, hillsides, weeds and trees is strictly prohibited (*Those who violate this by-law will be fined Sh. 10,000*).
- When cultivating, leave some reserve narrow strips of land along

boundaries, roadsides, livestock tracks, etc. (*Those who violate this by-law will be fined Sh. 5,000*).

- Whoever cuts down trees shall plant replacement trees (*Those who violate this by-law will be fined Sh. 5,000*).
- Every household shall cultivate fruits, such as avocados (*Those who violate this by-law will be fined Sh. 5,000*).
- Anyone who owns or rents land in another village should abide by the NRM by-laws obtaining in that village.

Note:

Village policy task forces (PTFs) should have representatives at LC3 (sub-county) level.

Approach 3—Socially-optimal watershed diagnosis to capture diverse "voices"

The third approach consisted of systematically capturing the perspectives of diverse social groups within the watershed, first through socially disaggregated focus group discussions and next through household surveys in which representatives of these different groups were purposively targeted. The approach consisted of the following key steps:

1. Contact local leaders to inform them of the project mandate and interest in supporting livelihoods and NRM in their areas of jurisdiction.
2. Conduct focus group discussions in each watershed village according to social categories likely to influence people's priorities in NRM, namely by gender, age, wealth, and landscape location (farmers with households and plots upslope and downslope, where relevant).[3] The following set of questions can be used as a guide for eliciting watershed problems:
   a) How have changes in the landscape and land use over time influenced your livelihood?
   b) Do your neighbors' on-farm management practices have any influence on your livelihood? How about the management of resources by neighboring communities?
   c) Are there any NRM problems that could benefit from collective action?
   d) Are there any problems associated with communal resources?
   e) Are there any conflicts associated with land or natural resource management (within or between villages)?

   Local leaders are singled out during this process and their views obtained through key informant interviews.

3. Generate a single list of identified watershed issues for the whole watershed.
4. Conduct participatory ranking of these issues according to disaggregated social categories (again, by gender, age, wealth, and landscape location), either in focus groups or through interviewing key informants from each village—ensuring that views are captured equally across all social categories.[4]
5. Analyze data in the office to generate average ranks by village, gender, age, wealth and—where relevant—landscape location, and highlight watershed issues of high priority across all social categories (Table 3.7).
6. Identify entry points based on Step 5, with attention to those key priorities that can bring the most immediate benefits to a majority of watershed residents to heighten their enthusiasm for future watershed innovations (Box 3.6).
7. Conduct a one-week planning session for research and development teams to explore the causal interactions among identified watershed themes and generate clusters of issues to be addressed through integrated solutions.[5] If research is to be conducted together with development interventions, research topics and protocols are also generated at this time.
8. Conduct participatory watershed planning involving all watershed residents. The process involves the following steps:
   a) Feedback of issues identified by the community, how they were prioritized differently by different social groups, which issues are ranked highly by all watershed residents.
   b) Presentation and discussion of constituted watershed "clusters," and the logic underpinning these groups.
   Note: In some sites, the teams subjected these ranks to community scrutiny and priorities emerging from socially disaggregated ranking caused issues of high priority by some groups to be subsumed in importance to issues considered more important by outspoken community members. We therefore recommend excluding corrections to identified priorities during these watershed planning fora.
   c) Solicitation of additional feedback, clarifications and inputs without letting the new feedback take precedence over the socially differentiated views captured beforehand.[6]
   d) Group work based on identified R&D clusters and related sub-themes to plan in detail for how to address the issues in an integrated manner.
   e) Group feedback in plenary.

When the watershed is large and it is therefore impractical to include all residents in planning, mechanisms for effective representation must be put into place. In Lushoto, for example, local school teachers and leaders, and male and female farmer representatives from all watershed villages, were called together to plan on behalf of others (see Plate 6).

Ensuring effective representation, however, goes far beyond simple selection of individuals to represent a particular interest group. Those individuals must be sensitized on the need to plan not for their own individual interests,

TABLE 3.7 Sample database illustrating socially disaggregated ranks at watershed level (Ginichi BMS)[a]

No.	*Watershed issue*	*Watershed priorities of each social group*					
		Men	*Women*	*Elders*	*Youth*	*High wealth*	*Low wealth*
1	Loss of seed, fertilizer, soil from excess run-off	6	6	**3**[b]	9	6	7
2	Water shortage for livestock and humans	11	9	11	8	7	4
3	**Poor water quality**	**2**	5	**2**	**3**	**1**	**1**
4	Conflict from lack of common drainage	15	12	15	15	16	16
5	Crop failure owing to drought	12	10	9	14	**3**	8
6	Soil fertility decline	**3**	4	5	7	4	**3**
7	Feed shortage	7	13	4	10	11	15
8	Shortage of oxen	13	3	10	5	8	10
9	Land shortage owing to high population pressure	5	2	6	2	5	5
10	Lack of improved crop varieties	9	15	13	11	10	13
11	Wood shortage	4	8	8	4	9	6
12	**Loss of indigenous tree species**	**1**	**1**	**1**	**1**	**2**	**2**
13	Effects of eucalyptus on soils and water	14	11	14	13	14	14
14	Theft of agricultural products	18	18	18	18	17	18
15	Conflict over paths and farm boundaries	17	17	16	16	18	17
16	Low productivity of animals	10	16	12	12	12	11
17	Lack of access to improved seeds	8	7	7	6	13	9
18	Conflict among villagers over watering points	16	14	17	17	15	12

Notes:

a These ranks were derived from averaging responses of all members of that social category across all watershed villages.

b Bold fonts denote the top three priorities of each social category. Rows with many bolded numbers represent issues of high priority to most watershed residents.

but on behalf of the group they are representing. Furthermore, decisions taken by this small group must be fed back to their villages or identified constituencies to solicit reactions and input from a broader group, and to foster broader buy-in to the work plan.

This approach helps to ensure that the priorities and perspectives of diverse community members are captured and adequately reflected in the prioritization and planning process. However, it has the disadvantage of minimizing direct participation by affected households, thus limiting the capacity to utilize the planning process as a critical step in community mobilization.

BOX 3.6 SELECTION OF ENTRY POINTS IN GALESSA

While loss of indigenous tree species ranked highest among most watershed residents at Ginchi BMS (Table 3.7), benefits to afforestation with indigenous tree species would only be derived in the medium term. The team therefore looked to the second and third priorities, and highlighted water quality as a problem that could be addressed in a period of several months through construction of concrete collection chambers around springs. Spring construction, with contributions of labor, materials, and money from watershed residents, was therefore selected as the entry point. At the same time, activities designed to address the loss of indigenous tree species were also initiated through negotiation support and nursery development.

Approach 4—Stakeholder-based planning

This approach to planning, while unique in its approach, is nevertheless embedded in one of the above planning processes to enable more intractable issues to be addressed. In this approach, specific landscape issues requiring collective solutions are analyzed with respect to the local interest groups who either affect or are affected by the issue. Planning is based around the integration of the views and interests of these different local stakeholders or interest groups, as follows:

1. Identification of landscape niches where the specific watershed problem is manifest.
2. Identification of local stakeholders to be involved in problem-solving, focusing on one of the following:
   a) Parties affected negatively, but in different ways, by the issue at hand;
   b) Those most and least affected by the problem, who have different levels of motivation for investing in NRM solutions; or
   c) Those affected and those perceived to be causing the problem (see Box 3.7 and Chapter 4 for a more detailed treatment of problem and stakeholder characteristics).

BOX 3.7 COMMON SCENARIOS REQUIRING THE NEEDS OF MULTIPLE STAKEHOLDERS TO BE MET

Scenario 1—Both parties are negatively affected by current practice

This can be illustrated by the case of Mt. Elgon National Park. Park rangers complained of illegal extraction of forest products, livestock grazing, and encroachment, while the indigenous Benet community complained of landlessness, loss of their traditional livelihoods following park establishment, and physical abuse by park officials. This had created a breakdown in communication between protected area officials and the communities surrounding the park.

Scenario 2—Collective solutions are required but one party has more to gain from the intervention than the other

This scenario is exemplified by the control of porcupine in Areka, where some farmers are much more affected than others owing to the particular nature of their landholdings and crops—yet collective action is required to solve the problem. It is also exhibited in some sites by upslope farmers who feel they have less to gain from labor-intensive run-off control measures on their fields, yet are being asked to allocate valuable land for these structures for the sake of negatively affected farmers residing downslope.

Scenario 3—One party is negatively affected by the actions of another party

This scenario is represented by landowners in all sites who were found to be using destructive land-use practices up to the edge of springs (grazing, cultivation, pesticide use, cultivation of "thirsty" trees), causing harm to spring users. It is also represented by farmers planting fast-growing trees on farm boundaries so as to minimize the trees' competition with their own cropland—thus intensifying competition between these trees and neighbors' cropland.

3. Consultation of individual stakeholder groups to identify their perceptions on the causes and consequences of the issue, possible opportunities for 'win–win' solutions, and the approaches they are comfortable with for entering into dialogue with the other stakeholder group(s)—including the selection of facilitators seen to be impartial and respected by each party. These consultations also help to demonstrate their external party's concern for their 'stakes' in the issue, and to reduce their fear of engagement (Box 3.8).

BOX 3.8 THE ROLE OF PRIOR STAKEHOLDER CONSULTATIONS IN MULTI-STAKEHOLDER ENGAGEMENT: THE CASE OF THE SAKHARANI BOUNDARY, LUSHOTO, TANZANIA

During the participatory watershed diagnosis in Lushoto, farmers identified negative effects of fast-growing boundary trees, particularly eucalypts, as a priority problem. One of the key stakeholders identified by farmers for improved boundary tree management was the Sakharani Mission. In 1946, the mission bought land and established high-value trees and crops. Eucalypts were planted in 1970 to secure the farm boundary from encroachment, and neighboring farmers had experienced negative effects of these trees on their cropland and low season spring flow. This was the main reason that multi-stakeholder negotiations were pursued between Sakharani and three neighboring villages.

The first step following participatory watershed diagnosis consisted of visiting the Mission to convey the concerns of farmers to the Mission's farm manager. This visit was instrumental in moving multi-stakeholder negotiations forward in several ways. First, watershed problems had only been diagnosed in the minds of smallholder farmers, failing to capture the views of other land users such as Sakharani. This preliminary meeting was therefore instrumental in highlighting concerns that the Mission had with regard to land-use practices of neighboring villages. These included the destruction of tree seedlings by freely grazing livestock and decline in the Mission's water supply from upstream land-use practices (see Plate 7). Owing to the impartiality demonstrated by the facilitators for the concerns of the Mission in addition to those already expressed by neighboring farmers, the farm manager began to view the dialogue as an opportunity rather than a threat.

A second outcome of this preliminary stakeholder consultation was to enable the farm manager to make suggestions on how the multi-stakeholder engagement itself would be facilitated. The farm manager was asked to contribute his suggestions on the date and venue for the meeting and the agenda. Contributions to the meeting's agenda included the inclusion of local leaders from neighboring villages and efforts to depolarize the concerns of each party. The latter led us to develop facilitation materials that emphasized the commonalities rather than the differences in the interests of each stakeholder, as illustrated in Table 3.8.

While the first two concerns were the main reasons for farmers to approach the Mission, the new concerns raised by the Mission were also included as farmers' concerns. This was justified by the fact that they were identified in the watershed exploration and therefore of concern to both parties. Furthermore, by emphasizing shared concerns rather than polarized interests, this helped set the stage for collaborative dialogue.

TABLE 3.8 Identification of concerns common to each stakeholder in the Sakharani boundary case

Problem	*Problem faced by:*	
	Farmers	*Sakharani*
Competition between boundary trees and neighboring crops	✓	
Eucalyptus depleting water in springs	✓	✓
Decline of rainfall	✓	✓
Depletion of water sources by catchment deforestation	✓	✓
Damage caused to crops and trees from free grazing	✓	✓

By accommodating the concerns and interests of the Mission, the proposed meeting for multi-stakeholder engagement was now seen as an opportunity by the farm manager to dialogue with his neighbors toward more optimal natural resource management for the benefit of both parties.

4. Facilitation of multi-stakeholder dialogue between the two parties, through the following steps:
   a) Provide feedback to participants on steps taken so far and their outcomes
   b) Jointly establish ground rules for dialogue, such as being respectful in listening fully to others and focusing on needs and interests rather than specific solutions when each stakeholder presents their perspective on the issue
   c) Ask each interest group to express their views using the ground rules
   d) Support the negotiation of socially-optimal solutions that meet the needs of each stakeholder group and which do not overly burden households who have little to benefit from the outcome
   e) Develop a detailed implementation plan with responsibilities and timeline (Box 3.9).

This approach makes divergent interests around any given issue explicit, and fosters "middle ground" solutions in which each party makes amicable concessions for the sake of harmony and the collective good.

In addition to using one of these four approaches, most sites used complementary diagnostic tools from the Participatory Rapid Appraisal methodology (Chambers, 1994; Rietbergen-McCracken and Narayan, 1998). For example, participatory resource mapping enabled the spatial identification of environmental

BOX 3.9 THE IMPORTANCE OF DETAILED ACTION PLANNING DURING MULTI-STAKEHOLDER NEGOTIATIONS: THE CASE OF AMEYA SPRING

Management of the Ameya spring had been the subject of ongoing conflict in Galessa watershed between the landowner and spring users. While the landowner was benefiting from the cultivation of Eucalypts near springs (growth rates being higher when water is more abundant), the spring users complained about the reduced water discharge and absence of alternative water sources to meet their basic needs.

During the first multi-stakeholder meeting, a heated discussion ensued focusing on each stakeholder's views: the spring users on problems resulting from Eucalypts, and the landowner on the need to protect his woodlot investment. The landowner eventually proposed a solution: if each spring user raises and plants a tree somewhere else on his farm, he would remove the Eucalypts from the spring. After some hesitation, one spring user stood up and said he would comply—with others eventually following suit. However, the meeting was closed with no detailed action plan (the "when," the "how" and the "who") on how the agreement was to be implemented.

The landowner ended up cutting down a small section of the woodlot as a gesture of cooperation. Yet Eucalypts coppice, requiring the trees to be uprooted. This is a very laborious exercise for the landowner who has no incentive to uproot. Furthermore, no plan for how replacement seedlings would be grown, or how the newly fenced woodlot would be established, was put in place. In a follow-up meeting, the landowner came with a host of additional demands which the community was unwilling to meet. These included financial compensation for trees uprooted, and community labor investment in uprooting trees and establishing a new woodlot. Had a detailed action plan been developed during the first meeting, many of these problems would have been avoided by moving directly into roles and responsibilities for implementation.

hotspots in the watershed (see Plate 3); current, seasonal, and extinct springs and waterways; and harmful tree lines and woodlots. Historical trends analysis with local elders also enabled the identification of causal factors behind major NRM degradation processes, and the magnitude of changes observed over time through matrix ranking of the degree of expression of identified variables (cover of indigenous and exotic trees, water flow, extinction of medicinal plants, etc.) during different time periods. Transect walks further complemented R&D teams' understanding of how watershed issues are manifest on the ground and raised awareness among community members about issues otherwise taken for granted.

Lessons learned

A cross-method comparison is useful in distilling the strengths and weakness of each approach based on a set of parameters of potential interest to project planners (Table 3.9). Interestingly, different approaches may be best suited to different purposes. The strengths of the first approach lie in efforts for widespread mobilization, articulation of farmer demands for support from development agencies, and being locally led. The merits of the second approach lie in the strong inclusion of local government agencies with ultimate responsibility for service provision and natural resource governance. The third approach, on the other hand, is beneficial for its efforts to explicitly capture the views or "voices" of diverse social categories, and the scientific validity of methods used to diagnose problems. The approach employing stakeholder-based planning is time-consuming, but is perhaps the only method for surfacing latent conflicts of interest and unlocking the potential for socially optimal (and thus politically and economically feasible) solutions.

General lessons learned from the development, testing and use of these methods in the field include:

- The selection of participatory planning processes effective in sensitizing and mobilizing the community at the planning stage can go a long way in setting the foundations for effective implementation.
- Local government and opinion leaders can play an instrumental role in mobilization, coordination, and strengthening buy-in at all levels.
- The need to ensure that outspoken community members, leaders or technical agents do not suppress the voice of less empowered actors at local level—either through socially disaggregated diagnostic activities or the use of skilled facilitators and disaggregated planning processes (such as by gender and ethnicity) in the context of large community planning fora.
- Opportunities to identify strategies for integrated and "win–win" solutions to complex landscape problems are often lost in the absence of multi-stakeholder processes and due to the emphasis on disciplinary planning.
- Participatory watershed diagnosis and planning should not be done with research teams alone; ideally, researchers should work in partnership with development agents experienced in community mobilization to bring complementary skills and mandates to the table.
- Communities are not homogeneous entities, but are often polarized by divergent interests or "stakes." Divergent interests should be understood, made explicit and cautiously but proactively reconciled if equitable solutions to watershed problems are to be identified.
- No single approach is "best." All approaches have unique strengths as well as shortcomings, and integration of their respective strengths into "hybrid" approaches is strongly encouraged.

TABLE 3.9 Relative strengths and weaknesses of approaches for participatory watershed diagnosis and planning

Aspect of Approach	*Approach 1*	*Approach 2*	*Approach 3*	*Approach 4*
Duration	Approx. 6½ wks	Approx. 4 months	Approx. 6 wks	Approx. 2 wks
Mobilization	Very Strong (emerging leaders, watershed committees leading the process, stakeholder engagement based on expressed demand)	Strong among leadership; medium among community members	Weakest in initial stages	Not ideal for mobilizing large numbers of people, but can unlock entrenched problems
Ability to capture diverse local perspectives	Strong among leadership; weak in ensuring socially differentiated views are effectively captured	Very strong among leadership; weak in ensuring socially differentiated views are effectively captured	Good in capturing the interests and priorities of diverse local groups and leaders	Very strong in reconciling divergent "political" interests on NRM
Topical coverage	Elicits most salient landscape and livelihood issues	Focused on conflict and areas of marked environmental degradation	Very broad (all system components; salient landscape and livelihood issues)	Applicable to many NRM issues, but used for specific niches or causes of conflict
Emphasis on integrated solutions to watershed issues	Medium (landscape approach helps to integrate)	Medium (landscape approach helps to integrate)	High (explicit effort to articulate linkages and plan by "cluster")	High (most issues involve landscape-level processes or boundary issues)
Involvement of support agencies	Strong and in response to local demand	Strong with local government, less strong for NGOs	Medium (agencies not directly involved in diagnosis are brought in only after plans have been developed)	Low (involvement can compromise the negotiation process if outside agencies are biased or lack conflict resolution skills)
Territorial coverage	Full coverage of few villages, but may be scaled up	Targeted to degradation "hotspots" and areas with high local initiative	Full coverage of a few villages, but may be scaled up	Targeted to specific landscape niches

Research and development team planning for landscape integration

Given that AHI's mandate included an explicit objective to develop research methods for participatory integrated watershed management, a lot of effort went into operationalizing the research component (questions, methods, ultimate application) within a participatory and integrated approach to solving real problems with farmers. To differentiate these approaches from the farm-level approaches described in Chapter 2 and to capture watershed issues that extend beyond the hydrological realm, we employ the term "landscape integration."

While research inputs were needed at diverse stages of the participatory watershed management process, this step of the watershed planning process was unique in involving primarily R&D teams. Iterative steps of planning and implementation in different benchmark sites were used to consolidate a single methodology for R&D team planning. This section is devoted to describing this unified approach.

Approach development

Following participatory identification of watershed problems by local residents, a lot of effort was devoted to answering the following two questions: (i) how to move from a "laundry list" of discrete problems to integrated solutions at landscape level; and (ii) how to operationalize the research component of participatory integrated watershed management or participatory landscape integration. A draft methodology was generated by the regional team, and a series of follow-up planning events was held at site level to test and improve upon the methodology. The methodology presented herein was a result of this iterative process of planning, application and lessons learning at site and regional levels (see also Stroud, 2003; German and Stroud, 2004).

Step 1—Creation of functional R&D clusters

The first step consisted of moving from a discrete list of concerns of watershed residents to functional "clusters" defined by strong causal relationships. The rationale for this was both to focus interventions on a few integrated objectives and interventions to facilitate implementation by addressing multiple problems simultaneously, and to structure interventions likely to foster positive synergies among diverse problems or components. Two criteria were utilized to develop an integrated intervention strategy from the list of identified watershed problems, one grounded in social principles and the other on ecological principles.

Principle 1: Focusing on watershed issues with high ranks from most social groups can enhance the likelihood of success

By focusing on the issues of greatest concern to most watershed residents, future R&D efforts are likely to have greater pay-offs as a function of the broad social support they receive within watershed communities. In each AHI benchmark site, a list of watershed issues was generated through systematic consultations with diverse social groups. Issues were solicited from various groups according to gender, wealth categories, physiographic location of plots or homesteads, and age. Once the issues were identified, the groups ranked them and identified the functional/causal linkages between diverse issues. By looking at the rankings given to these issues by different social groups, it is possible to prioritize those that have broad social support.

Principle 2: Focusing on watershed issues with strong functional relationships can enhance returns from any given investment

The second principle is to identify watershed issues that are functionally linked. The rationale behind this is twofold. First, it helps to identify issues that should be managed in an integrated manner to enable greater pay-offs from investments. Second, it makes the causal interactions and spin-offs (both positive and negative, at present and following alternative interventions) characterizing interactions between these issues explicit, enabling their management.

An example from the Ginchi site helps to illustrate how these principles are applied in practice. Thirty-nine watershed issues were identified by local residents in the Ginchi site and combined on the basis of their similarity into 18, namely:

1. Loss of water, soil, seeds, and fertilizers owing to excess run-off
2. Water shortage for livestock and human beings
3. Poor water quality
4. Problems associated with lack of common drainage
5. Crop failure from shortage of rains
6. Soil fertility decline and limited access to fertilizer
7. Feed shortage
8. Shortage of oxen
9. Land shortage owing to population pressure
10. Lack of improved crop varieties
11. Wood and fuel shortage
12. Loss of indigenous tree species
13. Effects of eucalyptus on soils, crops, and water
14. Theft of agricultural produce
15. Conflict over paths and farm boundaries

16. Low productivity of animals
17. Limited sharing of seed
18. Conflict between villages over watering points

These 18 issues were then ranked by different social groups in the watershed. The resulting ranks of the priority issues are presented in Table 3.10.

Several issues were considered either beyond the means of the R&D teams to address, or could only be addressed indirectly through other activities, for example addressing land shortages by intensifying crop and livestock systems or addressing drought through soil and water conservation. While the site teams decided to leave these issues out of subsequent clustering activities, this is something that should be reconsidered by others applying the methodology as opportunities for addressing more intractable problems might be lost by eliminating the issues from further discussion and analysis.

After applying the first principle—identification of watershed issues prioritized highly by most social groups, it was then necessary to apply the second principle; namely, identifying clusters of watershed issues with strong functional

TABLE 3.10 Rankings of watershed issues by social group, Ginchi benchmark site, Ethiopia

Watershed issues	*WS rank*[a]	*Social categories*					
		Men	*Women*	*Elder*	*Youth*	*High wealth*	*Low wealth*
Loss of indigenous tree species	1 (1.3)	1	1	1	1	2	2
Poor water quality	2 (2.3)	2	5	2	3	1	1
Land shortage[b]	3 (4.2)	5	2	6	2	5	5
Soil fertility decline	4 (4.3)	3	4	5	7	4	3
Loss of fertilizer and seed from run-off	5 (6.2)	6	6	3	9	6	7
Wood and fuel shortage	6 (6.2)	4	8		4	9	6
Limited access to improved seed	8 (7.8)	8	7	7	6	10	9
Shortage of oxen	7 (8.0)	12	3	10	5	8	10
Water shortage for livestock and humans	9 (8.3)	11	9	11	8	7	4
Crop failure from drought	10 (9.3)	12.5	10	9	14	3	8
Feed shortage	11 (10.0)	7	13	4	10	11	15

Notes:

a Watershed ranks were computed by taking the average of ranks given by each social group.

b Issues in italics are those the R&D team considered could only be addressed indirectly, through other activities.

relationships. This involved looking at the short list of issues emanating from the participatory ranking exercise, and trying to lump them into smaller clusters based on their functional relationships—as defined by a biophysical (nutrients, water), social (conflict and cooperation), economic (competition among components or users for scarce resources), or other logic. When the Ginchi site did this, they ended up with the following clusters based on what they knew about the system:

Cluster 1:

- Poor water quality and quantity (for humans and livestock)
- Loss of seed, fertilizer, and soil from excess run-off
- Loss of indigenous tree species
- (Crop failure owing to drought)[6]

The rationale for this clustering is based on the recognition that: (i) water quality is being affected by seed, fertilizer, and soil run-off from fields; (ii) substitution of indigenous trees with eucalyptus has caused the depletion of groundwater and the drying of springs; (iii) integration of appropriate trees and soil conservation structures on the landscape could enhance spring recharge (water quantity) and reduce the loss of seed, fertilizer, and soil from the landscape; and (iv) crop failure owing to drought could be ameliorated by reducing water loss from run-off through water harvesting. The common logic behind the perceived relationships caused the team to name it the "Soil and Water Management" cluster.

Cluster 2:

- Soil fertility decline
- Wood and fuel shortage
- Loss of indigenous tree species
- Limited access to improved seed
- Feed shortage
- (Land shortage owing to population pressure)

This clustering of issues was based on the following observations: (i) loss of indigenous tree species and fuel wood availability has exacerbated soil fertility decline through the increased use of dung and crop residues for fuel (and the former must be dealt with to ameliorate soil fertility decline); (ii) intensification of the system to reduce land pressure will require a balancing act so that increased agricultural production (crop, livestock, trees) does not further compromise the already ailing nutrient status of the system; (iii) "improved" seed often requires high soil fertility, and places demand on already limited nutrient resources; and (iv) the traditional practice of rotating between cropland and fallow (for grazing) between seasons and years means that interventions in the livestock system will have a direct impact on the cropping system, and vice versa. The common logic behind these perceived relationships caused the team to name this the "Integrated Production and Nutrient Management Cluster."[7]

These clusters are depicted graphically to illustrate the relationship between discrete problems and the integrated solution (Figures 3.4 and 3.5). The left-hand arrows in Figure 3.4 illustrate how solutions (middle of the diagram) do not address a single problem, but multiple problems simultaneously. In the same way, the three intermediate solutions can be further clustered into a single process of integrated (micro-) catchment management in which the whole is greater than the sum of its parts. For example, agroforestry practices should be able to add value to soil and water conservation objectives and water resource protection if the appropriate trees are selected for their functional role in addressing other watershed problems, as well as for the direct economic benefits they may bring. Alternatively, by addressing spring development as a high priority entry point, farmers may be more enthusiastic about trying out soil and water conservation measures or investing in the longer-term returns associated with the cultivation of tree species compatible with soil bunds, springs, and outfields.

In Figure 3.5, all issues identified in the cluster are represented with the exception of land shortage. As mentioned above, the R&D team decided that the land constraint would be addressed only indirectly, through the intensification of the crop, livestock, and tree components of the system. Our intention was not to suggest that such seemingly intractable issues should be marginalized up front; rather, we would encourage that such issues be fully explored to identify whether there are dimensions of the problem that can be taken on board by communities, the R&D team, or other actors. Limited availability of oxen was another issue identified by farmers but left out of the planning process by the team. One possibility put forward was to foster labor-saving technologies in other spheres to address the labor constraint implied by this concern, yet we found such linkage to be tenuous at best and instead constructed the diagram around the biophysical synergies we hoped to achieve. As farmers could very

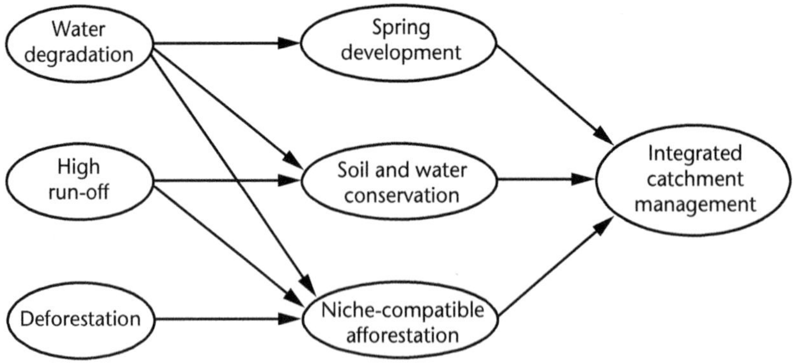

FIGURE 3.4 Soil and water management cluster

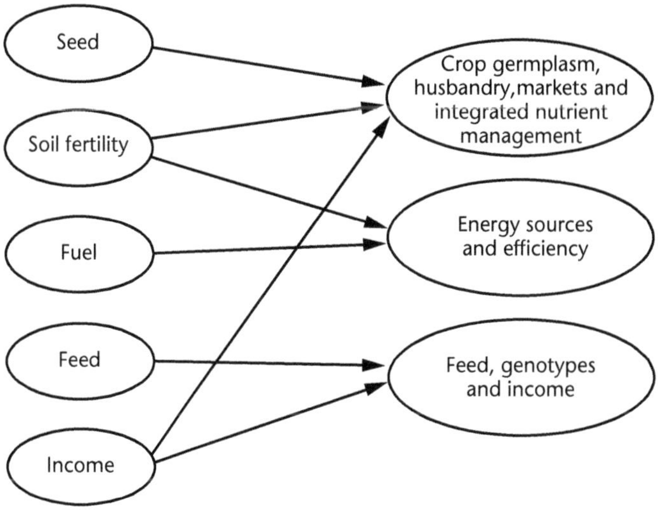

FIGURE 3.5 Integrated production and nutrient management cluster

well have prioritized labor saving over productivity gains, this decision represents a value judgment and should be duly questioned together with farmers before proceeding into participatory planning and implementation.

Step 2—Integrated planning

Once clusters have been identified, integrated research and community action protocols must be developed to articulate both a vision and an operational plan for bringing change within each cluster. The overall objective of the cluster is first articulated, followed by the objectives of each integrated solution (the middle or right-hand circles in Figures 3.4 and 3.5, respectively). The objectives must express "higher-level" goals that go beyond any given discipline or system component to an integrated target that involves optimizing returns to different system goals (i.e., crop production, livestock production, nutrient conservation) or understanding trade-offs that emerge when giving greater emphasis to one system goal over others (i.e., production over water conservation). Through this approach, interventions within each sub-cluster are aimed at addressing problems within that area as well as within other sub-clusters with which functional linkages are strongest.

The following sample objectives from the Ginchi site help to illustrate what higher-level targets look like:

> Objective 1 (Soil and water management cluster): To enhance the positive synergies between water, soil, and tree management in micro-catchments.

Specific objectives corresponding to each sub-cluster are:

- To improve the quantity and quality of water for both human and livestock use and enhance community enthusiasm for future watershed activities.
- To reduce run-off (loss of soil, seed, fertilizer, water), improve productivity (of crops, trees, fodder), and enhance infiltration and groundwater recharge.
- To increase the prevalence of trees in their appropriate niches to minimize run-off while increasing the availability of tree resources (fodder, fuel, income, timber).

Objective 2 (Integrated production and nutrient management cluster): To improve farmer incomes and system productivity (including crops, livestock and trees) while ensuring sustainable nutrient management in the system.

Specific objectives corresponding to each sub-cluster are:

- To improve farmer incomes from crops through improved crop husbandry (including varieties and management), integrated nutrient management, and marketing (while ensuring sustainable nutrient management in the system).
- To improve the availability and quality of feed resources (while ensuring sustainable nutrient management in the system).
- To enhance the availability of fuel and tree income (while contributing to the restoration of system nutrients).

As originally stated (without the phrases in brackets), these specific objectives are phrased in such a way that the integrated approach to managing the resource base for multiple outcomes could be easily lost. For example, sub-teams managing each specific objective began to focus on conventional research topics—namely, component-specific goals (livestock productivity, crop productivity, etc.) rather than on their integration or optimization. When testing new barley varieties, for example, it is important to monitor not only grain yield—illustrating a bias toward the crop component, but also biomass yield for feed, and the resulting impact on soil nutrient stocks. When exploring alternatives for improving the productivity of fallows, it is important not only to consider the yield of feed, but also the yield of subsequent crops in the same area and the quality of dung which will be recycled in the cropping system. It is for this reason that it is important to manage the entire cluster as a whole rather than according to its sub-components. It is also critical to ensure that farmers—natural systems thinkers seeking to optimize diverse benefits from any given innovation—have strong decision-making and oversight powers to determine what options or innovations are to be tested and the key parameters to be observed or measured for each.

At this point, integrated research and development work plans are developed around specified R&D targets or objectives. To assist in developing action plans toward the achievement of these targets, it is important to define two types of activities and their respective contributions to learning and change:

1. Community-led learning and change processes; and
2. Research contributions (social, biophysical, economic, policy) that can assist watershed residents or support institutions to make well-informed decisions.

Detailed planning for each is required at both community and R&D team levels. Yet planning also evolves as the learning process evolves—with new community-led change and research priorities emerging as critical knowledge gaps hindering informed decision-making emerge. Planning at the level of R&D teams at this stage requires: (i) articulation of the facilitation process to be used to help communities meet their own objectives; and (ii) articulation of research questions and methods, and how research results will feed back into decision-making processes at community or higher levels. A protocol was developed for the purpose of helping R&D teams to structure the planning of integrated research and development interventions (Box 3.10). Table 3.11 illustrates the relationship between cluster-level objectives and research questions and specific sub-components of these protocols (community facilitation, action research, and empirical research). As mentioned in Chapter 1, action research is different from empirical research in both the questions asked and the methods used—with action research emphasizing the "how" questions (to answer the question, "what works where and why?") and empirical research placing emphasis on the "what" questions (system characterization). Applications of empirical research in watershed management within AHI are summarized later in this chapter.

BOX 3.10 FORMAT FOR INTEGRATED R&D PROTOCOLS FOR EACH CLUSTER

1. Title
2. Background and justification
   - Problems in the cluster and why they are functionally linked
   - Why the problem persists despite community concern and the rationale for new types of interventions (including facilitation and research)
3. Cluster objective and primary research question (see Table 3.11)

4. Community facilitation process

    - How the overall change process will be facilitated for reaching the cluster objective (key steps, who to be involved and why, facilitators and facilitation process)
    - How action research and empirical research are sequenced in time with the facilitation process (i.e., how new knowledge will be used to inform decision-making)

5. Action research

    - Research question 1 (research methodology; research outputs and how they will be used)
    - Research question 2 (research methodology; research outputs and how they will be used)

6. Empirical research

    - Research question 1 (research methodology; research outputs and how they will be used)
    - Research question 2 (research methodology; research outputs and how they will be used)

7. Roles and responsibilities

    - Who will be responsible for leading each cluster, and for each research question and output?

Lessons learned

The following lessons were learned in efforts to apply and improve upon the methodology for research and development team planning for landscape integration:

- Planning for integrated research and development interventions at R&D team level was instrumental in reaching a common understanding of the complexity of management challenges facing farmers, and of how to organize the team to assist in navigating amidst this complexity.
- Planning at R&D team level must be iteratively validated and informed by watershed residents themselves, both as inputs to the planning process and as a means to raise awareness among farmers of the functional relationships being targeted by integrated interventions.
- Integrated planning is challenging, but staying integrated in practice is much more challenging. Researchers and practitioners trained within single disciplines or sectors will tend to sway toward conventional views and biases, forgetting to look at the system as a whole. For research, planning at

TABLE 3.11 Planning framework for integrating diverse learning approaches in research and development

Major activity/ step	*Objective*	*Facilitating participatory action learning and action research*	*Action research questions*	*Empirical research questions*
Watershed diagnosis	To identify major watershed problems from the perspective of local residents.	Primary Research Question: What are effective, equitable processes for participatory diagnosis and planning for watershed management?		
		1. Consultations with diverse social groups to identify key watershed problems, and opportunities and barriers to their resolution. 2. Development of participatory watershed action plans. 3. Program-level planning for integrated R&D interventions.	1. What is an effective approach for planning at local and program levels? 2. How can problem diagnosis be balanced with the need for immediate impact, so as to keep community interest high?	1. What are watershed priorities by gender, age, wealth, and landscape position? 2. What are key opportunities and barriers to addressing identified watershed problems? 3. How effective are current by-laws and natural resource governance?
Soil and water conservation (SWC) and management	To enhance the positive synergies between water, soil, and tree management in micro-catchments.	Primary Research Question: How can natural resource management innovations enhance agricultural productivity through decreased run-off (reduced loss of soil, seed, fertilizer, water) while enhancing spring recharge in the long term?		
		1. Spring development with spring management plans (responsibilities, rules, sanctions). 2. Negotiation support and local by-law reforms for spring maintenance, common drainage ways, investments in spring recharge, and greater niche compatibility in agroforestry. 3. Adaptive research on SWC structures and niche-compatible afforestation to control erosion, enhance water recharge and minimize loss of inputs.	1. If a high-priority entry point (spring development) is used, will outcomes of future R&D investments be greater? 2. What are the necessary conditions for people to invest in a shared resource? 3. What are effective approaches for reaching the overall cluster objective?	1. What is the impact of chosen SWC measures on run-off, soil and nutrient loss, and infiltration? 2. What are farmers' key indicators for SWC, and how do these change over time? 3. Which trees are compatible with different niches? How do prioritized tree species perform in different niches? 4. Who are the stakeholders for each issue, and how do they view the cause and solution?

Continued

TABLE 3.11 Continued

Major activity/step	*Objective*	*Facilitating participatory action learning and action research*	*Action research questions*	*Empirical research questions*
		Primary Research Question: How can income be improved through increased agricultural productivity (of crops, livestock, and trees) and marketing while maintaining or enhancing system nutrient stocks?		
Integrated production and nutrient management	To improve farmer incomes and system productivity (crops, livestock, trees) while enabling sustainable nutrient management.	1. Test alternative crop, feed and livestock husbandry practices and monitor effects on the system. 2. Raise awareness on fuel-nutrient dynamics; negotiate and test viable alternatives (fuel-efficient stoves, afforestation, regulations on dung collection from outfields). 3. Negotiation support for benefits sharing and collective investments in outfields (nutrient management, alternative fuel source).	1. What is an effective and sustainable approach for scaling out tested varieties and integrated nutrient management technologies? 2. What are effective approaches for improving livestock and feed production, minimizing system nutrient loss, and catalyzing collective investments in a sustainable fuel supply?	1. What is the effect of different varietal-nutrient management combinations on yield, income, plot fertility and system nutrient dynamics? 2. What is the effect of different feed and management innovations on income, livestock productivity, and system nutrient dynamics? 3. How much energy/fuel wood is needed to substitute unsustainable fuel sources? What is the "absorption capacity" of trees in different types of households and landscape niches?

the level of variables to be measured (for empirical research) or indicators to be monitored (for action research) helps to ensure multiple perspectives are considered. For further details on this methodology, see German (2006). For community facilitation, having cluster leaders within R&D teams to keep individual members focused on the bigger cluster goal and their activities aligned with this goal, and fostering integrated planning and monitoring at community level, are both instrumental.

Selecting entry points

The basic characteristics of good entry points were highlighted in Chapter 2. In AHI, two different types of entry points were used to bring early benefits within new watershed villages. A description of each approach and its underlying principles is provided below.

Approach development

Approach 1—Use of farm-level entry points

The first set of entry points builds upon prior work in the participatory farm-level innovation theme, because at the farm level is where new technologies are validated on farmers' fields before more widespread dissemination. The properties of these entry points should be in accordance with known principles specified in Chapter 2, namely: is of high priority (addresses felt needs of intended beneficiaries); able to bring quick benefits (often economic in nature); and has been previously tested (thus carrying low risk). A few case studies help to illustrate how farm-level entry points have been applied at early stages of watershed management within AHI benchmark sites (Box 3.11).

BOX 3.11 FARM-LEVEL ENTRY POINTS

Taro in Areka—Farmers in Gununo watershed quickly gained confidence in AHI interventions during watershed entry owing to the multiple benefits derived from the dissemination of a new taro variety called *Boloso-I*. The variety gives higher yield, requires less time and wood fuel to cook, has a good texture and lower concentration of oxalic acid, and generates more income compared to local varieties. This early success increased rates of repayment of in-kind loans (in the form of planting material) and increased community participation in subsequent meetings following the intervention. Therefore, technologies that bring quick benefits in the form of increased food security and income can serve as excellent entry points and improve the likelihood of community investment in future activities.

Tomato in Lushoto—During the PRA in Phase 2, farmers complained of low income from their enterprises. While exploring options for improving income, they mentioned the need for tomatoes that can withstand long-distance transport to distant markets in Dar es Salaam and Arusha without being damaged. New varieties were tested, and two were found to be much better performing in this regard, and therefore in great demand by intermediaries. The great success of this crop in bringing income quickly (within 4 months) at low risk (most households have access to valley bottom land and some irrigation water) made it a very successful entry point when expanding activities beyond Phase II villages to the watershed.

Approach 2—Use of watershed-level entry points

Landscape-level entry points have similar properties to farm-level entry points, but bring benefits at community rather than household level. Two cases from AHI help to illustrate this, and the importance of jointly considering diverse criteria (high priority, quick benefits) when selecting entry points (Box 3.12). In Ginchi, for example, spring development was chosen over the highest priority watershed issue (loss of indigenous tree species) owing to its ability to bring quick returns to the community.

BOX 3.12 LANDSCAPE-LEVEL ENTRY POINTS

Spring development in Ginchi—In Ginchi, the participatory ranking process identified in the participatory watershed planning session highlighted loss of indigenous tree species as the highest priority across most social groups (Table 3.12). However, as this entry point would defy a key principle of entry points (that it yields quick returns), the second priority across many social groups was selected. The community was mobilized to contribute labor, material (rocks, sand) and small sums of money for constructing cement structures around springs to protect water quality. Farmers were highly enthusiastic, with one individual exclaiming, "I had no idea the kind of water you buy in bottles can come from this spring."

TABLE 3.12 Final ranks of the top two watershed issues in Ginchi site

Watershed Issue	*Men*	*Women*	*Elder*	*Youth*	*High wealth*	*Low wealth*
Loss of indigenous tree species	1	1	1	1	2	2
Poor water quality	2	5	2	3	1	1

This entry point was also selected to create an incentive for community investment in ensuring long-term water supplies (a benefit with more delayed returns) through soil conservation structures designed to enhance infiltration and spring recharges as well as increase land productivity.

Controlling run-off in Kabale and Kapchorwa—For a long time, people in highland areas of eastern Africa have struggled with the challenge of effectively controlling excess run-off and landslides that destroy crops, fields, property, infrastructure and even lives. In the process of trying to overcome these challenges, different innovations have been tried with limited success. One of the most effective strategies in Kabale was the use of check dams together with local by-laws (an innovation of Africare). Cross-site visits to Igomanda Watershed by farmers from Rubaya and other sub-counties exposed farmers to these successful innovations and aroused their keen interest in working collectively toward adopting these best practices. The entry point, in this case, included several complementary elements: a cross-site visit, the technology, the by-laws and a few tools required for heavy digging. This led to overwhelming levels of community enthusiasm and collective action, as evidenced by 100 percent household participation.

In Kapchorwa, recent landslides led villages in the Tuikat Watershed to highlight this as their top priority. Cross-site visits were also used to learn from other farmers in Kaseko Parish who had been supported earlier on by the Kapchorwa District Landcare Chapter, mobilizing the community's commitment to apply these innovations to landscape areas most affected by landslides and excess run-off. Within two weeks, 50 percent of households in the watershed had adopted the innovation, banning free grazing to further control land degradation and ensure survival of tree seedlings planted on the contours.

Lessons learned

The following lessons were learned through our experience in testing diverse types of entry points for watershed management:

- The same principles apply for watershed entry points as for farm-level entry points (high priority, quick returns, previously tested), yet previous testing does not have to be from the same institution or project—as illustrated by the use of spring development as an entry point for stimulating innovations in agriculture and NRM.
- The scale of intervention (farm vs. landscape) does not have to be a determining factor in selecting entry points; proven farm-level entry points can mobilize community enthusiasm for future landscape-level innovations. Some watershed-level entry points may have an added biophysical

advantage over and above the social advantage conferred by ensuring quick benefits to communities. For example, it was hypothesized that spring development—an entry point that brings immediate improvements to water quality—might catalyze community interest in building soil and water conservation structures above the springs to help ensure a long-term water supply through groundwater replenishment.

- The success of entry points does not depend only on immediate livelihood improvements. Entry points can also be evaluated on the basis of the social capital built among community members and the effect this has on their confidence to engage in other collective endeavors in the future.

Empirical research inputs into decision-making

Action research is not a substitute for empirical research. The latter can be instrumental to decision-making for individual farmers as well as R&D teams and policy makers.

Approach development

Approach 1—Scientific data as inputs to decision-making

Scientific data can be instrumental in bolstering political commitment to a new approach (for example, impact assessments to illustrate the relative merits and demerits of new approaches for agricultural research and extension) or to new policies. The latter may be shown in research conducted in Lushoto to help legislators make tough decisions about whether and how to regulate eucalyptus cultivation in the district (Box 3.13). Scientific research can also make new indicators visible to farmers, raising awareness and mobilizing their interest in finding solutions. This is best illustrated by a case study from Ginchi, where soil erosion experiments helped to make visible the benefits of soil erosion control (Box 3.14). It has also been shown to empower communities to question the actions of more powerful actors, as illustrated by a case in Lushoto where scientific experiments were used to bolster support from external agencies in enforcing new boundary management practices less detrimental to farmers' livelihoods (Box 3.13). Finally, scientific data from watershed exploration and diagnostic work can help to ground interventions by identifying problems important to local residents, environmental hot spots (where these problems are most extreme), social conflicts, opportunities (i.e., local institutions respected by most parties) or other important guiding parameters.

BOX 3.13 SCIENTIFIC RESEARCH CAN HELP INFORM POLICY MAKERS AND LEGITIMIZE LOCAL STAKES VIS-À-VIS MORE POWERFUL ACTORS

During the participatory watershed diagnosis, farmers mentioned the incompatibility of eucalyptus with adjacent farmland as a multi-stakeholder problem among neighboring landowners. Since this problem can be partially addressed through policies regulating the location and density of eucalyptus on or near farm boundaries, empirical research was undertaken to assess soil chemistry, soil moisture, and maize yields near boundaries of eucalyptus and other species perceived by farmers to be harmful to crops. Identification of significant negative impacts on crop yields or thresholds (specific distance from tree lines at which negative effects rapidly decline), as illustrated in Figure 3.6(b), would both be useful for guiding policy. While the former would provide a justification for a policy intervention in the form of restrictions on species or planting locations, the latter would provide a clear design principle for such interventions (i.e., species X not to be planted within Y meters of farm boundaries).

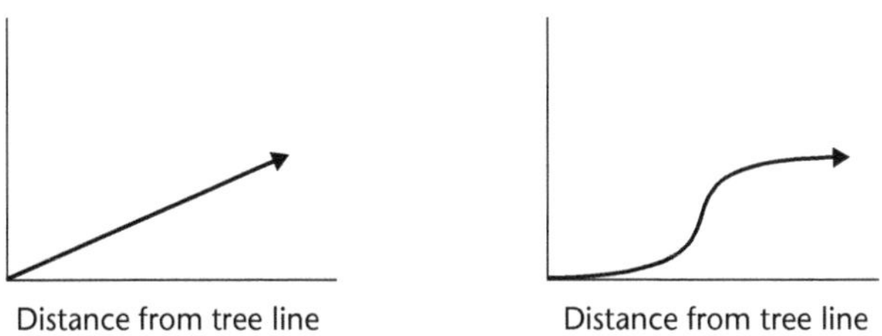

FIGURE 3.6 Hypothetical impact of boundary trees on the yield of adjacent crops in cases with (b) and without (a) thresholds

While this was the motive for conducting this research, one farmer living next to the Sakharani Mission and "hosting" an empirical research experiment used the clear visual evidence of reduced yields near the Sakharani boundary to support his interests. He requested the District Forest Officer to visit his field, see the outcomes of the experiment, and demand for land-use change by the Mission in the form of substitute species compatible with adjacent cropland. Clearly, such experiments can have both intended and unintended outcomes for livelihoods, learning, and social justice.

BOX 3.14 SCIENTIFIC RESEARCH CAN MAKE VISIBLE PROCESSES OTHERWISE DIFFICULT FOR FARMERS TO SEE

While gulley and rill erosion are highly visible to farmers, sheet erosion is less visible. Furthermore, farmers tend to focus on immediate economic needs over long-term sustainability, making them focus more on the damage caused by excess run-off over soil loss per se. Field demonstrations linked to scientific research proved to be instrumental in Ginchi for raising awareness by making new processes that are otherwise difficult to observe visible to farmers. An experiment in the Ginchi site was conducted using run-off plots and three treatments: (i) plots without conservation measures planted with barley; (ii) plots with soil bunds planted with barley; and (iii) fallowed plots. Water and sediment were collected at the bottom of each plot, with water color and the amount of sediment now visible indicators of soil loss in each of the three treatments (see Plate 9). Since the experiment was located near the main road, several farmers inside and outside the watershed were observing these indicators and became convinced of the importance of soil bunds in reducing both run-off and soil loss. During a farmers' field day conducted in October 2006, one farmer stated that, "seeing is believing."

Approach 2—Local knowledge as an input to decision-making

Systematic studies of local knowledge using social scientific methods can also help to make highly specialized or localized knowledge available to a broader community, creating opportunities for collective solutions to a shared problem. The "formalization" of local knowledge can be useful for a number of reasons. The first is the specialized nature of local knowledge, in which some members of a community may know much more about a topic than others owing to natural variations in individuals' interests and experience or incentives that keep that knowledge from being shared (Box 3.15). Formally documenting local knowledge can also help in multi-stakeholder negotiations, either by identifying inconsistencies in the knowledge systems of different stakeholders (and the need to reconcile these differences), or by feeding common local understandings on cause and effect into decision-making processes (Box 3.16). Finally, studies of local knowledge can help to target intervention strategies that are most strategic and to identify traditional practices already proven in addressing local social or environmental concerns (Box 3.17).

BOX 3.15 LOCAL KNOWLEDGE ON VERTEBRATE PESTS: SPECIALIZED KNOWLEDGE AND BARRIERS TO SPONTANEOUS SHARING

Research into local knowledge of porcupine and mole rat control in Areka was instrumental in finding a way to address the damage they cause to crops. The content of local knowledge included both control methods and the landscape locations where these are effective (Table 3.13). Yet this knowledge had not been effectively mobilized by the community to control the pests because they lacked institutions of collective action for doing this, and some of the more effective local control practices were not understood by the majority of community members. Some of the knowledge proved to be highly specialized and coveted, as the few farmers familiar with them earned income from controlling porcupines and mole rats on others' fields and benefited through secrecy. Making this knowledge more available to the broader community and using it to mobilize all residents to assist in control efforts were fundamental to addressing the problem.

TABLE 3.13 Characteristics of local control methods for porcupine

Control method	*Niches where applied*	*No. of knowledgeable farmers*
Wire body traps	Grassland; graveyards; forest; under eucalypts and bamboo	One
Deep digging at outlet of porcupine hole	Grassland; river beds and banks; under eucalypts and bamboo	Many
Circular ditch	Graveyards	Many

BOX 3.16 SOCIAL SCIENCE RESEARCH ON LOCAL KNOWLEDGE AS INPUTS TO MULTI-STAKEHOLDER NEGOTIATIONS

In Ginchi and Lushoto, local knowledge on the properties of different tree species and their suitability to different landscape niches was used as a first step in addressing problems related to incompatible trees on farm boundaries (competing with cropland), near springs (drying up springs and changing the taste of water), and on state land (where trees planted along roads and boundaries of protected areas compete with crops and cause the drying of springs). This knowledge was used for: (i) identification of niches where trees are or could be grown; (ii) identification of culturally important, harmful, and niche compatible (and incompatible) trees; and (iii) identification of the properties that make species compatible and incompatible with different niches. Participatory ranking was done to assess the degree to which different species exhibit different properties or "niche compatibility criteria" (German et al., 2006b). These data were then fed back to stakeholders during negotiation support processes to identify tree species exhibiting properties important to each stakeholder group, and species to be avoided in particular landscape niches (German et al., 2006a).

BOX 3.17 LOCAL KNOWLEDGE ON SPRINGS: IDENTIFICATION OF ENVIRONMENTAL "HOTSPOTS" AND TESTED SOLUTIONS

In Lushoto, social research on local knowledge of the causes and consequences of spring degradation helped to target appropriate solutions to this priority problem. First, identification of historical trends in spring degradation helped to identify springs for priority intervention based on: (i) the perceived threat to the resource (status of vegetation, historical changes in spring discharge); and (ii) the "social importance" of the spring (its importance as a function of distance to households, volume and seasonality of water, and number of users). Studies of local knowledge also helped to identify tree and grass species with conservation functions that researchers were unaware of, which were subsequently utilized to rehabilitate degraded springs. Finally, such research can help to validate traditional knowledge on environmental conservation that is coming under threat from changing belief systems. A traditional taboo forbidding the collection of crabs in springs, for example, may very well have an important conservation function as many crab species are known to manipulate water *quality* by removing detritus and circulating and oxygenating the water (Schubert et al., 1998).

Lessons learned

The following lessons have been learned in AHI attempts to integrate scientific research into development and natural resource management processes:

- Scientific knowledge is most useful when employed to make the (otherwise hard-to-see) consequences of local natural resource management practices visible to local actors and used to inform ongoing change processes.
- Value added from integration of scientific research into participatory watershed management may derive from a number of different things: its role in awareness creation (making visible previously invisible processes and unthinkable opportunities); its ability to help shed new light on cause-and-effect relations that need to be understood in the context of negotiated decision-making or policy design; supporting community advocacy vis-à-vis government agencies and more powerful actors; and for mobilizing the potential of local knowledge in local problem-solving.
- Local knowledge may be more effectively applied if combined with other forms of external support, including social science research to legitimate, systematize, or publicize it.
- Symbolic explanations for biophysical processes, often disrespectfully called "superstitions," should not be discredited because they are not explained

through scientific rationales. Not only do these explanatory frameworks often explain underlying biophysical processes in ways consistent with scientific explanations, but discrediting them may have negative effects on sustainability by eroding the natural resource management practices they help to sustain and encode.

Participatory monitoring and evaluation

Monitoring and evaluation is perhaps the most fundamental step to participatory integrated watershed management because without active monitoring at community and project levels, other investments of time and energy are likely to yield few returns. Monitoring helps to capture challenges early on so that they may be addressed before they lead to failure. It also enables opportunities to be effectively captured by identifying them and fostering agreement on how they can best be seized. By monitoring, participants can share their views on the challenges faced, generate "best bet" solutions, and agree on how these solutions will be put into practice through revised work plans and division of responsibilities.

Approach development

AHI has experimented with approaches to participatory M&E at both community and R&D team levels. This section profiles three distinctive approaches. The first two approaches emphasize participatory M&E at community level, the first drawing on local concerns or indicators alone, and the second employing exogenous indicators and/or scientific methods. The last approach illustrates participatory M&E at the level of research and development teams.

Approach 1—Participatory monitoring and evaluation (PM&E) with communities

Within AHI, approaches to participatory monitoring at community level, based on observations of local residents themselves, have included both informal and formal approaches. Case studies help to illustrate how feedback elicited from farmers during implementation helped to change the approach being used for improved impact, as lessons were learned through implementation.

(i) Informal PM&E at community level

The first strategy has been simply to ensure a continuous presence in the watershed, and to continuously ask watershed residents (both active and less active ones) how they view or perceive the effectiveness of ongoing activities. This can occur through active questioning or through sharing time together informally over a cup of tea or through other forms of socializing. It is often through such informal interactions that the most honest reflections are shared.

Case 1—Monitoring the "mood" of the community during watershed exploration to sustain community enthusiasm

Although watershed delineation and characterization were seen as a necessary step by the Lushoto team—providing baseline information and helping to identify constraints to livelihoods and improved natural resources management, these processes took time. Participatory diagnosis and prioritization of watershed issues and participatory watershed planning, while more engaging, also took time. Once planning was finalized, diverse sub-teams engaged farmers in more detailed planning meetings and training workshops on select themes. All of these activities placed demands on farmers' limited time, and diverted them from other important activities. While farmers were getting fatigued, frequent visits by the R&D team, together with open communication and good rapport between team members and farmers, enabled farmers' concerns to be expressed openly. Researchers were therefore in a position to respond and to brainstorm on ways to keep farmer enthusiasm high. Issues of highest concern by farmers, identified during early phases of characterization and diagnosis, became the subject of discussions on how to bring immediate impact and sustain farmers' trust and enthusiasm. It was planned that some efforts and resources should be invested in the rehabilitation of degraded water sources. Local residents were asked to identify water sources to be rehabilitated using jointly agreed upon criteria, including level of degradation and number of households depending on the water source. Several water sources were chosen and communally rehabilitated through contributions from local residents (stones, sand, labor) and the project (technical, cement, and financial). Water source sanitation and water levels were observed to immediately improve, restoring confidence of the community in the program as a whole.

(ii) Formal PM&E using local indicators

The most notable difference between informal and formal PM&E is the latter's explicit use of local indicators to monitor performance. Rather than following a complex typology of principles, criteria and indicators, here 'indicators' were loosely interpreted to include quantitative or qualitative statements that can be used unambiguously to describe desired situations and measure changes or trends over a period of time. Use of locally formulated indicators helps to foster a shared understanding of the objectives being sought, and what a successful change process should look like. It is important to note that the indicators of "success" for one individual or group may not be the same as for others. It is therefore necessary to either: (i) seek broad consensus on the indicators chosen, and to ensure that diverse views are captured; or (ii) carry out stakeholder-based M&E to ensure that the interests and concerns of diverse interest groups are adequately captured and monitored among groups sharing common interests.

Steps to formal PM&E using local indicators are summarized in Box 3.18. For stakeholder-based monitoring, stakeholder analysis would come as a first step, and different stakeholder groups would be encouraged to formulate their own views when monitoring the performance of local indicators (in Steps 4 and 5).

BOX 3.18 BASIC STEPS IN PARTICIPATORY M&E

1. Identify objectives of the activity to be carried out. This can be defined at the cluster level, or at the level of specific problems being addressed.
2. Identify local indicators. This can be done by asking the following questions: "If [activity X] is successful, what will be different in [6, 12, 24 months'] time? What changes will you see?"
3. Set a schedule for follow-up meetings to monitor progress.
4. Monitor progress according to established indicators. This can be done by asking the following questions: "You mentioned that if you are successful, you will see [e.g., more water discharge from springs]. Have you observed any changes yet in [spring discharge]?"
5. Revised work plans as needed to adjust activities so that the ultimate objectives are more likely to be achieved. This can be done indicator by indicator, for example by asking the question, "Is the observed change in [indicator X] enough, or does more need to be done to see [e.g., more water discharge from springs]?"

The following case studies help to illustrate what the use of local indicators and stakeholder-based monitoring looks like in practice. The first case study also demonstrates how monitoring of local indicators may be used to monitor successive stages of a single approach or different approaches tested over time.

Case 2—By-law reforms in Kabale

Two waves of participatory by-law reforms were carried out in Rubaya Sub-County over the course of several years. Formal participatory monitoring was done using a combination of indicators proposed by local communities and project facilitators to observe how these two approaches, implemented sequentially, compared to one another and relative to the pre-intervention period. Table 3.14 illustrates the nature of outputs from this type of monitoring.

These findings clearly show the incremental nature of local governance improvements, where villages with prior experience implementing NRM by-laws were more effective in utilizing by-law reform processes to catalyze collective action.

Case 3—Equitable technology dissemination in Areka

A case from Areka also helps to illustrate how participatory monitoring using local indicators may be done. Following the approach to equitable

TABLE 3.14 Performance of identified indicators by phase of intervention

Local indicator	*Prior to intervention*	*Phase I intervention*	*Phase II intervention*
Muguli and Kagyera (villages participating in Phase I and II interventions)			
Number of soil conservation structures	Limited use of natural resource management technologies (few trenches, no tree nurseries, only 5 farmers used bench terraces)	– 226 trenches – 3 check dams – 6 tree nursery beds – 31 bench terraces – terraces 5,500 *Calliandra, Grevillea*, and *Alunus* spp. planted	– 85 additional trenches – 19 additional bench terraces – 1 additional check dam
Number of NRM conflicts reported per month and mode of resolution (fines vs. consensus)	– No committee to resolve conflict – 15 cases of free grazing reported per month – Cases resolved with fines	– Policy Task Force resolves conflicts (16 cases from 2003-04 reported to LC1 Court referred back to PTF for resolution) – Prevalence of conflict reduced – Resolution of 8 cases of free grazing through consensus and without fines	– Further reduction in prevalence of conflict – No fines applied – Resolution of conflicts through the sub-county Committee
Local leadership support to by-laws	No support from political leaders, for fear of losing votes	Increased support from local leaders as a result of sensitization and committee membership	More extensive support by local leaders to the extent of participating in reviewing, monitoring and implementing by-laws

Local indicator	*Prior to intervention*	*Phase I intervention*	*Phase II intervention*
Change in behavior/ social relations	– Conflicts resolved in LC courts with fine – Increase in hatred and selfishness among conflicting parties	– Conflict resolution by consensus and not in courts – Collective action every Thursday to help those negatively affected by by-laws – Spirit of sharing (tree seedlings) and trusting one another while in meetings – Collective action in input supply (tools, seedlings)	– Conflict resolution by consensus and not in courts enhanced – Spirit of sharing and trusting one another further enhanced
Katambara and Mushanje Villages (villages participating in Phase II only)			
Number of soil conservation structures	None	N/A	– 70 new trenches in Kantambara – 56 new trenches in Mushanje – Collective action for trench digging every Thursday
Number of NRM conflicts reported per month and mode of resolution	– More than 20 cases reported on monthly basis – No committee to resolve conflicts or enforce by-laws	N/A	– Formed a committee of 9 people who monitor the performance of by-laws and resolve conflicts without fines – About 5 cases per month reported in Katambara and 7 in Mushanje, but resolved in harmony
Local leadership support to by-laws	No local leadership support	N/A	By-laws are working, but limited support from local leaders could undermine sustainability

technology dissemination described in Chapter 2, mixed groups of farmers from each village were called together to assess how different local indicators were performing. Groups were asked to do matrix ranking to compare the approach used by the formal extension service with the AHI approach. Participants were asked to discuss the relative performance of the two approaches for each indicator, and to use seeds to rank the two approaches based on their perceived performance (with more seeds meaning better performance). Results are presented by approach, with "before" representing formal government extension and "after" the AHI innovation, in Figure 3.7. While this approach to evaluation does not explicitly capture views of different stakeholders (e.g., by gender and wealth), participants were asked to jointly reflect on the effects of different approaches on women and poorer households. While this evaluation method has the benefit of generating a collective awareness of the performance of different extension approaches for different stakeholders, it may not be effective in ensuring the voice of those same groups are adequately captured. Subdividing the group of farmers by gender or assets (e.g., landholdings) for matrix ranking and then comparing the outcomes would ensure diverse perspectives are captured. The presentation of these gender- or wealth-disaggregated results would involve the generation of two figures (one for each group), or the inclusion of additional columns (to represent how different groups evaluated each approach).

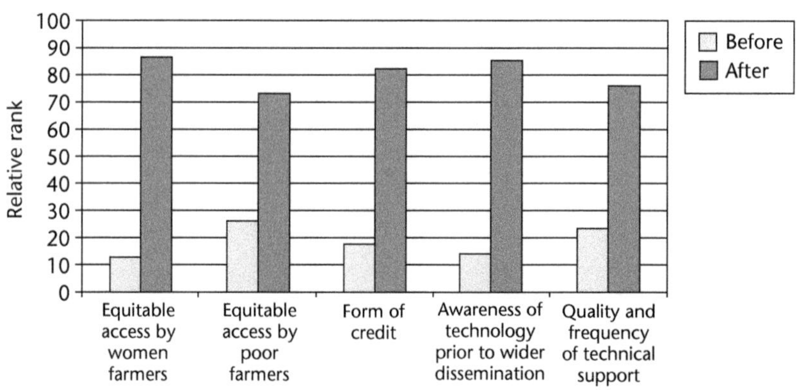

FIGURE 3.7 Farmers' perceptions of the relative equitability and benefits of the AHI/HARC approach as an alternative to that employed by the Government Extension Service, Areka, Ethiopia

Note: The AHI/HARC approach ("After") included negotiation support to agree on mechanisms and rules for equitable access; participatory by-law reforms to support local agreements; and in-kind credit.

Approach 2—Formal M&E using scientific indicators or methods

Formal M&E can draw on scientific principles or methods of data collection in a number of ways. One way is for local communities to identify indicators and for scientific methods to be used to gather data and monitor performance of the indicator. This approach consists of the following basic steps:

1. Identify objectives of an innovation process together with farmers representing different social or interest groups.
2. With each group, identify local indicators that will be used to monitor progress toward agreed objectives. This may be done by asking participants, "If you are successful in [achieving objective X], what changes will you see? What will be different in [2 months' time, 6 months' time, 2 years'] time?"
3. Researchers may also suggest indicators they think are important for monitoring, and explain the reasons why (e.g., they are complementary to farmers' indicators, help to capture outcomes of importance to the project—such as equity or sustainability, etc.).
4. Researchers and farmers agree which indicators will be monitored by whom, and how.
5. Researchers and farmers jointly develop an action plan to articulate the activities to be undertaken, who is responsible for each, the timeframe and plans for feedback of findings to the wider group (including for researchers to share any findings from the scientific or local indicators they are charged with monitoring).

This approach is again illustrated by the porcupine case from Areka, where scientific methods were used for unbiased sampling and systematic data collection in the monitoring of local indicators through formal household surveys, and the results shared back with farmers to enable them to take appropriate actions. Farmers were asked to identify local indicators for assessing performance of the activity. These included levels of crop loss for crops that are economically important and susceptible to attack, time spent and number of family members involved in guarding fields at night, and incidence of weather-related illness resulting from high levels of exposure to the elements when keeping watch of fields by night. These were measured by research across a representative number of households, and results compiled (Figure 3.8).

The most marked livelihood benefits were found to result from reduced crop damage, improved health, and labor savings. Levels of crop damage were reduced by 80 percent following intervention, while frequency of visits to health clinics as a result of weather-related illness also declined. Yet one of the most important successes in the minds of farmers was the reduction in efforts required to guard fields at night. Such an indicator could have been easily overlooked had scientists been the only ones to identify indicators to be monitored, yet it was the most critical success in the minds of farmers.

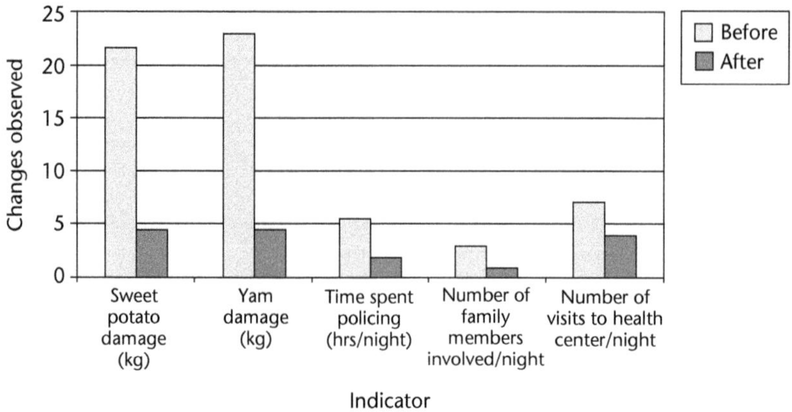

FIGURE 3.8 Observed impacts from collective action in porcupine control

Another example of formal monitoring, using a combination of local and scientific indicators to track spontaneous farmer-to-farmer spread of technologies, was presented in Chapter 2.

Approach 3—Participatory M&E at R&D team level

"Participatory" M&E or self-reflection at the level of R&D teams is also necessary for a number of reasons. First, it helps to align activities with specified objectives, community facilitation processes, and research protocols. Implementation is always easier in theory (during planning) than in practice; therefore, self-reflection among R&D teams is an important component to backstopping community-led efforts. It also helps to monitor contributions from different organizations and team members, to ensure that all team members are fulfilling their roles and responsibilities. Finally, and perhaps most important for AHI, it has served as a platform for team learning and innovation by fostering group reflection, cross-checking of assumptions, and fine-tuning how different disciplinary perspectives are articulated within community facilitation and research practice.

In AHI, "participatory" or self-monitoring was conducted largely through periodic meetings among R&D teams. These meetings consisted of reflections on progress made since the last meeting, discussions on how to improve team performance and innovate in line with program objectives and community priorities, and to adjust work plans accordingly. The content of reflections included both the technical content of watershed work (for example, to collectively assess the extent to which behaviors reflect principles of participation and integration) and the approach to teamwork itself. The latter might include, for example, joint reflection on the extent to which the team is simply planning together as opposed to learning together in the field. For reflections on how the team as a whole is interfacing with watershed communities to foster change, a process documentation methodology was used to facilitate systematic reflection on the same. This methodology is illustrated in Box 3.19.

BOX 3.19 FORMAT FOR PROCESS DOCUMENTATION

I. Prior to any activity or step (planning stage):

Objective: What are you trying to achieve in the particular community intervention being planned for?

Approach: What is it that you will do to achieve the objective? (What steps will be taken? Why did you choose these steps? Who will you involve, and why?)

II. Following any activity or step (reflection stage):

Approach: What did you actually do to achieve the objective? [The planning process for such community events is rarely complete, as there are always unforeseeable circumstances that affect the facilitation process. How was the approach modified during the event itself to accommodate these changes, and why?]

Successes and challenges: What went well? What did not go well? [This should include reflections of the successes and stumbling blocks faced by the communities or stakeholders since the last meeting, as well those faced in the facilitation process itself. It should also include observation of why these successes and challenges occurred, providing lessons for others wishing to learn from your experience.]

Findings: What did you learn that you did not know before? [These findings are generally derived from statements about reality made by farmers or other participants, which were new to the facilitators.]

Resolutions: What decisions were taken by participants? [These should include agreements reached by the participants about the principles behind the work, and about the way forward.]

Lessons: What lessons or insights can be derived from these experiences to share with others trying to carry out similar activities? [These should include things you were surprised to find out, both about the approach that was used to engage with communities and about the findings.]

Recommendations: Replanning. [Here the team should reflect on what they would do the same and differently next time to build upon successes and overcome challenges faced during implementation. This can be for the benefit of the team itself in terms of approving outcomes from similar activities to be conducted in other locales, or for purposes of sharing experiences with a wider audience.]

The following case studies help to illustrate the importance of frequent monitoring at the R&D team level. The first illustrates use of R&D team monitoring to learn lessons in the field, and the second the importance of continuous monitoring of team performance to achieve greater integration of disciplines, perspectives, and system components.

Case 1—Use of process documentation tool to reflect on methods for multi-stakeholder negotiation

The process documentation tool presented in Box 3.19 has proven to be an important tool for fostering learning at the level of R&D teams on approaches under development. It has been used as a means to operationalize action research—namely, to learn lessons "what works, where, and why." Such a tool can assist in advancing iterative, cumulative learning over time on any given case, or to learn lessons across cases on a particular set of approaches, such as the facilitation of multi-stakeholder negotiation processes. For the latter, cross-case comparison could contrast a set of facilitation approaches when applied in different landscape niches (springs, farm boundaries, and waterways), different topics (niche-compatible agroforestry, free grazing, soil and water conservation) or different contexts (districts, countries, agro-ecological zones). Ultimately, it assists R&D teams to reflect on learning and to decipher emerging patterns. These patterns might be seen through *iteration*—a sequential approach to learning in which different approaches are tested over time, evaluated and modified as needed to address weaknesses. In this approach, outcomes obtained at different stages of the innovation process are observed to distil key points in time—and elements of the approach being applied at that point in time—that brought the most profound change. They may also be deciphered through *comparison*—namely, trying a similar approach across different cases and observing how theme, stakeholder characteristics, or context influence the outcomes in each case.

An example of a process documentation output helps to illustrate how the methodology is used in practice (Box 3.20). This case documents a single meeting in which two stakeholders—the Sakharani Mission and neighboring farmers—were brought together to negotiate more socially optimal land-use practices to address latent conflict. The event followed preliminary activities to diagnose watershed problems from farmers' perspectives, ethnoscientific research to document local knowledge on the properties of different tree species and their positive and negative effects within different niches, and preliminary stakeholder consultations.

BOX 3.20 PROCESS DOCUMENTATION OF MULTI-STAKEHOLDER NEGOTIATIONS FOR THE SAKHARANI BOUNDARY CASE

I. Prior to activity/step

Objective: To advance multi-stakeholder dialogue and planning for improved management of the Sakarani boundary.

Approach (as planned):

1. Call together stakeholders (Sakharani farm manager, affected farmers, and local leaders from neighboring villages) for multi-stakeholder negotiations.
2. Share steps carried out so far (and findings): (i) Participatory watershed diagnosis (competition of eucalyptus planted on farm boundaries with crops and water); (ii) Research on local knowledge (niches needing improved management, species causing problems, niche compatibility criteria of farmers); (iii) Stakeholder consultations with Sakharani and neighboring villages.
3. Solicit reactions and clarification from participants.
4. Negotiate "binding" criteria for tree species selection by prioritizing the most important in the list of niche compatibility criteria mentioned by each stakeholder (*farmers:* improves soil fertility, produces few seeds, crop-compatible, small shade, does not dry soil; *Sakharani:* secures boundary, fast growing, coppices, few branches, good for fuel, lumber, and income).
5. Identify tree species that fit combined criteria.
6. Develop work plan with activities (*what?*), responsibilities (*who?*) and timeline (*when?*).

II. Following activity/step

Approach (as actually carried out):

- Rather than negotiate "binding" criteria, we went directly to the negotiation of tree species acceptable to both parties, as it was conceptually easier for both parties.
- We did not plan the "when" in work plans, owing to time limitations and the need to consult more people before making specific work plans.

Successes and challenges:

- [S]: The event successfully overcame the communication impasse and led to agreements to address the latent conflict.
- [S]: The outcome was favorable to both parties, as it addressed concerns of both.
- [C]: Representation of different hamlets was not good; the meeting dragged on for long.

Findings:

- Sakarani rejected peach as a boundary tree (because fruits would attract villagers), resulting in the addition of a new "binding" criterion to the list.
- *Mtalawanda* (*Markhamia obustifolia*) has drawbacks for both stakeholders (slow growing, produces many seeds), but advantages outweighed disadvantages (height, limited branching, and shade, compatibility, life span).
- *Agrocarpus* is not a good boundary tree because its roots invade farmland and compete with crops.

Resolutions:

- To replace eucalyptus with *Mtalawanda*.
- To hold a second meeting with all farmers bordering Sakharani to discuss a detailed plan for the felling of eucalyptus and managing tree seedlings, to be called by the Village Executive Officer.
- A host of technical and policy solutions for rehabilitation of springs and waterways (buffer zones, water-conserving vegetation, by-law reforms and enforcement).

Lessons/Insights:

- Terminology matters, either polarizing the issues (e.g., "stakeholder") or minimizing conflict ("party").
- Use of language to manage power dynamics is essential, for example acknowledging the property rights of the landowner by asking him whether he can accommodate the concerns of the other party.
- Crucial role of a third party to bring dialogue in situations of latent conflict, and the power of simple dialogue in unlocking deadlock.
- It is easier to discuss niche-compatible species than niche compatibility criteria.

Recommendations:

- Move directly from sharing the niche compatibility criteria of each stakeholder to negotiating species acceptable to both parties. Return to specific criteria only if solutions are not forthcoming.
- Divide approach into several steps: a) preliminary dialogue with tentative solutions, b) broader consultations (follow-up meetings with affected farmers), and c) development of final work plans.

By observing and documenting these experiences through *iteration* (sequential phases of innovation and learning on the same case), we learned a number of important lessons. The first is that it is essential to involve the right authorities in decision-making. Failure to involve Fathers higher up in the Benedictine order caused difficulties for the farm manager when it came to implementing agreements. A second lesson is the need for detailed planning, for example going beyond *criteria* to guide the sequence of tree felling (i.e., where trees posed a risk to households, followed by areas where trees posed a risk to cropland) to specific *locations*. Early efforts by the Mission to fell trees did little to address the safety concerns of one particular farmer whose house continues to be at risk from boundary trees. Finally, close follow-up monitoring by a neutral party or local authority is required to ensure agreements are implemented and operationalized. In addition to the aforementioned complaints about the location of felled trees, local residents and leaders experienced difficulty in holding the Mission accountable, given how they benefited from a host of services (schools, worship, etc.) provided by the Mission.

By observing and documenting these experiences through *comparison* with other cases (multi-stakeholder negotiations for managing other niches, topics, and contexts), some of the above lessons were confirmed and other new lessons learned. Lessons that were confirmed include the need for detailed plans of action and systems for follow-up monitoring. Other lessons consolidated through comparison are the fundamental importance of a respected neutral party to convene and facilitate multi-stakeholder events; the need to develop formal by-laws to back up resolutions involving high "stakes" (for example, in the case of lost income); and the importance of legal texts in supporting or undermining negotiations. Formal laws may support the inalienable rights of landowners, thus undermining any concessions agreed to by the landowner once he or she learns of these rights. Alternatively, formal laws on environmental protection can render illegal those land-use practices that undermine the provision of environmental services, thus bolstering the claims of parties negatively affected by these practices (as in the case of spring degradation).

Case 2—The importance of R&D Team monitoring and evaluation to strengthen integrated research

Multidisciplinary teamwork where different professionals from different institutions and personal backgrounds come together to address common issues is one of the new ways of working that has been adopted by AHI. This was necessitated by the reality that NRM issues confronting highland farmers in eastern Africa require holistic solutions that go beyond specific system components (crops, soil, trees, livestock) to integrated ecological processes, and beyond technologies to encompass collective action, marketing, policy reforms, and new forms of institutional behavior and cooperation. However, multidisciplinary team leaders faced many challenges when trying to foster collaborative efforts, including team members who were reluctant to learn or value other team members' disciplines, the tendency of scientists to pursue questions of interest within the confines of their own disciplines, and limited institutional support (Mowo et al., 2006; see also Pirrie et al., 1998). Other institutional bottlenecks included the lack of an incentive scheme that recognizes and rewards team work and team products. The imbalance in skills and experience among team members was also a challenge in efforts to foster collective understanding. In extreme cases, some individuals never believed in the potential of multidisciplinary research and pulled out of the team altogether to pursue more conventional forms of research.

With experience gained through time, and with the use of outcome mapping techniques (see www.idrc.ca) and facilitated M&E sessions at team level, changes were observed in a number of areas. Reduced antagonism among disciplines, increased leadership competence and willingness to explore more holistic research questions and methods were among the most notable changes observed. Therefore, project-level M&E to reflect on team performance in addition to the end goal (e.g., community engagement and related outcomes) is a crucial dimension of monitoring.

Lessons learned

The following lessons were derived from AHI experiences with participatory M&E:

- Participatory M&E at multiple levels (community, R&D team) is instrumental in ensuring any change process is successful, given the need to proactively reflect on challenges faced and the approaches being used to address these challenges or reach the agreed end goal.
- Stakeholder-based processes for participatory M&E at community level can be useful in capturing diverse perspectives on the effectiveness of approaches, and on the winners and losers of any intervention.

PLATE 1 Farmers in Kwalei village, Lushoto, load up their tomatoes for transport to Dar es Salaam

PLATE 2 Metallic hook used to trap mole rats in Areka

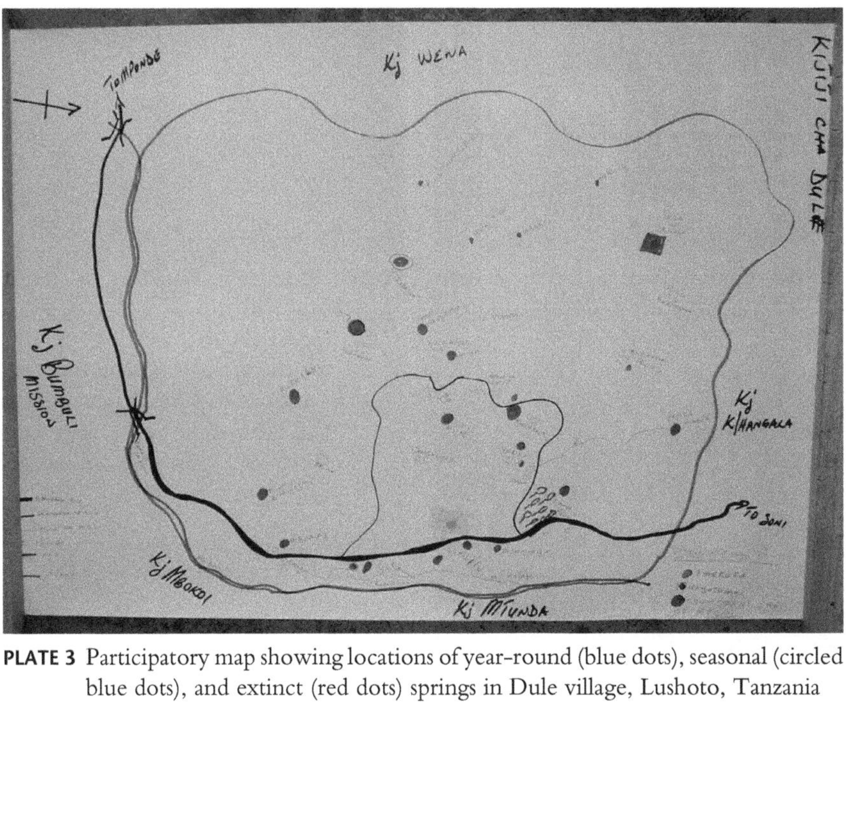

PLATE 3 Participatory map showing locations of year-round (blue dots), seasonal (circled blue dots), and extinct (red dots) springs in Dule village, Lushoto, Tanzania

PLATE 4 Spring in Kwekitui Village, Lushoto, which yields much less water today than in the past

PLATE 5 Tolil Watershed Committee in Kapchorwa, Uganda

PLATE 6 Village representatives involved in participatory watershed planning in Lushoto, Tanzania

PLATE 7 Progressive clearing of forest and absence of soil and water conservation activities in the catchment and riparian zone just upstream of the Sakharani Mission are believed to have caused sharp declines in the Mission's water supply in recent years

PLATE 8 Introduction to the watershed approach to farmers in Rwanda

PLATE 9 Seeing is believing: water and sediment collection chambers in Ginchi BMS make the extent of soil loss visible to farmers

PLATE 10 Ginchi landscape prior to soil conservation interventions

PLATE 11 Ginchi farmers exploring terraced landscape at Konso

PLATE 12 Farmers in Lushoto complain that eucalypts, such as those lining this tea estate boundary, lead to the drying of nearby springs

PLATE 13 Cultivation up to the edge of a spring in the Baga watershed, Lushoto

PLATE 14 Landscape with (bottom) and without (top) natural resource governance

- The frequency of community-level participatory M&E events must be adapted to the activity at hand (frequent for tree nurseries, for example, because poor group performance can within days lead to death of seedlings), the stage of implementation (more frequent at early stages, for example, when farmers and team members are beginning to learn how to work together) and the complexity of the challenge (Box 3.21).
- Regular monitoring of the performance of R&D teams with respect to the use of integrated and participatory approaches is essential for ensuring their continued use in practice, given high levels of specialization of disciplines, institutional mandates and mindsets.

BOX 3.21 "DON'T GIVE UP!": THE IMPORTANCE OF SUSTAINED MONITORING TO ADDRESS CHALLENGES

Perhaps the most challenging watershed activity in all AHI benchmark sites has been to facilitate outfield intensification in the Ginchi benchmark site. The reasons behind this relatively intractable challenge, as described in prior case studies, include low tenure security from histories of land reform and limitations on private property rights, seasons of restricted and open access to private farmland, and limited options for livestock feed. For such a challenging task, R&D teams must work closely with watershed communities to design and redesign innovation strategies, and to monitor their effectiveness in practice. Participatory planning and monitoring highlighted a host of challenges farmers face when struggling to innovate and intensify their outfields. If farmers were to fence outfield plots owned by the household, they would face an insurmountable barrier of sourcing alternative feed, since they would be restricted from accessing the fallowed plots of other farmers and the cultivation of fodder crops would compete with food crops. Collective action in reducing free grazing was also seen as inviable given that many livestock come from distant villages and fostering collective action at that scale is nearly impossible. Farmers finally agreed to construct soil bunds, plant soil stabilizers along them and manually protect them from livestock damage by guarding the fields and fencing seedlings. This was tested in the 2005 growing season, but since this approach was labor intensive, most of the seedlings were destroyed from grazing livestock (Table 3.15). In the 2006 season, follow-up monitoring and negotiations led farmers to agree to test local by-laws to restrict free grazing in selected sub-catchment areas until conservation bunds and trees were established, and then shift these areas to gradually cover the entire watershed. The agreement also included the testing of new tree species along bunds. While the approach worked during the rainy season (when outfield areas are under cropland or restricted access grazing), it broke down during the dry season when open access grazing is practiced. Follow-up monitoring elucidated farmers' concern about equity issues—namely, that only those landowners falling within the protected sub-catchment benefiting in the short term from restricted grazing, as well as a strong reluctance to follow through with agreements.

TABLE 3.15 Seedling performance under diverse outfield intensification strategies

		% seedling survival by species	
Season	*Species*	*End of rainy season*	*End of dry season*
Jun 05–May 06	*Chamacytisus palmensis*	83	2.3
	Acacia decurrens	68	1.8
Jun 06–May 07	*Chamacytisus palmensis*	92	0.5
	Vetiveria zizanioides	98	10
	Pennisetum purpureum	89	0

Under such extreme challenges, what options remain? One could simply give up, in which case a problem affecting huge portions of Ethiopia where some of the poorest in the world live would remain largely unresolved. Alternatively, one could utilize empirical data showing the linkage between tenure security and levels of farmer investments in sustainable NRM to advocate for structural changes such as improved property rights. While this was something to be considered, it also posed major challenges and would require complementary innovations in the free grazing system in order to bring about changes in land use. The team therefore decided to make one more attempt to solve the problem locally, namely through use of an economic "pull" that might induce outfield innovation. They decided to introduce high-value trees (apple) as an incentive for outfield intensification. The problem faced at the time of writing was that the team failed to make technology delivery conditional on certain types of management (e.g., that they be planted in outfields) and farmers have largely planted seedlings in infield plots. So the question remains as to whether this will be a viable option for inducing outfield innovation in the Ginchi site, or whether other options such as strong enforcement of local resolutions might also work. The "take home" messages from this case study are: (i) ensure and build on early successes so that farmer engagement with tough challenges can be sustained; (ii) frequent monitoring of local and scientific indicators can be useful for guiding new approaches; and (iii) "don't give up" when addressing tough watershed challenges!

Addressing implementation challenges

Timely monitoring at all levels can go a long way in increasing the chances of success in watershed management. However, it is also useful to acknowledge that some challenges are likely to arise under the best of monitoring systems, and to explore what AHI has learned in our efforts to address them. A few key challenges stand out and merit specific attention here.

Approach development

In this section, approaches are presented according to the nature of key challenges faced.

Challenge 1—Overcoming historical and "structural" constraints

A number of challenges faced in participatory landscape-level innovation emerge from historical or higher-level "structural" constraints. Historical constraints have included legacies from the colonial era in British East Africa, where conservation methods were enforced from above through violent means. This entrenched a very negative attitude in people's minds towards soil conservation. In an act of defiance, all structures were systematically destroyed at independence. In Ethiopia, histories of land reform and shifting land tenure policies through feudal, socialist, and contemporary eras have instilled a sense of tenure insecurity in the minds of farmers, causing them to resist any conservation investments in the less secure outfield areas. While the government promises no future land reforms, the experience of shifting governance systems overrides any security farmers may feel over land rights. Land insecurity resulting from de-gazettement of protected areas for resettlement of prior indigenous residents has created similar problems of tenure insecurity in the Kapchorwa site, undermining land investments. Approaches to break through the cognitive and psychological barriers to innovation are often needed to help overcome such entrenched, historically grounded attitudes posing barriers to innovation (Box 3.22).

BOX 3.22 SEEING IS BELIEVING: THE IMPORTANCE OF CROSS-SITE VISITS IN EXPANDING FARMERS' MENTAL MODELS ON "WHAT IS POSSIBLE"

At times, farmers' understanding of "what could be" is largely constrained by "what is." The landscape in Ginchi and much of highland Ethiopia is almost devoid of trees, and biomass and enterprise diversity is extremely low (see Plate 10). For farmers to imagine a conserved landscape with controlled grazing and diverse crop and tree enterprises requires a very big leap of the imagination. Field visits to Konso, where terraced landscapes and a diversity of enterprises at plot and farm level are the norm, opened farmers' eyes to "what could be" (see Plate 11). As stated by Atu Yirga Tafu, a Ginchi farmer, "If I had not been to this place I would not have believed human beings can construct the whole district in such an artistic manner." Others expressed their surprise at the number of crops they were unfamiliar with and the number of different crops growing in a single plot.

Other historical constraints have included failed development efforts from past administrations or projects, making farmers reluctant to trust outside actors. Hand-outs used by past or concurrent projects and histories of aid (food and cash for work programs) have inculcated a "dependency syndrome" in some sites (Areka, Kabale), undermining voluntary contributions by farmers and causing farmers to focus on "quick fixes" rather than more comprehensive development strategies. In some cases, traditional norms and beliefs hinder effective solutions, as in the aforementioned case of communal grazing areas in Areka or the case of male-dominated decision-making, land tenure and management of household finances in much of the region.

Structural constraints are those factors largely beyond the control of communities which nevertheless influence the extent to which problems may be readily solved. Such structural constraints included national policies (i.e., government land tenure in Ethiopia, which has a similar effect on farmers' willingness to invest in their land as prior government appropriation of land during land reform programs), poor market opportunities and infrastructure which make agriculture less profitable than other livelihood options, and institutional practice which undermines equitable and effective development. Examples of such institutional practices included a bias toward wealthy male farmers in many agricultural extension programs, failure of research and development organizations and different sectors to work in partnership, and the demonization of indigenous knowledge, beliefs, and behaviors through modernization and its institutions (religious, educational, agricultural, etc.). A final institutional constraint emerged from efforts to step outside standard institutional mandates to embrace broader dimensions of NRM based on the "integration" concept, as observed by the Holetta Agricultural Research Centre (HARC) when using spring development as a watershed entry point (Box 3.23).

BOX 3.23 ADMINISTRATIVE HURDLES FACED WHEN STEPPING OUTSIDE SECTORAL MANDATES

The participatory approach to watershed diagnosis challenged AHI partners—largely agricultural research and extension organizations—to step outside their normal institutional mandates. The compartmentalized nature of institutional mandates caused a host of bureaucratic challenges. When proposing to use spring construction as an entry point for watershed interventions—to raise farmer confidence in AHI and catalyze their interest in other water-conserving catchment management practices—research center officials (administrators, finance heads, and auditors) expressed concern. Their main reservation was that HARC has no mandate to carry out such construction activities without the knowledge and permission of the engineering department within their umbrella organization, the Ethiopian Institute of Agricultural Research (EIAR).

Because of this, HARC requested EIAR for their endorsement of and support to spring construction. Being the first time EIAR and its engineering department had receive such a request, they were unsure how to respond. They raised questions such as:

1. Do construction works for spring management fall within the mandate of agricultural research?
2. Since the springs along with the construction materials will be handed over to the community and Bureau of Water Resources, how will they be registered? Will this create a problem for official transfer of property and for internal audits?

These complexities forced the site coordinator to make frequent trips (about 10 in all) to Addis Ababa to convince officials of the need to support this activity, and to follow through with the diverse administrative procedures. Finally, by presenting coherent arguments for the logic of spring construction in the context of integrated watershed management, Deputy Director Generals for Research and Administration approved the request.

While a more logical approach would be to reach out to ministries with a relevant mandate, this was ineffective as the selected watershed site fell outside their priority areas of operation (largely urban). This case study therefore shows that agricultural research and development agencies must be in a position to reach out to district partners and expand their institutional mandates when engaging in a holistic approach to landscape-level NRM.

Many of these constraints have not been overcome, making progress slow in addressing certain locally felt NRM concerns. However, progress has been made in some areas to either minimize the extent to which these problems hinder solutions, or to address problems despite these hindrances. One prominent example comes from the Ginchi site. Histories of land reforms and public land ownership in the country and traditions of open access grazing in the site have jointly undermined farmer investment in outfields. However, cross-site visits to areas that have been intensified despite the odds have opened farmers' eyes to what is possible, and increased their enthusiasm for working within these broader constraints to improve their livelihoods by making better use of existing resources. Following feedback meetings to share these discoveries with other watershed residents, enthusiasm was much higher for developing and testing collective solutions to the outfield dilemma. Another example comes from efforts to counter the dependency and apathy resulting from a prolonged history of hand-outs from development actors and top-down decision-making from government. Continuous efforts to *facilitate* farmers to think for themselves and solve their own problems have changed farmers' attitudes, boosted their confidence in working collectively on their problems, increased their willingness to experiment and, ultimately, improved their livelihoods. Challenges remain, however, in finding a suitable institutional model for sustaining such heavy facilitation efforts. Closer partnerships between research and development institutions and training of community facilitators at the local level are promising options.

Challenge 2—Managing complexity

Managing landscape-level processes in an integrated and participatory manner is a complex task. AHI's approach went beyond soil and water management to encompass agroforestry, crop and livestock production, water management for domestic purposes (water harvesting, spring protection), energy, markets, and the social and governance dimensions of each of these. Since farmers had an interest in each of these areas, many activities were ongoing at any given time. Some activities also had to be aligned with the seasons, requiring rigid implementation schedules. This was at times hindered by slow administrative procedures within R&D organizations and limited availability of required materials (i.e., seedlings of certain tree species), among other factors. Social, economic, and political life outside agriculture is also rich in rural communities, with many activities competing for farmers' time. Learning how to sequence, coordinate, and harmonize activities in time and space given each of these factors is challenging. Very detailed and consultative planning at the outset, including activities to be conducted, their timing and well-defined roles and responsibilities can assist in this regard, as can frequent replanning to adjust actions with emerging realities. One challenge which is difficult to overcome was the tendency for development processes to be embedded within projects of short duration and overly influenced by external institutional mandates. While landscape-level NRM takes time, periods of donor funding were limited to a few years at a time. The implication is that it is important to set realistic plans with donors in terms of the time required for any given change process to unfold and to bring significant impact. Time horizons must be set on the basis of whether methods are to be adopted from elsewhere and simply applied, or whether the project aims to engage in a process of action research and methodological innovation—which takes considerably longer. If the former, we estimate that 3 to 5 years may be sufficient if facilitators are adequately experienced and thus able to quickly gain rapport with farmers, and if they receive prior training in INRM approaches.

The participatory nature of the watershed management approach used by AHI also required strategic balancing of attention to activities bringing short- and long-term benefits to community members. First, in contexts characterized by high levels of rural poverty, NRM strategies needed to be grounded in immediate livelihood concerns such as food security. Thus, efforts to address NRM challenges bringing only medium- or long-term benefits required that interventions be accompanied by strategies to address the immediate needs of farmers. Second, no extended periods of time should pass without farmers seeing any benefit from watershed management activities so as to sustain their interest. If approaches used for watershed exploration took time, for example, it was necessary to identify and apply entry points based on identified priorities of farmers to sustain community interest and build trust. Since some activities take many years for benefits to be seen (e.g., soil fertility improvements through reduced erosion, groundwater recharge from conservation structures and niche-compatible trees), the concurrent implementation of activities with short-, medium- and long-term benefits is required. In AHI sites, activities with short-term benefits (e.g., spring protection, dissemination of proven varieties) were conducted alongside activities with benefits derived over the medium-term (e.g., technology and by-laws for niche-compatible agroforestry) and long-term (e.g., soil and water conservation).

Another dimension of complexity emanated from efforts to work in teams with people from different institutional, sectoral, and disciplinary backgrounds through team work and partnerships. Different disciplines working in R&D teams, for example, had different views on the meaning of watershed management and strategies to be used. The tendency to "disintegrate" into areas of disciplinary expertise was strong and had to be continuously reflected upon as a team and addressed for teams to come together toward integrated solutions. This extended down into the specific research questions and variables to be tracked by researchers, who tended to focus on research questions, methods and variables from within their own areas of expertise rather than integrated research protocols. This problem extended to the participatory nature of research, which should ensure farmers' priority variables (which tend to cut across disciplinary boundaries) are brought on board within formal and action research. Strategies that helped to overcome these barriers to more effective collaboration included regular review and planning meetings to reflect on the approach being used, participatory monitoring with farmers, and an official "policy" within AHI to work through partnerships and interdisciplinary teams.

Challenge 3—Social justice and equity

A final area that presented substantial challenges was in managing equitable approaches to participatory landscape-level innovation. This challenges stems from both the realities on the ground and the approaches used to bring change. Highly polarized local interests in NRM make natural resource management a political process in terms of who wins and who loses from current land-use practices and related innovations. The challenge lies in bringing solutions that benefit multiple parties at the same time, or maximize the benefits for most land users while minimizing the cost to any given party. External institutions often have a role in favoring some local land users over others owing to their failure to consider equity and monitor effects on different local groups. These issues will be discussed in greater detail in Chapter 4.

Lessons learned

The following lessons were distilled from efforts to address challenges that arise through implementation:

- Constraints emanating from historical and "structural" influences at higher levels may at times represent significant constraints to INRM.
- The dependency syndrome and negative attitudes stemming from past experience can be overcome, but it requires continuous sensitization (most notably, through cross-site visits, see Box 3.22), dialogue and observing concrete improvements among early innovators.
- Multidisciplinary teams do not ensure interdisciplinary approaches. It is easier for diverse research disciplines to work toward disciplinary aims when on multidisciplinary teams than to work toward integrated solutions. Sensitization and frequent reflection meetings at the level of R&D teams are necessary (but not sufficient) to overcome disciplinary barriers to interdisciplinary team work and integrated approaches. Political support and performance review systems that reward multidisciplinary approaches and results are also important.
- Agricultural research mandates focusing more on research than on development concerns, and having a narrow productivity focus, hinder action research and integrated approaches to addressing landscape-level problems.
- Achieving equitable solutions to landscape-level NRM requires fostering synergies between governance and technological interventions. It also requires behavioral change among external R&D actors in the way they interface with rural communities (to proactively avoid elite capture and foster equitable benefits capture) and monitor outcomes (so as to capture socially differentiated effects). Additional details on the empirical basis for this lesson may be found in Chapter 4.
- Frequent monitoring and replanning at local and R&D team levels is of fundamental importance to adaptive learning in addressing complex NRM challenges.

Missing links

Addressing landscape-level livelihood and natural resource management concerns of farmers is a challenging task. Substantial progress has been made by AHI in the eastern African highlands in identifying approaches to operationalize participatory landscape-level innovation, including methods for participatory problem diagnosis and prioritization, participatory planning, and participatory management of change (including monitoring and adjustment). Scientific methods to support these processes have also been articulated and refined, including setting baselines for subsequent impact assessment, delineating and characterizing watersheds (methods developed largely outside of AHI but refined internally), embedding scientific research in locally owned change processes, and supporting local change through facilitation and support in addressing common implementation challenges. However, there are a number of methodological gaps for coming full circle in our efforts to operationalize participatory landscape-level innovation. These gaps highlight a number of priorities for future research and methodological innovation in AHI and the region at large.

1. *"Minimalist" approach to watershed characterization.* While all of the data collected during the watershed characterization was useful to researchers, who may generate ample material for publications, not all of it is directly relevant to planning, monitoring, or impact assessment. This approach can be simplified so that the minimum data needed to identify opportunities, develop strategies, and monitor performance is collected at this time. This can minimize farmer "fatigue" and make more efficient use of financial and human resources.
2. *Hybrid approach to participatory watershed diagnosis and planning.* Different approaches employed for participatory watershed diagnosis, prioritization, and planning each had their respective strengths and weaknesses. It is likely that "hybrid" approaches building on the strengths of each will be better than any of these approaches in isolation, and efforts should be dedicated to testing such approaches in practice.
3. *Sequencing of steps in R&D team planning for landscape integration.* As integrated R&D protocols are developed at the level of R&D teams, they need to be continuously informed by participatory decision-making processes involving the intended beneficiaries. This is done not only through participatory planning around previously identified watershed problems, but also by continuously cross-checking assumptions about the most relevant causal processes to serve as the organizing logic for clustering, the most appropriate approaches for community facilitation and the most critical research questions. For example, while biophysical scientists and extension practitioners may emphasize a biophysical logic for clustering, farmers may use different rationales for clustering focusing on social or economic processes. Local residents may also perceive different research priorities than researchers, based on what they know to be critical gaps in their knowledge base. Beneficiary groups and communities should increasingly assume decision authority as the range of possible meanings and uses of "research" come to light and as local capacities to design and monitor change processes are improved. More research is needed on how to effectively sequence participatory, community-level planning with planning at the level of R&D teams. Ultimately, the latter should build upon (during the design phase) and support (during implementation and synthesis of lessons) community-level objectives and decision-making.
4. *Participatory approach to generating functional R&D clusters.* Another knowledge gap in the planning stage is the extent to which the creation of functional

R&D clusters can be made a fully participatory activity with communities. This would go a long way in addressing the aforementioned sequencing issues in R&D team planning. To what extent can farmers develop, with minimal external assistance from research and development actors, fully integrated action plans at watershed levels that foster "win–wins" in livelihood and environment, and optimal returns to different social actors (based on gender, wealth, ethnicity, or specific sets of interests vis-à-vis what is being planned)? What is the optimal distribution of planning responsibilities between communities and R&D teams? These questions should form the basis of future research and methodological innovation in this area.

5. *Testing of integrated solutions to difficult landscape-level challenges*. The means to address the multiple challenges of intensification in Ethiopian outfields remains a challenge. Innovative or multifaceted solutions should be explored, including the use of incentives (e.g., conditional delivery of high-value crops and trees, payments for environmental services, among others), regulations (e.g., policies to control livestock movement, support the implementation of local agreements, or link incentives to specific problem niches), and institutional innovations (e.g., fostering collective action at higher levels, privatization of tenure conditional on good land management, innovations within support agencies). Multidisciplinary and multi-institutional teams with a strong sense of dedication to the process are an essential component, as are "systems thinkers" (from both the community and the facilitation team) who can help to think outside the box.
6. *Adaptive testing of proven approaches in other highland sites and agro-ecological zones*. While methods already developed within AHI have only been field-tested in the eastern African highlands, they are likely to be of wider applicability as they were generated in multiple contexts and address challenges that are widespread. This suggests that the methods are likely to be relevant to other highland areas throughout the tropics and, for some issues, perhaps also other eco-regions (e.g., densely settled lowlands). A missing link which is likely to yield high returns with limited effort is therefore the adaptive testing and modification of these methods in new settings. We welcome opportunities to partner with other R&D actors within the region and elsewhere to expand the learning process in this regard.

Conclusions

This chapter illustrates a sequential series of methodologies for facilitating a process of participatory watershed entry, diagnosis, management, and governance. While the process is complex and challenging, it also yields rich rewards for rural livelihoods, sustainable natural resource management, and more harmonious relationships within densely settled highland communities. While the approaches presented here are ready for uptake by other organizations, there is need for more experimentation with the various approaches presented—so that strategies may be refined to meet the unique circumstances of different countries and localities. And, as always with any sort of methodological or institutional innovation, *pilot first and scale up later*!! Only with such experimentation can the necessary lessons be learned that will enable programs to avoid propagating errors and instead disseminate methods proven to work in a variety of settings.

Notes

1 These included farmers with prior involvement in NRM Task Forces supported through prior efforts of CIAT and AHI in Rubaya Sub-County, Watershed Management Committees who had worked with Africare in Hamurwa Sub-County, soil conservation groups who had worked with NEMA in Bubaare and individuals from new villages with severe environmental problems.
2 In two sub-counties where villages were closer to one another, the village-level NRMPCs decided to form higher-level coalitions to work jointly on common NRM challenges, electing smaller committees to coordinate work across several villages.
3 In some AHI benchmark sites, there is no such distinction between upslope and downslope households, either because most farmers hold land in different parts of the landscape or because the landholdings of different households are arranged in strips from the hilltop to valley bottom. In villages where some farmers hold land in multiple landscape locations while others hold land in only one locations (e.g., upslope or downslope), farmers whose perspectives are likely to be most different from other households—in this case, with plots restricted to single landscape locations—are called together for consultation.
4 Focus group discussions are less time consuming than household surveys for ranking watershed priorities. However, individual ranking ensures that diverse views are better captured, as dominant individuals will always influence what final number is put on paper.
5 A key gap in AHI methods development is in the testing of such "clustering" methods at community level to see to what extent fully integrated watershed action plans can be generated with minimal outside assistance. This should constitute a priority for future R&D interventions.
6 Issues denoted by parentheses are those that would be only partially addressed through interventions focused on the cluster of issues, because they are partially the result of issues beyond local control.
7 Clearly, the identification of such functional clusters requires a relatively intimate knowledge of the system. It is important to note that this knowledge can be provided by either farmers or researchers who have been working in the system in a participatory manner for some time. We would encourage the exploration of both options when applying this methodology in new sites.

References

Ashby, J.A., J.A. Sanz, E.B. Knapp and A. Imbach (1999) CIAT's Research on Hillside Environments in Central America. *Mountain Research and Development* 19(3): 2–18.

Bellamy, J.A., G.T. McDonald and G.J. Syme (1998) Evaluating Integrated Resource Management. *Society & Natural Resources* 12: 337–53.

Chambers, R. (1994) Participatory rural appraisal: Analysis of experience. *World Development* 22(9): 1253–1268.

Constantz, G. (2000) Grassroots-based Watershed Conservation in Central Appalachia. *Mountain Research and Development* 20(2):122–25.

Datta, S.K. and K.J. Virgo (1998) Towards Sustainable Watershed Development through People's Participation: Lessons from the Lesser Himalaya, Uttar Pradesh, India. *Mountain Research and Development* 18(3): 213–33. FAO-UNESCO (1987) *Soils of the World*. Food and Agriculture Organization and United Nations Educational, Scientific and Cultural Organization, New York: Elsevier Science.

De, R. and S.P. Singh (1999) *Watershed Development Imperatives for Boosting Rainfed Agricultural Productivity – An Indian Experience*. New Delhi: Indian Council of Agricultural Research.

German, L. (2006) Moving beyond component research in mountain regions: Operationalizing systems integration at farm and landscape scale. *Journal of Mountain Science* 3(4): 287–304.

German, L. and A. Stroud (2004) Social learning in regional R&D programs. *AHI Brief* D2.

German, L., S. Charamila, and T. Tolera (2006a) Managing trade-offs in agroforestry: From conflict to collaboration in natural resource management. *AHI Working Papers* No. 10.

German, L., B. Kidane, and R. Shemdoe (2006b) Social and environmental trade-offs in tree species selection: A methodology for identifying niche incompatibilities in agroforestry. *Environment, Development and Sustainability* 8: 535–552.

German, L., S. Ayele, W. Mazengia, M. Tsegaye, K. Abera, K. Bedane, E. Geta, T. Tolera, and H. Taye (2008) Institutional Foundations of Agricultural Development in Ethiopia: Drawing Lessons from Current Practice for Agricultural R&D. *Quarterly Journal of International Agriculture* 47(3): 191–216.

Knox, A., B. Swallow and N. Johnson (2001) Conceptual and Methodological Lessons for Improving Watershed Management and Research. *CAPRi Policy Brief* 3: 1–4.

Johnson, N., H.M. Ravnborg, O. Westermann and K. Probst (2001) User Participation in Watershed Management and Research. CAPRi Working Paper 19: 1–25.

Mowo, J.G., L. German and T. Tolera (2006) Local institutions and their role in INRM: A regional synthesis. In: T. Amede, L. German, C. Opondo, S. Rao and A. Stroud (eds), *Integrated Natural Resources Management in Practice. Enabling communities to improve mountain livelihoods and landscapes*. Proceedings of a conference held at ICRAF HQ, Nairobi Kenya October 12–15 2004, African Highlands Initiative; Kampala Uganda.

Pirrie, A., V. Wilson, J. Elsegood, J. Hall, S. Hamilton, R. Harden, D. Lee and J. Stead (1998) *Evaluating Multidisciplinary Education in Health Care*. Edinburgh: SCRE.

Ravnborg, H.M. and J.A. Ashby (1996) Organising for Local-Level Watershed Management: Lessons from Rio Cabuyal Watershed, Colombia. *AgREN Network Paper* 65, 14.

Rhoades, R. (2000) The Participatory Multipurpose Watershed Project: Nature's Salvation or Schumacher's Nightmare? In: L. Rattan (ed.) *Integrated Watershed Management in the Global Ecosystem*. London: CRC Press, pp. 327–43.

Rietbergen-McCracken, J. and D. Narayan, comps. (1998) *Participation and Social Assessment: Tools and Techniques*. Washington, D.C.: World Bank.

Schubert, W.D., Klukas, O., Saenger, W., Witt, H.T., Fromme, P., and Krauss, N. (1998) A common ancestor for oxygenic and anoxygenic photosynthetic systems: a comparison based on the structural model of photosystem I. *J Mol Biol* 280: 297–314.

Shah, A. (1998) Watershed Development Programmes in India: Emerging Issues for Environment-Development Perspectives. *Economic and Political Weekly*, June 27, 1998, A66–A78.

Stroud, A. (2003) Combining science with participation: Learning locally and generalizing regionally. *AHI Brief* B3.

Turton, C. and J. Farrington (1998) Enhancing Rural Livelihoods through Participatory Watershed Development in India. *ODI Natural Resource Perspectives* 34: 1–4.

Van Horen, B. (2001) Developing Community-Based Watershed Management in Greater São Paulo: The Case of Santo André. *Environment & Urbanization* 13(1): 209–22.

4

PARTICIPATORY LANDSCAPE GOVERNANCE

Laura German, Waga Mazengia, Simon Nyangas, Joel Meliyo, Zenebe Adimassu, Berhanu Bekele, and Wilberforce Tirwomwe

Context and rationale

This chapter builds on AHI experiences in participatory watershed management, but focuses on the social and institutional dimensions of natural resource management challenges at landscape level. As mentioned in Chapter 3, time–space interactions between plots and common-pool resources, lateral flows of materials (water, nutrients, pests), and interdependence between users in terms of resource access and management, require decision-making and intervention strategies beyond the farm level (Johnson et al., 2001; Knox et al., 2001; Ravnborg and Ashby, 1996). Therefore, in addition to emphasizing effective participation and integrated decision-making to acknowledge linkages among diverse system components and users, processes for governing biophysical processes connecting different land users and interest groups are sorely needed. This is particularly true where demographic, economic, and ecological dynamics have outpaced the ability of customary systems of natural resource management to cope with change. Efforts to foster collective action and govern natural resource decision-making should therefore be considered a fundamental component of any watershed management process—particularly where local motivations (e.g., issues that are of deep concern to at least some land users) do not translate into effective solutions in the absence of intervention.

Social and political dimensions of NRM are very poorly addressed in NRM research and development programs. These include a complex social fabric within communities based on differences of gender, kinship, tribe, wealth status, religion, and politico-administrative divisions. They also include internal polarization around "appropriate" land-use practices based on economic or other interests. The divergent "stakes" that lend a political dimension to landscape management generally go unrecognized. The intractability of many natural resource management challenges—which manifest as inherently biophysical on the surface—may in fact result from underlying epistemological, cultural, and political factors (German et al., 2010; Leach et al., 1999). These include divergent interests associated with either a mismatch between efforts required of individuals to implement an innovation and the benefits they anticipate from them, or from land-use practices for which benefits accrue to some land users and costs to others. At times, the prevalence of practices that carry negative consequences for other land users may be owing to limited awareness of the consequences of current behavior on others or on environmental services of local importance (e.g., water). However, as this chapter shows, it is often owing to the fact that important benefits accrue to those households continuing practices viewed as detrimental by others. These divergent interests often create latent conflict that leads to a breakdown in communication among those who most need to plan collectively to address the problem.

Such divergent interests need to be taken into consideration in the approaches used by NRM research and development organizations. The tendency, however, is to focus on technological solutions to NRM problems and on individual decision-making on which solutions to test. Who participates and who benefits are questions that are largely left unaddressed, as are those issues that require collective or negotiated decisions in order to be effectively addressed. Such deficiencies may create further inequities, undermine the effective resolution of landscape-level NRM problems, or result in lost synergies between social capital development, technological innovation, and natural resource governance.

As this chapter and the wider literature illustrates, collective action and participatory governance processes are required to regulate rights and responsibilities to common property resources and public goods such as water, communal grazing lands, and community forests (Gaspart et al., 1998; Gebremedhin et al., 2002; Munk Ravnborg and Ashby, 1996; Ostrom, 1990; Scott et al., 2001). Collective solutions are also required to manage biophysical processes that do not respect farm boundaries, such as control of pests and excess run-off, minimizing damage caused from free grazing, or managing the effects of boundary vegetation on adjacent farms (Munk Ravnborg et al., 2000). Collective action is likewise necessary to negotiate joint investments and technological innovations for enhanced productivity or income, for example to enable the sharing of transaction costs of organizing or marketing, and to regulate benefits capture from outside interventions (Meinzen-Dick et al., 2002).

When developing and testing social and institutional innovations to be tested in this arena, it was necessary to ground the design of interventions on established theoretical understandings of the foundational elements to effective local governance of natural resources. For this, we drew heavily on the work of Eleanor Ostrom (1990). Ostrom's work was instrumental in countering the theory laid out in Hardin's *The Tragedy of the Commons* (1968), where it is posited that open access and unrestricted demand for a finite good in common pool resources will inevitably cause over-exploitation and resource degradation—thus requiring enclosure or privatization of the commons. The Ostrom tradition has clarified how groups of users can create institutions to fulfill a set of functions required for managing resources sustainably—namely, exclusion, allocation among users, and establishing conditions of transfer. By studying a large number of case studies from traditional common property regimes across the world, they were able to distill a set of features common to institutions that have proven effective in managing common pool resources sustainably (Ostrom, 1990; see also Pandey and Yadama, 1990; Wittayapak and Dearden, 1999). These include:

1. A clearly defined community of resource users and clearly defined resource (both of manageable size).
2. The presence of a set of clearly defined "collective choice rules" developed voluntarily by users to clarify the rules of the game (e.g., what is permissible, what contributions are required in exchange for use rights, and sanctions for non-compliance) and which help to balance the costs of collective action with the benefits derived from it.
3. Sanctions that are "graduated" or matched to the level of the offense.
4. Systems for monitoring the status of the resource and for adaptive management (to enable rules to be modified as need arises).
5. Conflict resolution systems.

Yet despite the rich body of research and accumulated wisdom on customary systems of natural resource governance emanating from the Ostrom tradition, efforts to apply these principles to contemporary natural resource management challenges are hard to come by. To what extent can negotiation support processes enable local interest groups to see the value of collective action or to overcome the silences that characterize latent conflicts and thus contribute to inaction? To what extent are the principles of Ostrom's "self-governing institutions" relevant to contemporary problems where institutions are insufficient for addressing natural resource management problems of local concern? And are they effective in building solutions that lie beyond common pool resources? To what extent can external research and development institutions move beyond generic forms of support to undifferentiated communities towards "more explicit partiality" (Leach et al., 1999)? These are the questions that the body of work presented in this chapter attempted to address.

The following section presents a typology of natural resource management issues, constructed based on experience in supporting local land users to identify and address natural resource management issues requiring collective solutions. The three sections that follow describe different approaches employed to strengthen local governance of these issues, and the closing sections synthesize lessons learned and remaining challenges.

Toward a typology of issues requiring improved landscape governance

In AHI watershed sites, two broad scenarios were found that require efforts to foster collective decision-making among divergent local interest groups. The first involves NRM issues that remain unresolved owing to inadequate collective action among community members. The second and more intractable scenario involves local interest groups with divergent views and interests around the issue. The nature of these issues determines the type of strategies most effective for improving equitable landscape governance. For a brief illustration of the two scenarios as they apply to vertebrate pest control, see Box 4.1.

BOX 4.1 CONTROLLING PORCUPINE IN AREKA: TWO DETERRENTS TO COLLECTIVE ACTION

Crested porcupine was the most important vertebrate pest identified during the problem diagnosis stage of AHI in Gununo watershed. Porcupines cause tremendous crop losses for Areka farmers. Porcupines eat primarily maize cobs, followed by roots of sweet potato, leaves of cabbage, roots and tubers of yams, potato, cassava, haricot bean, and seeds of field pea. As mentioned in Chapter 3, farmers spend their nights keeping watch over their fields against porcupines during the growing season, causing loss of sleep and frequent visits to the health center for weather-induced disease. The porcupine case is an interesting collective action challenge because it applies to each of the two scenarios, as follows:

A. The problem remains unresolved owing to inadequate collective action

As porcupines travel up to 14 km at night in search of food, crossing many farm boundaries, efforts to kill or trap them by individual farmers are largely ineffective. Its management must extend beyond farm and watershed boundaries, even up to district level, to minimize the effect of re-infestation from adjacent villages or Peasant Associations (PAs). The large areas over which re-infestation may occur makes individualized efforts at porcupine control largely ineffective.

B. Divergent interests of local stakeholder groups hinder the quest for easy solutions

There is a second challenge to catalyzing collective action for porcupine control, namely, the presence of local interest groups here defined by the level at which different households are affected by porcupine. Farmers growing crops vulnerable to porcupine damage are more eager to engage in collective solutions than those growing crops less susceptible to attack (for example, teff, wheat, barley, and enset). For this purpose, tools for awareness creation were used to encourage high levels of collective action across several PAs. Community meetings were called to raise awareness of the fact that households less affected today may be susceptible in the future if they shift to crops eaten by porcupine. By-laws were also developed through participatory dialogue among the different interest groups to hold individual households accountable to collective interests.

In recognition of these barriers to collective action, combined strategies were used both to mobilize the overall community (megaphones, local music, and awareness creation) and to foster equitable solutions between the local interest groups (negotiation support, participatory by-law reforms). When applied concurrently, these strategies were instrumental in reducing crop losses, labor burden, and illness resulting from long nights spent policing fields against the pest among PA residents.

We now delve into each of the two scenarios in greater detail as a means of illustrating the diversity of issues that falls within each, as well as the sub-classes into which each of these issues may be further differentiated.

Scenario 1: Issues remain unresolved owing to inadequate collective action

In this scenario, either the solution is not fully effective when carried out by individuals, or in the absence of collective action the issue simply cannot be solved. The following types of NRM problems are *less effective* when done on an individual basis:

- Control of many pest and weed species that easily spread across farm boundaries (as in Box 4.1).
- Controlling run-off and soil erosion, for which greater levels of collective action imply more effective solutions, owing to "aggregate effects" of many households implementing soil conservation structures (Box 4.2).
- Nursery management, where "free riders" (who fail to invest time according to agreements) undermine incentives of others to engage in collective action (Box 4.3).

BOX 4.2 CONTROLLING RUN-OFF IN KAPCHORWA DISTRICT, UGANDA: FROM "LONE RANGER" TO COLLECTIVE ACTION

Mr. Akiti Alfred of Tolil village in the Benet Sub-County has, in recent years, been constructing soil and water conservation structures in an attempt to control the run-off in his fields. However, his fields continued to be affected by the ever-increasing run-off from his upslope neighbor's fields. He approached one of his neighbors from the adjacent village, Mr. Kissa Peter, and told him about the continued run-off affecting his fields. Mr. Kissa said that he was also experiencing similar problems of soil degradation and declining crop yields and that his crop yields of maize had reduced by about 60 percent. Mr. Kissa further explained that there were other farmers in other villages experiencing similar problems, despite the fact that they had adopted soil conservation structures in their fields. From this experience, Mr. Akiti approached two other neighbors about the problem and discovered that they were equally concerned. Mr. Kissa and these other neighbors advised him to call for an urgent village meeting to share with other farmers ideas on how they could deal with the run-off affecting livelihoods of the entire community. A meeting was convened in the Tolil village to discuss strategies for controlling run-off.

In the village-level meeting, it was resolved that a broader meeting should be held among four of the most heavily affected villages. In that meeting, residents of the four villages resolved to form Village Watershed Committees to take responsibility for common NRM problems. New by-laws governing common NRM issues were then formulated, and soil and water conservation technologies used to implement agreed by-laws. The Kapchorwa District Landcare Chapter continues to serve as a multi-stakeholder platform to backstop and support communities in their articulation and resolution of this and other common NRM concerns.

BOX 4.3 NURSERY MANAGEMENT IN GINCHI: LEARNING THROUGH ITERATIVE PHASES OF IMPLEMENTATION AND ADJUSTMENT

Extensive forest clearing for cultivation and over-grazing, and unregulated exploitation of forests for fuelwood and construction materials in the absence of reforestation efforts at household or community level, has led to the depletion of forest resources in the watershed. As an integral part of integrated watershed management, introducing multi-purpose tree species in Galessa watershed was seen as a priority of both farmers and the site team. An action research approach was employed to develop, test, and improve upon the approach over time.

Testing of approaches

Approach 1—A meeting was organized and farmers were reminded about the concern they ranked very highly during the participatory watershed diagnosis—namely, loss of indigenous tree species. Farmers were asked to identify the most preferred tree species in their locality. One tree nursery was established in the watershed for the whole watershed community in 2004/2005. The nursery was to be managed collectively by the entire watershed community.

Outcome 1—The performance and survival rate of seedlings in the nursery was poor (54.7 percent) owing to poor nursery management. This was in turn owing to lack of agreements on communal work (e.g., how responsibilities and benefits would be shared), lack of knowledge on raising seedlings, and lack of nursery tools.

Approach 2—In 2005/2006, those farmers with an interest in raising seedlings in each village were organized in groups and a single nursery was established at Legbatebo village with subdivisions into blocks corresponding to different villages, to clarify ownership. Collective choice rules were developed to specify how responsibilities for nursery maintenance would be shared and seedlings distributed. Trainings were given and continuous follow-up was made to reinforce local agreements.

Outcome 2—While the number of participating farmers declined (from 86 to 36), the performance and survival rate of seedlings was very good (97 percent) owing to the manageable size of the group and the local governance arrangements put into place.

The lessons learned from this experience included: (i) the need to have a manageable number of farmers to work together for nursery management; (ii) the importance of developing collective choice rules through the full participation of all participants; and (iii) the importance of ensuring these rules are enforced. Each of these is a key to sustaining collective action, as they assist in clarifying both responsibilities and the distribution of benefits.

Examples of natural resource management problems that generally *cannot be solved* in the absence of collective action include the following:

- Extensive use of outfields, in which free grazing traditions (including seasons of restricted and/or open access grazing) will subject any innovation to collective agreement.
- Extensive use of outfields, in which traditional beliefs governing the use of the common property resource prohibit any innovation (Box 4.4).
- Controlling extreme run-off, which requires trenches across the entire landscape and agreement on the location of common waterways which must pass through farmers' fields (to divert excess water from fields).

BOX 4.4 LOCAL BELIEFS GOVERNING THE USE OF COMMUNAL GRAZING AREAS IN AREKA

The communal grazing area in Areka covers approximately 60 ha. Residents of Gununo watershed say that the land was once privately owned but transferred to the community for grazing purposes. At that time, the landowner called a community meeting for the purpose of handing over the land to the community and about 100 cattle were slaughtered for the celebration. On this occasion, the owner made the community promise not to utilize the land for any purpose other than communal grazing. At that time, the land was productive owing to the low livestock population. However, now the land is utilized unproductively, scarcely supporting the large livestock population. When exploring options for intensification, the community strongly resisted touching the grazing area. One man whose farmland had encroached onto the communal grazing land in the past died, making the community believe that they will be cursed if they do the same. These beliefs may have an adaptive logic, such as ensuring access to pasture irrespective of household landholdings and supporting social safety nets (through the complex livestock sharing mechanisms mentioned in Chapter 3). However, its productive value is strongly undermined through overgrazing – suggesting the need for some form of collective action.

Scenario 2: Divergent interests of local interest groups hinder easy solutions

The issues that fall within this second category remain unresolved either because collective action requires that some individuals contribute or sacrifice more than they are likely to benefit from collective action, or because one party is benefiting while another party(ies) is harmed by the status quo. Problems stemming from the latter often involve latent or overt conflict and a resulting breakdown in communication.

Local interest groups or stakeholders for Scenario 2 may be defined in much the same way:

(i) Some households are more affected than others, and the motivation to participate in collective action varies among most and least affected households. Examples include the following:

- Controlling excess run-off, where upslope farmers benefit less from soil conservation structures because they are less affected by the damage caused by excess run-off from upslope.
- Crop destruction from porcupine, since some households grow crops that lure this pest (e.g., sweet potato, maize, haricot and faba bean), while others do not grow crops attractive to porcupine (please refer to Box 4.1).
- Loss of soil fertility from excess erosion under the following situations:
  - *When eroded soil is fertile*: Upslope farmers are negatively affected by loss of fertile topsoil, while downslope farmers benefit from the deposition of this same soil on their land.
  - *When eroded soil is infertile*: Downslope and valley bottoms are negatively affected by deposition of infertile soil over their more fertile

soil, while upslope farmers are losing only infertile soil and are less affected compared to downslope farmers (Figure 4.1).

– *Irrespective of the fertility of eroded soil*: Households with steep slopes are more affected by soil fertility lost to erosion than households with flatter land.

(ii) Land-use practices of some households (interest group 1) have a negative effect on other households (interest group 2). Examples include the following, which are common to most AHI sites:

- Fast-growing trees (most notably, eucalypts) planted on farm boundaries, which have a negative effect on adjacent farmers' fields owing to competition for nutrients, water and light, and to allelopathic effects.
- Spring degradation from land-use practices of landowners with springs on or near their land, owing to the cultivation of "thirsty" trees (which tend to grow better with improved water uptake), and to the loss of protective vegetation and contamination associated with the cultivation of crops up to the edge of springs (where land owners gain from bringing a larger land area under cultivation) (see Plates 12 and 13).
- Crop loss from free grazing, where households have very divergent livestock holdings and incentives to reduce free grazing only exist among households with low livestock endowments or deriving significant benefits from livestock sharing arrangements (Figure 4.2).

In the text that follows, AHI efforts to develop and test approaches for improving landscape governance—with an aim of enhancing both equity and sustainability—are described. The text is organized according to key methodological innovations that were tested for this purpose, including approaches

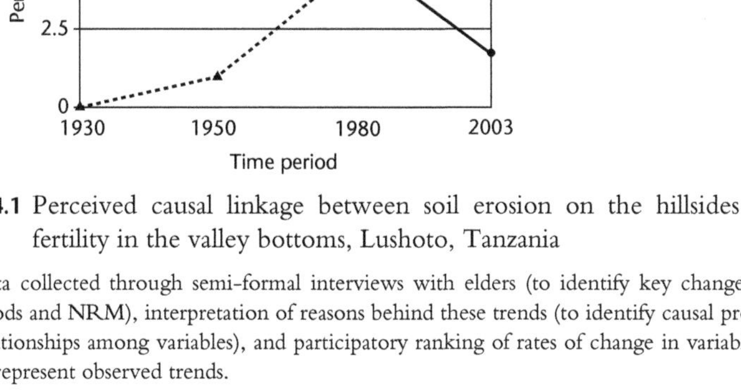

FIGURE 4.1 Perceived causal linkage between soil erosion on the hillsides and soil fertility in the valley bottoms, Lushoto, Tanzania

Note: Data collected through semi-formal interviews with elders (to identify key changes in livelihoods and NRM), interpretation of reasons behind these trends (to identify causal processes and relationships among variables), and participatory ranking of rates of change in variables selected to represent observed trends.

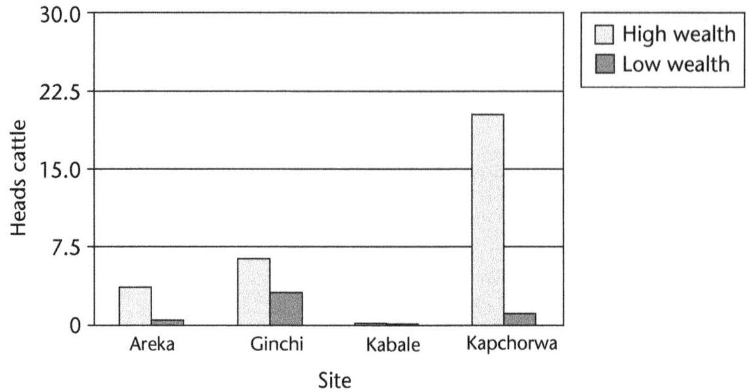

FIGURE 4.2 Livestock holdings by wealth category in four AHI benchmark sites

for supporting negotiations among local stakeholders with particular interests vis-à-vis the identified natural resource management challenges; methods for catalyzing collective action; and methods for participatory by-law reforms. Negotiation support was applied independently of the scenario under consideration (inadequate collective action, or divergent interests). The last two approaches, however, were explicitly targeted to consider the unique features of the issues being addressed. Strategies to mobilize collective action were applied largely to problems defined by Scenario 1, while participatory by-law reforms were applied largely to problems defined by Scenario 2. The reason for this differentiation is that formally endorsed by-laws are generally required to ensure negotiated agreements are implemented in practice, given the divergent interests characterizing the latter set of issues.

Negotiation support

The first strategy tested by AHI to address landscape governance challenges (insufficient collective action or divergent interests) was to support negotiations among affected parties. This helped to raise awareness of the need to act collectively, to reconcile latent conflicts among divergent local interest groups, and to devise strategies to hold external agencies (i.e., local government, extension, conservation agencies) accountable to locally felt needs. In AHI, several broad negotiation support strategies may be distilled, based both on the level at which negotiation support was carried out and the extent to which stakeholder interests are aligned or divergent. Strategies which may be defined by the level of intervention include support to negotiations among local stakeholders within the watershed area itself on the one hand, and support to negotiations between local communities and external stakeholders on the other. The second set of strategies may be differentiated according to whether the problem affects all stakeholders equally (and thus requires simple resolution of differences of opinion on whether and how to address it), or involves divergent interests and "stakes" (and thus interests which may be advanced or undermined through the negotiation process).

Approach development

Irrespective of the nature of the issue, the negotiation support strategy employed started with the following two steps:

1. Identification of specific landscape niches where the watershed problem is manifest.
2. Stakeholder identification, to explore which of the following scenarios the problem may be best characterized by:
   - Scenario 1—Different households are equally affected by the problem, but the collective action required to effectively address the problem is lacking;
   - Scenario 2—Different households are negatively affected by the problem but by different degrees—thereby causing them to have different levels of motivation for investing in a collective solution; or
   - Scenario 3—The issue may be characterized by two distinct interest groups, those perceived to be causing the problem and those affected by it.

At this point, once it is clear whether the problem is characterized by stakeholder groups with divergent interests, the approach diverges. If there is no such differentiation, one proceeds with the "undifferentiated" approach; if stakeholders with divergent interests are identified (Scenarios 2 and 3), one proceeds with the second approach—negotiation support involving stakeholders with divergent interests.

Approach 1—"Undifferentiated" negotiation support

If the problem is characterized by Scenario 1, the undifferentiated approach follows in which the following steps are followed with all involved parties present:

1. Provide feedback to participants on the steps taken so far and their outcomes (e.g., problem diagnosis, niche and stakeholder identification), and solicit reactions to the same.
2. Facilitate a discussion of the role of collective action in addressing the identified issue—confirming whether it is needed and why, discussing why it has been ineffective to date (or might have been more effective in the past), and agreeing on implications for the way forward.
3. Negotiate solutions that are acceptable to all parties present and implicated in one way or another by the proposed action.
4. Develop a detailed implementation plan with responsibilities and timeline.
5. Participatory monitoring and evaluation, and adjustment of work plans to address problems that arise during implementation.

As collective action involves a trial-and-error process that may not be effective the first time around, regular participatory monitoring and evaluation are required to identify deficiencies in collective action and solutions for overcoming these. The monitoring may reduce in frequency as participants become increasingly adept at working together towards a solution, but only stops once the underlying problem is solved.

Approach 2—Negotiation support involving stakeholders with divergent interests ("multi-stakeholder" negotiations)

If the problem is characterized by Scenarios 2 (households affected by different degrees) or 3 (losers and winners), the second approach is followed—which makes an explicit attempt to reconcile the interests and perspectives of different stakeholders. This approach aligns with the "stakeholder-based planning" approach described in the participatory diagnosis and planning section of Chapter 3. Following the two initial steps described above, it proceeds as follows:

1. Identification of appropriate mediators. Prior to the negotiation, an appropriate mediator should be identified—particularly for the more entrenched conflicts for which one or more parties are reluctant to enter into dialogue. This person should be someone well known and respected by both parties, knowledgeable about the technical and social aspects of the conflict, and neutral with regard to the outcome and the interests of each party. If the issue is not overly polarized, this facilitator could include project personnel, but more often local elders and opinion leaders, local administrative leaders, or spiritual leaders can be engaged as mediators, with support from project personnel.[1]
2. Consultation of individual stakeholder groups to identify their perceptions on the causes and consequences of the issue, possible opportunities for "win–win" solutions and approaches they are comfortable with for entering into dialogue with the other stakeholder group. These consultations also help to demonstrate the external party's concern for their "stakes" in the issue, and to reduce their fear of engagement (for fear of what they might lose). In cases of entrenched conflict or highly divergent interests, this step is often essential in bringing the two parties closer to dialogue and may involve a series of meetings (Box 4.5).

BOX 4.5 CASE STUDY ON CONFLICT: THE ROLE OF SEQUENTIAL NEGOTIATIONS

Farmers ranked spring degradation as the top watershed issue in Lushoto. Springs are communally owned according to national laws, even when located in people's fields or plots. However, individuals refused to abide by by-laws aimed to conserve the resource. Dialogue between spring owners and users was therefore necessary to avert conflict and address the problem. A series of multi-stakeholder dialogues were convened by AHI, bringing the negatively affected spring users and the landowners together to discuss how costs and benefits are distributed among local interest groups. While losses were occurring to both groups (through reduced access to water by spring users, and latent conflict for spring owners), benefits were only accruing to spring owners (e.g., from the expansion of cropping area or rapid growth of woodlots in the presence of water). Solutions were needed that acknowledged the stakes involved for both parties. In most such meetings, participants were able to agree and strike an acceptable balance. However, certain spring owners were initially reluctant to change, and often missed such meetings altogether. More targeted follow-up negotiations between local leaders and land users were effective in encouraging most of these landowners to protect the springs falling within their land. The few individuals who continued to protest—and even destroy investments made in spring protection by other community members—were eventually taken to court. Informal negotiations should be seen as complementary to formal law enforcement, given the ability of the former approach to avert longstanding conflict between families. The latter is, however, needed in some cases.

In the case of informal consultation of specific interest groups, it is also necessary to show compassion or empathy for the interests and concerns of *each* party. If the mediator is perceived at this time as being biased toward one party or having an interest in a particular outcome, it will jeopardize their ability to bring the two parties to the negotiating table. This should also include joint formulation of the agenda to be followed during the first negotiation, which will help diffuse tension and create a more comfortable and harmonious atmosphere for dialogue. Even the language that is used has a crucial role in either further polarizing the two parties or bringing them closer to the negotiating table at this time (Box 4.6).

BOX 4.6 PRINCIPLES OF MULTI-STAKEHOLDER NEGOTIATION: THE CASE OF THE SAKHARANI MISSION

The Sakharani Mission boundary case study described above illustrates some additional principles in multi-stakeholder negotiation. These include the following:

- *Showing empathy*. Having diagnosed watershed problems through the minds of farmers alone during the watershed exploration phase in effect marginalized a host of issues faced by Sakharani in relation to neighboring villages. These issues—including deforestation and its perceived effect on rainfall and water supply, and damage caused to tree seedlings from free grazing by neighboring farmers—were promptly brought to our attention in the first meeting (stakeholder consultation). By expressing empathy and concern for these problems in addition to those raised by neighboring smallholders, the farm manager perceived AHI to be a neutral and unbiased party and became more open to engaging in a negotiation process—as it was seen as a potential opportunity for addressing long-standing concerns of the Mission as well.

- *Use of language.* During our preliminary meeting with the Sakharani farm manager, one of the team members introduced the problem voiced by farmers—namely the negative impact of Sakharani boundary trees on neighboring cropland and springs. Use of language that unnecessarily polarized the interests of the two parties ("stakeholder") and presupposed compromise on behalf of the landowner ("negotiation") provoked an understandably defensive reaction in the mind of the farm manager. Careful choice of words to avoid further polarizing the issue is essential in early stages of stakeholder consultation and negotiation support. Words such as "party" and "dialogue," for example, are less threatening than words such as "stakeholder" and "negotiation."
- *Importance of balanced concessions.* The last principle relates to the first, in that deadlocks to constructive engagement can rarely be solved without each party "giving up" something for the collective good. In this case, the Sakharani farm manager agreed to change the boundary tree species from eucalyptus spp. to *Mtalawanda* (*Markhamia obtusifolia*), provided neighboring farmers kept their livestock from grazing within Mission boundaries and they both agreed to work together to recuperate degraded waterways.

3. Facilitation of multi-stakeholder dialogue between the two parties, through the following steps:
   - Provide feedback to participants on steps taken so far and their outcomes.
   - Jointly establish ground rules for dialogue, such as being respectful in listening fully to others and focusing on needs and interests rather than specific solutions when each stakeholder presents their perspective on the issue.
   - Ask each interest group to express their views using the ground rules.
   - Support the negotiation of socially optimal solutions that meet the needs of each stakeholder group and that do not overly burden households who have little to benefit from the outcome.
   - Formulate by-laws to support the agreements reached by the negotiating parties (see below and Chapter 5 for additional details).

- Develop a detailed implementation plan stating what is to be done (activities), who is to do it (responsibilities), when (timeline), where (in which particular areas), and how (Box 4.7). Written agreements with the signature of all participating parties bring greater legitimacy to agreements and ensure accountability by each party.

4. Formal endorsement of by-laws.
5. Participatory monitoring and evaluation, and adjustment of work plans to address problems that arise during implementation

BOX 4.7 THE IMPORTANCE OF DETAILED PLANNING FOR THE IMPLEMENTATION OF AGREEMENTS

Two cases illustrate the fundamental importance of detailed planning during multi-stakeholder negotiations.

Case 1—Sakharani boundary
Participants to the Sakharani boundary negotiations agreed on the criteria to be used for selecting priority areas for replacing eucalyptus with *Markhamia* species. However, the identification of specific boundary areas meeting those criteria was not done, leaving the most urgent case (where a neighboring farmer's home was at greatest risk of being destroyed from tree fall) unaddressed. Furthermore, the schedule of eucalyptus tree replacement was not specified, causing subsequent misunderstandings owing to divergent (unspoken) expectations. Therefore, the early trees to be removed were removed in areas of little consequence to adjacent landowners and subsequent actions were slow to materialize, undermining the spirit of agreements reached.

Case 2—Ameya spring
In the case of Ameya spring (Ginchi benchmark site), the landowner agreed during negotiations to remove eucalypts around the spring as long as other households contribute seedlings for re-establishment of the woodlot in other areas of his farm. Other details were left open. Thus the "who" was identified, but not the when, where, and how. A section of the woodlot was subsequently cut down as an expression of compliance on the part of the landowner, but the trees were not uprooted (enabling them to coppice and re-grow) and the other households did not contribute seedlings as agreed. While negotiations were ongoing at the time of writing, it is therefore clear that important opportunities are lost if concrete action plans are not developed in the first multi-stakeholder dialogue.

It is worth taking some time to reflect on individual steps in this process in greater depth. Before initiating negotiations, in addition to identifying different stakeholder groups in their aggregate it is important to identify the appropriate

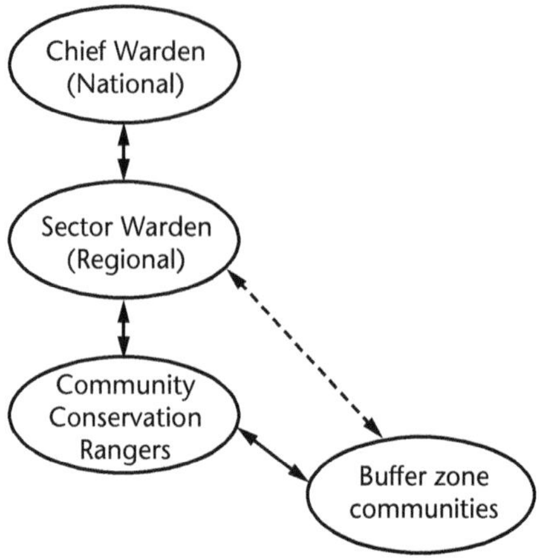

FIGURE 4.3 UWA communication and decision-making channels on co-management

avenues and levels of decision-making for each stakeholder. These authority figures can be brought directly into the negotiation process, or can be regularly updated as the dialogue progresses so as to give their blessing to the resolutions and to keep them informed. Whether or not local leaders are directly involved in the particular conflict or niche in question, for example, they should generally be present at the negotiations or be kept informed to lend legitimacy to the dialogue and help align their actions with the process and its aims. Formal institutions involved as a party to negotiations may also have established hierarchies. In Lushoto, for example, failure to involve authorities higher up in the Benedictine Order undermined the ability of the Sakharani Mission farm manager to follow through with some of his commitments on boundary management. Had the appropriate authorities been engaged in the first negotiation process, this problem could have been avoided. In the case of co-management of the Mount Elgon buffer zone (Kapchorwa site), efforts to circumvent standard communication pathways (solid lines in Figure 4.3) by taking community concerns directly to the Sector Warden helped to move beyond conflict to reconciliation, as corrupt local level officials (rangers) had more to lose from reconciliation than higher level officials.

It is also important to acknowledge the legitimate rights and authorities of each party under the law. For example, the landowner (with title or usufruct rights) may have greater authority over the use of his/her land than the affected party, as in the case of trees planted just inside farm boundaries. These rights must be acknowledged in the way the dialogue is mediated. Language matters here. For example, it is better to say to the landowner, "are there any alternative tree species that also meet your needs but minimize any negative effect on your neighbors?" than to ask the neighbors what tree species should be grown on another person's property. Similarly, supporting negotiations among parties with unequal power and authority may require a complex balancing act to acknowledge established hierarchies while also pursuing more balanced or equitable outcomes (Box 4.8).

BOX 4.8 NEGOTIATION SUPPORT IN THE BAGA FOREST BOUNDARY: MANAGING DELICATE POWER DYNAMICS

Tanzania has had a co-management policy since the late 1990s. However, these policies have yet to be operationalized in many parts of the country. In the watershed site, farmers complained that the eucalyptus—planted by the forest department to secure the forest boundary—was competing with cropland and reducing yields and contributing to the drying of springs. The site team brought the District Forestry Officer together with farmers to negotiate alternatives to current boundary management practices under the co-management umbrella.

At one point in the negotiations, the forest officer became visibly uncomfortable with the process being led by the external facilitator, as it departed from the standard approach used by the forestry department. Sensing we were losing one stakeholder's buy-in to the negotiation process, we quickly decided to hand over the facilitation role to the officer. This quickly brought him back to the table, but also tended to put much decision-making authority in the hands of only one party. Fearing we might in turn lose the community's commitment to negotiation, the team continued to play a guiding role in the negotiations, for example by pressuring the forest officer to commit to concrete deadlines for following through with agreed resolutions.

A number of additional observations may be made about the negotiation process itself. First, it is important to give each party an opportunity to express their respective views. Ground rules, such as "listening to the perspectives of others before intervening," can be either established openly through a facilitated dialogue or integrated into the process implicitly through a skilled facilitator. For example, if one party attacks the other when expressing his or her concerns and views, the facilitator needs to intervene and impress upon people the need to fully hear out the other party and acknowledge the legitimacy of one another's concerns.

It is also important to identify solutions that can ensure the interests of both parties are met. Such opportunities can be identified through detailed exploration of the main interests of each party, to see how they might come together to resolve the concerns not of one party but of both. While the benefit of reduced conflict with one's neighbors may in some cases serve as an important factor motivating the acceptance of solutions that are otherwise undesirable, strong economic rationales often underlie the more intractable problems—requiring a solution that addresses these constraints. Effective strategies can therefore involve efforts to minimize the cost or increase the benefits associated with alternatives (German et al., 2009). And while some solutions may take the form of "win–win" outcomes, the majority will involve balanced concessions from each party (Box 4.6). Ideally in such situations, each party concedes something while also securing certain benefits. Yet parties to a negotiation will seldom offer a concession for nothing; there is often need for reciprocity in such concessions to enable a "middle ground" to be met. This is more easily done through an emphasis on "bottom lines" than by trying to ensure that each and every concern of each party is adequately met. "Bottom lines" emphasize the basic interests that must be met for each party to continue participating in the dialogue (Box 4.9).

BOX 4.9 ENSURING THAT "BOTTOM LINES" ARE MET TO SUSTAIN STAKEHOLDER COMMITMENT TO A NEGOTIATION PROCESS

Case 1—Meeting the Uganda Wildlife Authority's bottom line of biodiversity conservation

The Mount Elgon co-management case study can help to illustrate how ensuring the "bottom lines" of certain stakeholder groups can help to keep the negotiating parties committed to dialogue. Co-management was undergoing implementation in other parts of Mt. Elgon National Park, but had thus far excluded the Benet owing to the history of conflict and the perception that the Benet were fundamentally against biodiversity conservation. Identification of this bottom line of UWA helped to keep them committed to dialogue and to advance the reconciliation process. This went a long way in fostering dialogue and a commitment to shared custodianship of the Park's resources.

Case 2—Meeting landowners' bottom line of livelihood security in spring negotiations

In the case of Ameya spring (Ginchi site), the landowner was strongly reluctant to enter into dialogue owing to his fear that he would lose a substantial investment and "safety net" if forced to remove his eucalyptus woodlot from the spring. Only when local elders expressed support for his interests did he agree to come to the negotiation table. Once there, only when the community agreed to meet his demand of bearing the costs of relocating his woodlot to another part of his farm would he agree to remove any trees from the existing location.

Resolutions reached through multi-stakeholder dialogue will also require frequent follow-up, particularly in early stages, to ensure effective implementation. When this is not done, stakeholders often identify opportunities to further their interests outside the scope of the agreement. This may be done either by failing to implement resolutions perceived to be less than beneficial to their interests, or by placing more conditionalities on their continuing cooperation than they earlier articulated to exploit gaps in the implementation plan. For example, the Ameya spring owner raised many new demands following failure of the initial agreement to specify the details of implementation. While the initial demand included only contributions of single seedlings per household, subsequent demands included community contributions to land preparation, transplanting, and fencing of the new woodlots. Furthermore, he specified that the uprooting of eucalyptus would be gradual over time, in accordance with the rate of biomass accumulation in the new woodlot. These new demands, if not covered in the original dialogue, can further polarize the issue in the minds of the other party—who by now perceives the relationship as one of exploitation rather than collaboration. Formulation of by-laws to enforce resolutions reached in early negotiations is often necessary to ensure that each party follows through with what is agreed. These proposed by-laws must often be endorsed officially by the relevant local government authorities to be effective, as described in the next section.

Finally, challenges often arise in implementing agreements owing to interactions between watershed and non-watershed communities. Two scenarios where this has occurred may be identified. In the first, parties not directly involved in the negotiations may influence the ability to effectively implement agreements. For example, non-watershed residents whose livestock freely graze in the Ginchi site during the dry season will create a burden on the efforts to police conservation structures. This was found to be particularly true in Tiro village, where the main road passes, thus concentrating livestock movement in adjacent farmers' fields. In the second scenario, farmers residing outside pilot villages feel resentment from being excluded from highly beneficial activities such as spring development or high-value enterprises. Such was the Ginchi case highlighted in Chapter 3, where the PA leader residing outside the watershed sabotaged village meetings by calling mandatory meetings at PA level during days when watershed activities were planned. This problem was addressed by involving him in his official capacity, thereby giving recognition to his importance in addressing problems within the watershed. This has been effective in dissipating the tension between watershed and non-watershed residents. It is therefore important to either ensure all relevant decision-makers are brought on board, or—should the problems persist—to consider ways to expand some of the benefits beyond watershed boundaries.

A strategy unique to Ethiopian sites has combined sensitization with persuasion to deal with certain problems that have persisted for long periods of time. This strategy entailed regular meetings with the concerned parties at diverse levels (Woreda, PA, sub-PA, watershed, village, and individual households) to raise awareness on the issues emerging from the communities themselves and to try to bolster commitments to collective efforts to solve a particular problem. This strategy has been necessary in several specific cases, namely soil and water conservation in Areka and Ginchi (to mobilize greater participation); eucalyptus management in Areka (to reconcile divergent interests); and outfield intensification in Ginchi (to progressively engage wider sets of stakeholders). In the case of soil and water conservation in Ginchi, this strategy has also been necessary to enable landowners to agree to collective drainage canals to pass through their farms, formerly resisted owing to the space it occupies and the potential for crops to be destroyed from channeling greater volumes of water through their fields. In some sites, by-laws were formulated to help consolidate agreements reached through informal means, which has helped to ensure individuals abide by established agreements.

The approach is, however, limited in addressing problems with highly divergent interests. It was ineffective in addressing the problem of eucalyptus on farm boundaries in Areka and springs in both Areka and Ginchi, owing to the cost (of both eucalyptus removal and foregone revenue streams). Landowners therefore actively resisted change. This resistance is actually backed up by national laws mandating that a landowner must be duly compensated for any loss of property. The only viable solution in this case would be for an external actor to bear the cost of woodlot removal (given the prohibitive cost of payment in cash), or for local communities to repay the farmer in kind (e.g., through contributions to moving woodlots to new locations). In Ginchi, the approach was effective in enabling early agreements on approaches to outfield intensification, but these agreements were not ambitious enough to actually solve the problem. For example, agreements did not involve curtailing free movement of livestock in locations where trees and structures were being established, but rather the active policing of outfield areas from livestock damage. The high cost of this activity to households meant that policing was ineffective in practice and most seedlings were damaged through livestock trampling or browsing. Reluctance to engage in more far-reaching innovations was likely due to the tenure insecurity in outfields as well as the open access nature of dry season free grazing, where the users are not well defined—complicating efforts to agree on collective rules for curtailing grazing.

Lessons learned

The following lessons may be distilled from the *negotiation support* experiences of AHI:

- It is critical to "get it right" the first time, to avoid the additional burdens inherent in follow-up negotiations.
- Effective negotiations require detailed action plans on how to implement resolutions (specifying what, who, where, when, and how), and in cases of divergent interests, ensuring sufficient weight is given to agreements through signed documents, close follow-up and the formulation and endorsement of formal by-laws. Signed documents can assist in making proper follow-up to agreements and minimize the emergence of new demands from both parties.

- For successful negotiations, it is important to identify and involve the appropriate actors throughout the process, including both mediators and stakeholders.
- Negotiation support is the only approach used by AHI that has enabled identification of solutions that effectively reconcile the interests of two divergent interest groups, as it fosters mutual understanding and concessions for the sake of the collective good. Generic watershed planning processes are generally ineffective in this regard, leaving contentious issues unaddressed.

Mobilizing collective action for common NRM problems

Some NRM problems that require collective action to be effectively addressed but do not involve divergent interests require simply mobilizing groups of people to work toward common goals that cannot be achieved through individual efforts. There is no single approach used for this, but rather a sub-set of approaches that differ slightly in their steps and aims.

Approach development

Approach 1—Working through local institutions effective in mass mobilization

The first approach consisted of the identification of local institutions known by local residents to be effective in mass mobilization, and facilitating their efforts to call people to action around a locally identified concern. It consisted of the following steps:

1. R&D team assists local residents to identify NRM issues of high priority through a watershed exploration exercise or stakeholder meetings.
2. R&D team facilitates a process whereby watershed residents identify local forms of collective action (CA) most effective for mass mobilization.
3. R&D team members and/or expert farmers train leaders (from identified forms of CA) on technical aspects of addressing identified NRM problems (based on scientific or local knowledge or both).
4. R&D team or other chosen party facilitates agreement on the roles of identified CA institutions in mobilizing the community around shared concerns (e.g., using megaphones or traditional methods of calling the community to action).
5. Set specific days convenient for all for mass mobilization initiatives.

The approach was very effective in mobilizing collective action at a large scale owing to the involvement of multiple collective action structures, each one operating in a small area and with few households but together covering a large area. If these local institutions of collective action also have the mandate to develop and enforce by-laws, they can be even more effective in mobilizing collective efforts (Box 4.10).

BOX 4.10 MOBILIZING COLLECTIVE ACTION FOR COMMON NRM PROBLEMS: THE PORCUPINE CASE

This box describes in detail how the challenges described in Box 4.1 related to crop damage from porcupine were addressed in practice. As mentioned previously, porcupines travel long distances at night, resulting in very low returns on control efforts applied by individual households. To mobilize collective action over a wide area, individuals or groups effective in mobilizing the community and highly respected by all are needed. Farmers were therefore asked to identify an individual or entity that could be effective in this regard. Farmers selected the "developmental unit" (DU) (local administrative units consisting of 25–30 households) as a local institution most capable of mobilizing farmers. In addition to the DU being of manageable size, DU leaders were thought to be capable of enforcing local by-laws developed for this purpose and effectively monitoring implementation.

Approaches

By-laws specifying contributions to be made by each household to porcupine control were developed and approved by local leaders. Local knowledge on porcupine control was studied, and control methods appropriate to different landscape niches were agreed upon. The DU leader coordinated the mass mobilization and each family in the DU agreed to devote one or two "development days" per week for collective control efforts. DU leaders then mobilized farmers on the designated development day using megaphones and a local instrument called a *tirumba* and farmers applied the agreed-upon control methods for the relevant niches.

Outcomes and lessons learned

Nearly 1,000 porcupines were captured or killed in the peak porcupine season through these collective efforts. Farmers selected DUs to enable implementation owing to its proximity to communities (small administrative units with few households) and the ease with which they can closely monitor activities in their area. Development of by-laws also enabled farmers to negotiate and clarify ahead of time who would be responsible for what activities, and to ensure those agreements were respected.

Approach 2—Working with self-mobilized local institutions

The second approach seeks to capitalize upon and support self-mobilized local institutions in supporting the evolution of stronger institutions of collective action in support of NRM. This approach consisted of the following steps:

1. Community members organize spontaneously around shared NRM concerns.
2. External development actors (NGOs, research, local government) identify existing "nodes" of collective action to support, and facilitate the forma tion of a higher-level watershed committee.
3. External facilitators encourage newly formed watershed committee to create awareness among CBOs, NGOs, and local government on the need to collectively address the issue(s) of concern.
4. Local CA institutions conduct village-level planning, and express demand for support from external development actors.
5. External development actors call sub-county and district-level planning meeting to articulate the roles of different actors in addressing local level concerns.

This approach to community mobilization was highly effective as it was grounded in emerging forms of self-mobilized collective action and systematic efforts to provide external support to these emerging initiatives. It further catalyzed interest in new forms of collective action as a result of its effectiveness in addressing the issue and bolstering support from outside development agents or service providers.

Approach 3—Mobilizing CA through local government and NRM champions

The third and last approach consisted in the identification of existing NRM champions and supporting their efforts to mobilize complementary interventions by local government and watershed residents. Basic steps in this approach include the following:

1. External R&D team assists local residents to identify NRM issues of high priority through watershed exploration or stakeholder meetings.
2. Local NRM structures are formed (by election) or strengthened (where existing NRM and leadership structures exist) to spearhead solutions to issues identified in Step 1.
3. Local NRM structures drive a process involving local government to address shared NRM concerns (Box 4.11).

BOX 4.11 VARIATIONS ON THE APPROACH FOR MOBILIZING COLLECTIVE ACTION THROUGH LOCAL GOVERNMENT AND NRM COMMITTEES

Case 1—Controlling run-off in Kabale

The approach utilized in Kabale began with the identification of farmers from communities observed to have "zeal" to find a way to address their own NRM problems. Second, meetings were held with sub-county stakeholders to sensitize newly elected local government representatives given the recent election, and to form sub-county NRM committees. Third, sensitization was carried out at village level by selected members of the sub-county NRM committee with the assistance of LC1 leaders, faith-based organizations, radio announcements, whistles, and word of mouth. LC1s, NRMPCs and AHI worked collaboratively from this point forward, assisting in putting checks and balances on LC1s to encourage them to respond to felt needs of the community. Villages then selected convenient days of the week to hold meetings and collective action activities to avoid clashes with other important activities (market days, hangover days, days of prayer, community development days). The mobilization concluded with action planning at sub-county and village levels, cross-site visits to observe successful strategies to control extreme run-off, and training of farmers on technologies for controlling run-off.

Case 2—Spring development in Ginchi

As already mentioned, decline in water quality was identified as a priority watershed issue in the Ginchi site. A watershed committee composed of representatives of each watershed village was formed. They called on the village-level government (*Gare Misoma*) to call a meeting on the need to formulate a plan for spring protection, who in turn mobilized the community. In addition to watershed residents and researchers, district-level ministries (the Bureaus of Water Resources, Health and Agriculture, and Rural Development) were called by the *Gare Misoma* to the meeting. A series of meetings with diverse local stakeholders was held to agree on how rights (to water use) and responsibilities (materials, labor, and cash for spring construction and maintenance) would be allocated both within the watershed and with adjacent villages. Contributions to be made from external stakeholders (funds from AHI, technical assistance from the Ministries of Water Resources and Health) were also agreed upon at this time. After the spring was constructed and officially "gifted" to the community, a Water Committee was established and trainings given on spring maintenance and governance. While spring users contribute small cash payments for maintenance, the Ministry of Water Resources retains the mandate to provide additional technical assistance as required.

This approach is highly effective because collective action was mobilized to address an acute problem facing the community. In those cases where newly formed or reconstituted NRM structures worked hand in hand with local government (spring protection in Areka and Ginchi, and run-off control in Kabale), complementary roles were played by these two institutions, further contributing to the success of mobilization efforts. Local government has a role in mobilization of development efforts in Ethiopia, as well as by-law formulation and enforcement functions in all countries. However, government actors had not adequately addressed common NRM problems owing to lack of capacity (financial, technical), greater emphasis on government-set development agendas than locally felt priorities, or local political interests (getting votes). In a study in Uganda (Sanginga, et al., 2004), ineffective by-laws were found to be the result of weak enforcement by local leaders, lack of awareness on by-laws, outdated regulations, legislative conflicts, small plot size, absence of extension facilities, and the desire to avoid confrontations within and among households and with the local leadership. Newly formed or reconstituted NRM structures emanating from the community itself therefore assumed a complementary "civil society" function of pressuring local government units to work on local NRM priorities. The downward accountability of local NRM structures combined with the authority of local government structures proved to be more effective than either actor working in isolation. In the Kabale case, additional success factors included the use of multiple strategies for mass mobilization and the network of local institutional structures (NRMPCs) to support the mobilization effort.

Lessons learned

Approaches tested for *mobilizing collective action* around common NRM problems have taught us that:

- Different actors (local government, community-based NRM structures, faith-based institutions, NGOs, CBOs) and strategies (radio, traditional methods, and megaphones) play complementary roles in the mobilization process. When several actors and strategies are engaged simultaneously, mobilization is likely to be more effective.
- Mobilization is easiest when building upon existing or emerging local institutions and collective action initiatives.
- Networks of local institutions are highly effective in mobilizing collective action because they combine a "personalized" approach (e.g., going door to door) with coverage of large areas (by working through multiple local institutions spread throughout the landscape).
- Early successes in mobilizing collective action around specific NRM issues can catalyze community confidence to address new challenges.
- Farmers are willing to invest in NRM on other people's land, provided that: (i) the benefits are for the majority; (ii) the problem cannot be effectively solved individually; and (iii) the gestation of benefits is short term.

Participatory by-law reforms

In addition to informal negotiation support, participatory governance was furthered through participatory by-law reform processes at village and higher levels. The process of negotiating rules and seeking their formal endorsement helped both to clarify aims among diverse parties, as well as to bolster commitment to putting into practice what was agreed upon in informal negotiations.

Three approaches are presented here. The first two, both applicable at village level, were selected based on the issue at hand and the implications for those who should be present during negotiations and/or by-law reform processes. The third approach is best differentiated from the first in its efforts to strengthen the capacity of local institutions and local government to facilitate the participatory by-law reform process, and may therefore be seen as a way to institutionalize the first two approaches.

Approach development

Approach 1—By-law reforms through village-level negotiations

The strategy used in this approach was to hold village-level meetings specifically focused on natural resource governance. While these meetings do not differentiate among local interest groups with divergent 'stakes' in the issue, equitable participation by gender and lower-level administrative units (i.e., hamlets in Lushoto site) must be ensured when identifying participants to be called to the meeting. Local leaders charged with proposing by-law reforms and with by-law enforcement should also be present, to help inform and guide negotiations. These meetings employed the following steps:

1. Use of graphical representations of landscapes with and without rules governing NRM, to foster a collective understanding of the role of governance (see Plate 14).
2. Feedback of watershed problems identified by local residents.
3. Introduction of meeting objectives (namely, to identify the need for governance solutions to address identified watershed problems) and identification of types of shortcomings that might exist in policies, norms and by-laws (namely, poor enforcement, gaps in coverage for certain watershed problems, and poor design undermining its utility in addressing the problem even if enforced).
4. Identification of existing policies, the extent to which they are enforced (and enforceable under local conditions), and how effective they are in addressing identified problems when enforced.
5. Discussion of whether any new by-laws are required to address identified watershed problems, and development of revised by-laws (where existing by-laws are deficient for addressing identified watershed problems) or new by-laws (in cases where no current by-laws exist to address identified problems).
6. By-law endorsement, implementation and monitoring.

Approach 2—By-law reforms in the context of multi-stakeholder negotiations

The second approach to by-law reforms is one and the same with the approach to negotiation support described above. It differs from the first in its explicit attempt to identify and engage in negotiations groups with divergent interests, so as to overcome the impasse that tends to characterize many collective action challenges. It therefore differs by the nature of participants, with this process involving only those with personal interests or stakes in the issue at hand. It is important to note, however, that when negotiation support leads to formal by-law reforms, it will be important to find a means to "scale up" agreements to the lowest administrative level at which by-laws may be formulated. This is because once approved, by-laws apply to all residents within an area; their buy-in is therefore critical.

Approach 3—By-law reforms embedded in local government structures

The third approach to participatory by-law reforms is embedded in local government structures established under decentralization reforms, as presented in Chapter 3 (Approach 2 of participatory watershed diagnosis and planning). Here, by-law formulation is integrated into a more general strategy of working through community organizations and local government structures to support participatory INRM. In this approach, these existing structures are themselves empowered with facilitation skills to support participatory INRM and by-law reforms within their respective local administrative structures.

A comparison of the three approaches gives an idea of their relative strengths and drawbacks (Table 4.1). The unique features and purposes of the three approaches, however, suggest that with adequate funding they could be best employed in tandem—with Approach 3 used to build the capacity of local institutions and administrators to implement the specific negotiation support processes outlined in Approaches 1 and 2. A comprehensive approach for linking actors at different levels in the by-law reform process, which could serve as the backbone to the multilevel governance reform process, will be presented in Chapter 5.

It is important to reflect on the stages after negotiations are concluded. It is clear that by-law endorsement by higher-level officials is a necessary condition for their effective enforcement. Yet even with this endorsement, by-laws can be ignored. The following reasons have been identified by farmers for poor enforcement of existing or new by-laws:

- The difficulty of holding certain community members (e.g., traditional healers, wealthy farmers, relatives of the leadership) accountable to by-laws, as they are feared for their status within the community or relationship with local leaders.
- Non-compliance of certain local government leaders with by-laws, which serves as a strong deterrent to others abiding by the by-laws.
- Negative livelihood consequences of enforcement for some households (Box 4.12).
- Failure to provide livelihood alternatives for activities forbidden or curtailed through by-law enforcement (such as disseminating fodder trees in exchange for restrictions on free grazing), thus undermining enforceability.
- Failure of government officials to apply sanctions when offenders are reported, owing to corruption or favoritism.

TABLE 4.1 Comparison of approaches to participatory by-law reforms

Characteristic of the approach	*Approach 1: By-law reforms through village-level negotiations*	*Approach 2: By-law reforms in the context of multi-stakeholder negotiations*	*Approach 3: By-law reforms through government and community NRM structures*
Entry point for by-law reform	Village, ensuring effective representation by gender and lower level administrative units	Interest groups involved in relatively "intractable" NRM issues involving divergent stakes	Sub-county government
Facilitation	R&D teams (but other entities can also facilitate the process)	R&D teams (but other entities can also facilitate the process)	Community-based organizations or sub-county government
Community participation in by-law revision and formulation	High (conducted first at village level; ensures representation of lower government structures)	Medium (involves only those with direct stakes in the issue at hand), but explicit targeting of those directly concerned makes it highly "equitable"	High (conducted first at village level)
Ease of by-law endorsement	Medium (government structures not brought on board initially, but can readily take up proposals once informed)	Medium (government structures not brought on board initially, but can readily take up proposals once informed) to Low (if negotiations target specific niches rather than all local residents with a stake)	High (local government at diverse levels giving oversight to the process)
Awareness of legal statutes	Medium to Low (existing by-laws and NRM policies are reviewed by the community, but awareness is incomplete)	Medium to Low (existing by-laws and NRM policies are reviewed by the community, but awareness is incomplete)	High (by-laws previously formulated are brought into the meeting by a knowledgeable party)

Characteristic of the approach	*Approach 1: By-law reforms through village-level negotiations*	*Approach 2: By-law reforms in the context of multi-stakeholder negotiations*	*Approach 3: By-law reforms through government and community NRM structures*
Duration	1 day mobilization; half day per village to formulate by-laws	Mobilization can take anywhere from 2 days (visits to each interest group) to several months (if certain parties refuse to enter into dialogue); half day per niche to negotiate solutions and by-laws	6 months
Outcomes	Community enthusiasm high; by-laws easily formulated for most NRM challenges	Stakeholder enthusiasm high if agreements reached are acceptable to both sides; by-laws more likely to "work" for the more intractable NRM issues involving divergent interests if this approach is followed	Community enthusiasm high; by-laws easily formulated for most NRM challenges; by-laws readily endorsed at sub-county level

BOX 4.12 THE LIVELIHOOD COSTS OF IMPROVED GOVERNANCE

Some by-laws proposed by community members themselves may carry detrimental effects for certain households. For example, by-laws to protect springs and waterways restrict the land area available for cultivation and grazing, in particular for those households that have springs, streams, or irrigation canals on or passing through their farms. Regulations on free grazing of livestock will have consequences on livestock productivity, in particular for households with larger livestock endowments or relying more on free than on zero grazing. By-laws regulating the distance at which certain tree species (those perceived to be overly "thirsty" or harmful to crops) may be grown relative to farm boundaries or springs restricts land-use options and revenue streams for those households practicing these activities. Governance must ultimately balance the social and environmental costs of the status quo (i.e., declining water resources, negative effects of boundary trees on neighbors, conflict) with the costs of solving these problems for the collective good. Alternative technologies (e.g., fodder, crop-compatible trees) can also go a long way in minimizing the livelihood costs of more equitable land management practices (German et al., 2009).

The livelihood costs of improved governance were a very crucial finding, suggesting there are often significant economic deterrents to more equitable, sustainable NRM. Two different possibilities exist for minimizing these costs—one technological and one social. As illustrated in Box 4.12 and in Table 4.2, technologies can play an important role in minimizing the cost of by-law enforcement to those households whose livelihood options would be curtailed in the process. It is interesting to note that during participatory by-law reform processes, complementary governance and technological solutions are almost always spontaneously proposed by participants. This has an important bearing on the sequencing of implementation. By-law *formulation* must come first, as new by-laws highlight technologies that may be introduced to help minimize the livelihood costs—and enhance the effectiveness—of by-law enforcement (e.g., fodder species providing a feed alternative to free grazing). Awareness of the by-law and its date of enforcement must also be effectively carried out far enough in advance of enforcement to enable households to adopt alternative practices to substitute those that will be curtailed through by-law enforcement. Only then must technologies be made available to all households, as awareness creation on upcoming by-law enforcement will affect demand for technologies and adoption levels. Finally, after livelihood alternatives are in place (i.e., fodder is now available), by-laws may begin to be *enforced*.

Interactions between watershed and non-watershed residents also have a bearing on effective by-law implementation. For example, a ban on free

TABLE 4.2 Proposed solutions to identified NRM problems in Ginchi benchmark site

Problem	*Technological solutions*	*Governance solutions*
1. Water quantity and quality in springs	(i) Spring development (concrete structures); (ii) Physical and vegetative structures to enhance infiltration and spring recharge.	(i) By-law specifying which tree species may be planted within a specific distance of springs (100m upslope, 25m downslope). (ii) By-laws to balance benefits with contributions to maintenance. (iii) [*Following negotiations at Ameya spring*] Each spring user to compensate landowner by planting 1 eucalyptus tree elsewhere on his farm.
2. Incompatible trees on farm boundaries	(i) Substitute species for farm boundaries that have most of the beneficial characteristics of incompatible species but carry minimal costs.	(i) Minimum 10m barrier between eucalyptus and cultivated land; (ii) Payment of reparations if policy is ignored; (iii) By-law specifying acceptable locations for eucalyptus (i.e., degraded areas).
3. Soil erosion	(i) Technologies for erosion control, drainage, gulley stabilization (physical and vegetative structures)	(i) Non-conserving farmers will compensate for losses to downslope farmers; (ii) By-laws governing drainage and gulley management.

grazing in the Tuikat Watershed of Kapchorwa has been effective only in controlling free grazing by watershed residents, but not by farmers living outside the watershed. A case from Kabale also illustrates this challenge (Box 4.13).

BOX 4.13 NRM BY-LAWS SHOULD EMBODY FAIRNESS IF THEY ARE TO BE UPHELD AND WIDELY ADOPTED

In Kabale District, AHI support to the formulation of local NRM by-laws raised questions of equity, owing to the initial emphasis on implementing them only in pilot villages. The consequences of NRM by-laws applied in one area but not in others are numerous and often controversial. Villages where the by-laws do not apply regard them as "alien" or "AHI" by-laws, and often resented or worked to actively undermine them. Owing to the prevalent practice of land fragmentation—where individual households own land plots in several landscapes and administrative units (i.e., villages, parishes, and sub-counties)—residents often complained that by-laws were unfair, and hence ineffective, since they only applied in particular locations. Consequently, some farmers were disturbed by the fact that they could freely graze their livestock in some areas, but were denied the right to graze in other "AHI" areas. At times such site-specific variations in by-laws aligned with areas under different administrative units. In a bid to partially redress this inequity brought about by the uneven application of NRM by-laws, NRM Protection Committees resolved to hasten the process of lobbying and convincing the different leadership structures at sub-county level to harmonize the diverse NRM by-laws emanating from different villages, and endorse and publicize these harmonized rules.

Lessons learned

The following lessons were learned from efforts to facilitate *participatory by-law reforms* in AHI benchmark sites:

- Informal resolutions are generally ineffective in ensuring agreed-upon rules are respected, requiring local government involvement in by-law endorsement and enforcement.
- The importance of creating awareness of the possible benefits of improved governance of natural resources, and of existing policies and by-laws, in the process of by-law review and formulation. Graphical representations of landscapes with and without by-laws can go a long way in stimulating awareness and interest in good governance during participatory by-law reforms.
- Corruption in different levels of government is a strong determinant of poor by-law enforcement, and must be addressed in efforts to improve natural resource governance in the region. This holds from village level (largely owing to interpersonal reasons and self-interest) up to district level (owing to financial and material gain from non-enforcement). It is of fundamental importance that local government leaders govern by example, and that these local leaders be sensitized in the consequences of their actions in this regard.
- By-laws must be applied uniformly in order to avoid negative transboundary effects; local government has an essential role to play in harmonizing by-laws across villages with a high degree of interdependence in their natural resource management practices.
- There is an urgent need to integrate livelihood considerations into landscape governance efforts to enhance their social and economic feasibility. For example, those negotiated agreements that create livelihoods costs to at least one party require livelihood options to minimize those costs. These

options often involve technologies that can substitute for the functions of foregone land uses (e.g., fodder in exchange for free grazing).

Missing links

Many natural resource management issues remain unaddressed *despite widespread local concern*, owing to the complex social and institutional dynamics that underlie them. These include ineffective or non-existent institutions for collective action, as well as divergent interests of local stakeholder groups that are difficult to reconcile. Substantial progress has been made by AHI in designing and testing approaches for participatory landscape governance which seek to address existing inequities in natural resource management as well as unlock the potential for solutions. These include methods for supporting local stakeholders to negotiate "socially optimal" outcomes, methods for mobilizing collective action and approaches for facilitating participatory by-law reforms. However, there are a number of methodological gaps that remain that would provide fertile ground for further methodological innovation:

1. *Exploiting synergies among different by-law reform processes*. In this chapter we profiled three distinctive by-law reform processes that were applied in different sites and for addressing a particular suite of challenges. Yet as illustrated in the comparative assessment of these approaches, there is a lot of potential to exploit synergies between these approaches. These include:
   a) *Efforts to link multi-stakeholder negotiations focused on specific landscape niches to village level by-law reforms*. Some challenges will remain unresolved in the absence of explicit strategies to identify and engage stakeholders with divergent interests. Efforts to engage specific sets of stakeholders in specific "problem niches" (e.g., farm boundary X, spring Y) are therefore essential. Yet in order for identified solutions to be supported through formal by-laws (thus making them more enforceable), they need to be officially endorsed. And for official endorsement and more widespread buy-in to identified reforms, these should be negotiated and vetted among all those with similar sets of "stakes" within local government jurisdictions (e.g., all farmers with eucalypt boundaries with all affected parties in village Z). Approaches for linking the sorely needed interest-based negotiations with wider endorsement processes remain to be developed.
   b) *Nesting village-level by-law reforms in multilevel by-law reform processes*. Similarly, detailed processes for negotiating by-law reforms at village level, presented in this chapter, have yet to be fully linked with formal processes of by-law reform and endorsement in AHI host countries (to be treated in Chapter 5). Yet there is considerable scope for refining and improving upon the ways in which this legislation is implemented in practice to achieve both equity and sustainability aims.

2. *Identification of an appropriate "institutional model" for mainstreaming participatory landscape-level innovation and governance in the region.* Most agricultural research and development organizations still focus on the farm level, where decisions on farm management are made by individual farmers or households. Organizations working at landscape level, on the other hand, tend to focus on conservation in isolation from livelihood concerns and livelihood barriers to conservation. Agricultural researchers tend to work independently from development agencies such as NGOs, local government and sometimes even extension. Other actors with a fundamental role to play, such as local courts and law enforcement agencies, generally do not view agriculture and NRM as their responsibility. Future innovation should seek to develop and test new forms of partnership and collaboration among such diverse actors to learn lessons on how to build upon complementary skills and institutional mandates in operationalizing integrated research and development agendas and fostering greater equity and sustainability in natural resource management.

Conclusions

Many natural resource management concerns of local land users require a collective approach. This may be because individualized actions are either ineffective (cannot solve the problem) or inefficient (requiring efforts that are greater than the benefits they yield in return), or because local stakeholders have divergent interests that hinder easy solutions. Issues that are particularly challenging either involve complex institutional challenges (e.g., open access resources, as observed in the Ginchi outfields) or require losses to be incurred by at least one party for the issue to be solved. This chapter provides both a typology of issues and processes identified by highland residents that require "participatory landscape governance," as well as a set of methodologies for identifying and negotiating solutions that balance the interests of multiple actors and for ensuring that commitments made through these planning processes are actually delivered on—without creating lasting enmities among neighboring land users.

Examples where previously intractable NRM issues were addressed illustrate the potential of drawing on Ostrom's principles in addressing modern-day NRM challenges where local institutions of collective action and local governance are deficient. This is seen most clearly in the benefits of a clearly defined set of users (observed in efforts to successfully identify and engage actors with specific yet divergent sets of interests in an issue, and unsuccessful efforts at intensifying open access outfields in Ginchi where users are many and often unknown), in the negotiation of voluntary collective choice rules (as seen in the context of negotiation support and participatory by-law reforms), and in the nature of those rules (including appropriate sanctions). Yet in addition to these principles, efforts to balance informal (follow-up persuasion with users to enhance compliance) with formal enforcement processes, and to marry local self-governance with efforts to engage outside institutions (local government, religious institutions, development agencies) in natural resource governance, are essential.

Note

1 For example, negotiations on spring management in Areka site were mediated by local administrative leaders at higher levels when more than one Peasant Association was involved, and both elders and PA leaders when consulting individual landowners. This approach enabled the landowner to tentatively agree to eliminate eucalyptus within a radius of 20m around springs, subject to subsequent community-level negotiations to agree on a plan for tree removal. In Ginchi, the project was unable to convince the landowner to come to the negotiation table, owing to the highly polarized and entrenched nature of the conflict, until local elders were brought on board.

References

Gaspart, F., M. Jabbar, C. Melard, and J.P. Platteau (1998) Participation in the construction of a local public good with indivisibilities: An application to watershed development in Ethiopia. *Journal of African Economies* 7(2): 157–184.

Gebremedhin, B., J. Pender, and G. Tesfay (2002) Collective action for grazing land management in mixed crop–livestock systems in the highlands of northern Ethiopia. *ILRI Socio-Economics and Policy Research Working Paper* No. 42.

German, L., W. Mazengia, H. Taye, M. Tsegaye, S. Charamila, and J. Wickama (2009) Minimizing the livelihood trade-offs of natural resource management in the eastern African highlands: Policy implications of a project in "creative governance." *Human Ecology* 38(1): 31–47.

German, L., J. Ramisch and R. Verma (2010) *Beyond the Biophysical: Knowledge, Culture and Power in Agriculture and Natural Resource Management*. Dordrecht: Springer.

Johnson, N., H.M. Ravnborg, O. Westermann and K. Probst (2001) User Participation in Watershed Management and Research. CAPRi Working Paper 19: 1–25.

Knox, A., B. Swallow and N. Johnson (2001) Conceptual and Methodological Lessons for Improving Watershed Management and Research. *CAPRi Policy Brief* 3:1–4.

Leach, M., R. Mearns, and I. Scoones (1999) Environmental entitlements: Dynamics and institutions in community-based natural resource management. *World Development* 27(2): 225–47.

Meinzen-Dick, R., A. Knox, B. Swallow, and F. Place (2002) Introduction. In: R. Meinzen-Dick, A. Knox, F. Place and B. Swallow (eds.) *Innovation in Natural Resource Management: The Role of Property Rights and Collective Action in Developing Countries*. Baltimore, MD: Johns Hopkins University Press, pp. 1–11.

Munk Ravnborg, H. and J.A. Ashby (1996) Organising for local-level watershed management: Lessons from Rio Cabuyal watershed, Colombia. *AgREN Network Paper* 65: 1–14.

Munk Ravnborg, H., A.M. de la Cruz, M. del Pilar Guerrero, and O. Westermann (2000) Collective action in ant control. *CAPRi Working Paper* No. 7.

Ostrom, E. (1990) *Governing the Commons: The Evolution of Institutions for Collective Action*. Cambridge: Cambridge University Press.

Pandey, S. and G.N. Yadama (1990) Conditions for local level community forestry action: A theoretical explanation. *Mountain Research and Development* 10(1): 88–95.
Ravnborg, H.M. and J.A. Ashby (1996) Organising for Local-Level Watershed Management: Lessons from Rio Cabuyal Watershed, Colombia. *AgREN Network Paper* 65, 14.
Sanginga, P.C., R. Kamugisha, A. Martin, A. Kakuru and A. Stroud (2004) Facilitating participatory processes for policy change in natural resource management: lessons from the highlands of Southwestern Uganda. *Uganda Journal of Agricultural Sciences* 9: 958–70.
Scott, C.A. and P. Silva-Ochoa (2001) Collective action for water harvesting irrigation in the Lerma-Chapala Basin, Mexico. *Water Policy* 3: 555–572.
Wittayapak, C. and P. Dearden (1999) Decision-making arrangements in community-based watershed management in northern Thailand. *Society & Natural Resources* 12: 673–691.

5

DISTRICT INSTITUTIONAL AND POLICY INNOVATIONS

Joseph Tanui, Pascal Sanginga, Laura German, Kenneth Masuki, Hussein Mansoor, and Shenkut Ayele

Context and rationale

In eastern Africa and indeed across much of the developing world, local government is increasingly being seen as a crucial nexus for rural development planning and implementation (IULA, 1993; Perret, 2004). While the move toward local government reforms and decentralized governance is strongly supported by multilateral development agencies (Khan, 2006), the structural adjustment programs of the 1980s also generated greater awareness among government officials of the potentially productive role of local government in development (Smoke, 1993). Being the arm of government closest to the people, it is argued that local government is best positioned to support democratization of political processes and enhance the quality and efficiency of rural development through grassroots participation.

Yet local governments face a host of challenges in meeting these expectations. Some of these are related to limited capacity and skills. These include limited experience with financial and human resource management, coordination, and planning; limited downward accountability; and lack of capacity to foster local participation (Perret, 2004). Other challenges are political and financial in nature, such as limited funding, high levels of dependence on central governments for resources, lack of motivation and political interference by central governments unwilling to relinquish control (Khan, 2006; Ribot, 2003). The need for institutional development and institution building is often acknowledged, but underfunded (Galvin, 1999). Khan (2006) argues that in order for local government to be an effective instrument of change, it must be supported financially and "backed up by consistent political will (by the state) and active society (people's participation)"—including the ability to serve heterogeneous village demands. Thus, institution building must be about both enhancing the capacity of local government to support democratic process, and the capacity of the grassroots to penetrate political and deliberative spaces and demand greater accountability.

Such challenges are compounded by historical antecedents. In most African countries, colonial governments strengthened their control over local populations through one of two means of domination: direct and indirect rule (Mamdani, 1996). With indirect rule, widespread in the governance of rural areas, the District Commissioner or "chief" served as the sole local authority, customary leadership and institutions were co-opted to serve the needs of the colonial rulers, and "local systems through which people were able to take collective action were neglected, distorted and sometimes destroyed" (Wunsch and Olowu, 1990: 27, cited in Galvin, 1999). Ironically, in the system of indirect rule, the district (in British colonies) and the *cercle* (in French colonies) was the seat of "customary" authority through which the centralized "civil" power of the state was leveraged. Thus, "decentralized" governance and district level government were tools of the central state to control its subjects (Mamdani, 1996). Newly independent governments maintained these systems for a short period, and then re-centralized government at the national level as a means to enhance central control.

Decentralization in Africa is therefore not new; however, the democratic principles driving the recent wave of decentralization represent a significant break from both the "decentralized despotism" of the colonial era (Mamdani, 1996) and the centralized control of early post-colonial states. Motives behind the current wave of decentralization include the desire to achieve administrative efficiency (owing to local decision making and coordination), enhance procedural and distributional equity, improve service delivery, deepen participation, and consolidate national unity (Ribot, 2003).[1] By placing decisions within the local sphere, it is argued that decision making will be faster and more responsive to local needs, transaction costs will be reduced by making decisions locally, and service delivery will be improved through better matching of supply and demand. These laudable goals create real challenges for operationalizing decentralized governance, as the very nature of the state must be transformed. Furthermore, as with South Africa, "the primary level of rural local government has not existed previously and thus requires support in the form of training, technical assistance and additional funding to begin to function effectively" (Galvin, 1999: 99).

Despite these challenges, the district does perform critical functions in the development agendas of nations in the eastern African region. It is usually the hub for commercial activity and social services such as hospitals and courts of law. The role of the district in providing social services, infrastructure, and other amenities is appreciated by community members who have in a number of cases petitioned their governments to create more districts, with the premise that it would bring services closer to the people. This suggests that district-level institutions and functions hold enormous potential as an engine for rural development. One of the key challenges that currently undermines more effective manifestation of this potential is the poor coordination among development initiatives and agencies, leading to duplication of efforts, missed opportunity for synergy and lack of sustainability. Various development actors (including government agencies, non-governmental organizations, and the private sector) working in the same locale often lack a consultative culture, and rarely discuss possible collaboration or coordination of efforts to capitalize upon their comparative advantage. Other challenges include weak planning processes; limited accountability; limited capacity and incentives to support participatory processes; and inefficient use of limited resources.

AHI's work at district level consisted of the design and testing of district institutional and policy innovations to explore means to capitalize upon the district's potential as a nexus for development planning and implementation and decentralized governance. This work centered around three core functions of local government in rural development: policy formulation and implementation; service delivery; and fostering democratic process or political representation (see also Galvin, 1999; Perret, 2004).

Democratic process

> Democracies are characterized by transparent decision-making and open, inclusive policy-reform processes. They provide for strong state-society links—the essence of democracy—at all tiers of government, and multiple communication channels between government leaders and citizens, affording opportunities for people to share their concerns with officials and to influence government decisions and actions. In well-functioning democracies … the availability and use of multiple forms of inclusion help ensure that citizen voices reach decision-makers and are acted on by government. These channels help citizens realize their rights and support the institutionalization of fundamental democratic principles such as transparency, responsiveness, and accountability.
>
> *Veit et al., 2008*

With the move toward decentralized governance, districts have become a key to democratic process. Via democratically elected local leaders, people gain a voice in policy matters—a voice that should be enhanced through the transfer of powers to decentralized local government. Through the proximity of local government and civil society, people can voice their development concerns directly to those charged with representing their interests. Where decentralization has increased the financial resources and discretionary powers of local government, the government should be more empowered to respond.

This is not to say that decentralization has always enhanced downward accountability or representative decision making. In decentralization of natural resource management, inequitable local decision-making and benefit distribution are frequently observed (Ribot, 2002a,b). Weak governance creates opportunities for local elites and vested interest groups to manipulate the opportunities created by decentralization for their own benefit (Tacconi, 2007). Local elites may be prejudiced toward the poor and dominant ethnic groups can use their new powers to take advantage of weaker ones (Ribot, 2002a,b). Therefore, choosing representative and accountable local institutions is a key for both equity and efficiency.

It is important to note that participation, an informal form of popular representation, also has its pitfalls. According to Veit and colleagues,

> [While] providing opportunities to directly engage in government matters, promoting the will of the people and giving voice to minorities while reinforcing majority positions, it can be time-consuming and expensive; is susceptible to rushed, uninformed decision-making; and often favours the most organized and powerful groups in society.
>
> *Veit et al., 2008*

The difference between democratic and undemocratic process—whether formal or informal—has a lot to do with accountability. Accountability refers to both the obligation to provide information and explanations concerning decisions and actions taken on behalf of others, and the ability to enforce rules and apply sanctions (Brinkerhoff, 2001, cited by Ribot, 2003). Accountability may be either upward or downward. While the latter is the essence of more democratic institutions, examples are rife of both unaccountable and upwardly accountable local institutions leading to misappropriation of funds intended for local communities (Brockington, 2007; Oyono, 2005).

Two issues related to democratic process were explored within AHI. The first concerns the development or strengthening of social *infrastructure* through which to articulate local development priorities. A number of authors warn against the creation of parallel local institutions for the implementation of development programs, given its effect on weakening democratically elected authorities and its potential to favor the most organized and powerful groups over majority interests (Ribot, 2002a). Yet some form of hybrid may be needed where government actors are too weak to fulfill their functions. The second concerns the development of social and institutional *processes* through which local voices are to be heard in setting local development priorities and enhancing rural governance.

Policy formulation and implementation

New approaches to natural resource management such as integrated natural resource management (INRM), integrated agricultural research for development (IAR4D) and sustainable livelihood approaches have emphasized the need to move beyond technologies to getting social and institutional innovations to work synergistically with technological innovations in addressing natural resource management challenges (Sanginga, 2004). Recent experience with more integrated approaches to natural resource management have illustrated the fundamental role participatory governance (and particularly the development of collective choice rules) has to play in addressing natural resource management concerns of local communities (German et al., 2008, 2010).

Districts have a fundamental role to play in policy formulation and implementation. Historically, this role was largely restricted to implementation of state-mandated policies. A system of local by-laws was first implemented by the British as a means to control the rural population, and utilized to enforce land management practices believed to be essential to environmental protection (soil conservation, forest protection, bans on burning, etc.). Imposed from without rather than developed through collective choice, these by-laws served to further the interests of colonial powers—creating a situation of resentment toward "modern" laws and the natural resource management practices enforced by them. By-laws have also been implemented as mechanisms for central state control in the post-colonial era, as illustrated by the use of agricultural by-laws during Nyerere's rule in Tanzania as a means to coerce farmers to produce more food (Sheridan, 2004). Until recently, by-laws were largely drafted by district or national governments and used as a means to promote national interests (Wily and Dewees, 2008).

While districts continue to have a fundamental role to play in the implementation and enforcement of principal laws formulated at national level, in many places they are also increasingly playing a role in fostering more participatory forms of rural governance. With the move toward decentralized governance, by-laws have been increasingly recognized as a means for rural self-governance. Today, registered villages in Tanzania and Uganda have the right to make by-laws in respect to any village matter,[2] provided they are consistent with the provisions of national laws. Ironically, the system of local by-laws established during the colonial era to further central government control may now offer an opportunity for more empowering forms of governance through the presence and political legitimacy of local collective choice rules. AHI work in this area involved the generation of approaches for mobilizing the latent potential of local government to bridge technological and governance innovations in addressing NRM concerns of local communities through participatory by-law reforms.

Service delivery

The district has a fundamental role to play in the delivery of public services, including health care, agricultural extension, infrastructure, education and—increasingly—information. Public service providers throughout the region have faced increasing pressure to demonstrate their relevance. This is largely owing to the predominant tendency toward supply- rather than demand-driven approaches. Agricultural extension, for example, has come under increased scrutiny owing to a host of institutional weaknesses that have limited its effectiveness, among these its highly centralized and bureaucratic structure; exogenous, donor-driven, non-participatory planning; lack of efficiency and accountability of financing and service delivery mechanisms; lack of motivated service providers responsive to farmers' needs; and decreasing public sector funding (Nahdy, 2004; Rivera and Alex, 2004). Key rationales behind resulting reforms have been the need to strengthen client demand for services through participatory approaches, and to enhance the role of the private sector in service delivery (Nahdy, 2004; Rivera and Alex, 2004).

One of the newer services to be delivered to rural populations is information. Ever since the concept of "information society" came to the fore in the 1970s, the correlation between access to information and poverty alleviation has been widely acknowledged (Flor, 2001). The main proposition was that information leads to resources and to opportunities to generate resources (ibid.). "Leaders in the World Bank, European Union, United Nations, and G-8 have highlighted the problem of exclusion from the knowledge economy, where know-how replaces land and capital as the basic building blocks of growth" (Norris, 2001: 6). While a few large-scale, commercial farmers on the continent have used some of the decision support tools that information and communication technologies (ICTs) are providing, relatively little attention has been paid to the potential benefits of the broader use of ICTs in the (largely informal) agricultural sector, one of the few in which women often predominate (ACACIA, 2006). The National Strategy for Growth and Reduction of Poverty and the Poverty Eradication Action Plan (2004/5–2007/8) of Uganda (Government of Uganda, 2004, 2005) assert that reasons for limited access to ICTs include low literacy rates, low incomes, and the limited number of ICT service providers.

At the district level, access to information by various actors continues to be a challenge and a deficit area. Any coordination among R&D actors is ad hoc, and coherent communications strategies at district level are lacking throughout the region. Information which is either highly specialized (and therefore found only in isolated pockets) or not readily available at district level often includes: market prices, seasonality and traders; the location of expert farmers and service providers; and agronomic information for non-traditional crops. Given limited coordination, the act of seeking information by R&D practitioners and providing it in usable forms to stakeholders is inefficiently handled on a case-by-case basis, as each entity is doing its own thing in its own location and according to its own sectoral interests. Such high transaction costs for limited returns (e.g., delivery to few farmers) could be addressed through a system of coordinated information access and delivery at district level. AHI took up this challenge by employing an action research approach to methodological innovation in information delivery.

This chapter summarizes experiences gained by AHI in evolving district level institutional and policy innovations for natural resource management in each of the arenas mentioned earlier. Through specific case studies, the chapter highlights lessons on systems for democratic decision making in NRM, district level governance of natural resources, and demand-driven service delivery.

Methods to foster democratic process and vertical stakeholder collaboration

The implementation and sustainability of landscape level natural resource management interventions require the participation and support from a variety of stakeholders at various levels, and ability to accommodate various points of view. Achieving effective representation of local level actors in decision making about development and natural resource management issues that affect them is a challenging task, given the number of actors and interests at the local level in any given district. For each of these reasons, effective approaches for achieving democratic decision making in efforts to link development actors at multiple levels are sorely needed.

The establishment of linkages between actors at different levels is required to address many problems given their unique and potentially complementary mandates and contributions in identifying constraints, implementing solutions, and in ensuring feedback from the grassroots to relevant government actors and service providers. In this regard there is need for an institutional arrangement at the district level that can: (i) systematically support the articulation of multiple local "voices" in development planning, (ii) help to reconcile alternative visions of land use and development, and (iii) foster local self-sufficiency while prioritizing issues requiring external support, and ensuring the responsiveness of the relevant actors through their integration into district-level planning. While many arguments have been put forward on the merits and demerits of "top-down" and "bottom-up" approaches to development, there is a dearth of information on effective processes for linking levels of decision making and action in rural development and NRM.

Approach development

Two basic approaches for linking levels of decision making and action have been tested in AHI. The first is more "ad hoc" in nature, and seeks to minimize the transaction costs of vertical linkages through the application of the "subsidiarity" principle—namely, that matters ought to be handled by the smallest, lowest, or least centralized competent authority. Only when approaches fail at this level are linkages with outside actors forged to help resolve conflicts or bring in resources required to unlock the potential for change. The second approach is more systematic in nature, establishing an institutional infrastructure for representative democracy in district-level development planning and implementation.

Approach 1—Vertical integration on demand

The more ad hoc approach to vertical stakeholder collaboration consists of the following basic steps:

1. During the participatory diagnosis and planning process, an institutional mapping is done of stakeholders at district and lower levels (the research and extension system, other government ministries, NGOs, farmer associations, government leaders at diverse levels, local courts and faith-based organizations, among others). All relevant[3] stakeholders are listed and their interests and mandates noted.
2. Initial consultative meetings are held to familiarize different actors with the initiative. This early involvement of other stakeholders can be vital for the targeting of actions, mobilization of resources and—importantly—for mobilizing their support at a later stage.
3. The project is implemented as planned through support to local level action planning and mobilization.
4. When a problem arises, communities and project representatives first discuss possible means to address the problem with existing stakeholders and using local resources.
5. If the problem cannot be resolved effectively at this level (either after agreeing that higher level intervention is required or after testing local level solutions and failing to find a solution), agreements are made on the nature of outside support that can help to address the problem and how to mobilize those actors to assist (Box 5.1).

BOX 5.1 VERTICAL INTEGRATION FOR WATER SOURCE PROTECTION IN GALESSA, ETHIOPIA

The participatory diagnosis of landscape-level NRM problems highlighted a serious shortage of water and high level of water contamination in the Galessa Watershed. With encouragement and guidance from the research and extension system, watershed residents showed a high level of interest in addressing the problem through community-level collective action. Through a process of stakeholder identification and negotiation support, local residents agreed on actions that would help to rehabilitate the springs and manage them properly and formulated by-laws to help support these agreements. Despite early successes—including farmer contributions of labor, money, and materials, and agreements between stakeholders on the removal of fast-growing ("thirsty") trees planted near springs (and actual cutting of a portion of a woodlot)—the process encountered some difficulties in implementation. Addressing these problems required inputs from district-level stakeholders.

Challenges in enforcing the by-law with neighboring villages

Owing to the commitment of watershed residents, concrete structures were installed around springs with financial, labor, and material contributions from watershed residents and were being well maintained with small monetary contributions from users. By-laws were being enforced and spring users from watershed villages expressed their willingness to accept sanctions for failure to abide by these local level agreements. The first challenge came from neighboring villages that wished to gain access to the springs but did not wish to pay the required fees or abide by the by-laws. They claimed that the developed spring was subsidized by the government and watershed communities therefore have no right to ask for labor or financial contributions. Watershed residents refused them access until they contribute in labor and money what they failed to contribute during spring construction. District level stakeholders were called in to assist in resolving the impasse. The point raised by neighboring villages was that they too had a right to government assistance in protecting the spring in their own village. If this support were provided, they expressed their willingness to make similar contributions to the collective good. As their position became clear, the district stakeholders assumed the responsibility in guiding these communities in developing their own spring.

Dealing with emergent conflicts

Following agreements on the removal of eucalypts around Ameya spring and strong pressure from the community, the woodlot owners cut down the portion of the woodlot closest to the spring. After some time, however, the eucalypts began to coppice and the spring owner refused to take any further action—instead placing increasingly stringent conditions on his compliance (requesting compensation, reducing the scope of earlier commitments). The watershed committee was unable to enforce the agreement. Village residents themselves were going to take it upon themselves to enforce the agreement, but were also unsuccessful. Therefore, district level stakeholders were called upon to assist in resolving the conflict, and district and PA-level government representatives assumed responsibility for finding a resolution to the conflict by resorting to the law. The conflict was ultimately resolved in support of the landowners, who would have to receive cash compensation for the eucalypts if they were to proceed with removal and village members could not generate sufficient compensation. Importantly, however, the process of dialogue and decisive intervention by government enabled the latent conflict to dissipate and neighboring farmers to continue living amicably despite the water problem remaining unaddressed.

Approach 2—A systematic approach to farmer representation in district-level development planning and decision making

The second approach to vertical stakeholder collaboration involves working through an (existing or new) institutional infrastructure for representation in district-level development planning and implementation. The limited institutional and financial resources for supporting rural development make such coordination an important part of any government-funded or exogenous development effort by ensuring that service delivery supports the most important concerns of communities throughout the district, and refocuses disconnected development or NRM efforts by leveraging synergies between actors and interventions. The following are generic steps in the process of ensuring representative decision making in district development planning:

1. Create awareness around the topic of concern (e.g., integrated development and conservation).
2. Develop a team of "champions" on the topic who are willing to volunteer during subsequent steps in planning, including farmers and other community members as well as local government officials at various levels.
3. Carry out a facilitators' training, during which skills in facilitating participatory processes are covered and a methodology for facilitating subsequent steps in participatory planning at diverse levels is agreed upon.
4. Carry out representative planning processes building upon the appreciative inquiry approach (to embed this planning in local level skills and resources), starting at the local level (villages or farmer groups) (Box 5.2).
5. For NRM interventions, carry out a middle level of planning at the landscape level together with village representatives. This process builds upon local level action plans but incorporates new activities that require landscape-level action.
6. Facilitate the identification of local-level innovations (including farmer learning processes and forms of indigenous knowledge-in-use) that can be built upon in addressing farmers' articulated needs.
7. Collate lower-level action plans from the appreciative inquiry process and the identified local innovations at district level, and hold a planning meeting with local government and different service providers to agree on how priorities articulated at the local level can be best supported with limited human and financial resources. Plans for mobilizing the recently trained facilitators in supporting development actions at various levels are developed at this time.

BOX 5.2 USE OF FARMER LEARNING CYCLES TO ARTICULATE DEVELOPMENT PRIORITIES AND INITIATE DIALOGUE WITH DISTRICT LEVEL ACTORS

In Kapchorwa District, Uganda, facilitators were trained to support grassroots structures (in this case, farmer groups) in strengthening their own learning and planning processes, as well as to make proactive linkages with district-level actors to leverage support for local development priorities. An approach called farmer learning cycles (or "reflect cycles") was employed, which is essentially a grassroot's planning and review process. At village level, farmers congregate to identify common problems, plan and implement agreed action plans. The process is facilitated by volunteers (from member organizations) as well as extension staff. Depending on the type and extent of resources required (financial and technical), these groups are able to implement their action plans by mobilizing their own resources. In situations where they are unable to raise the necessary resources, they pass on an action plan with a request for support through their facilitators to village or sub-county government. These requests are collated at sub-county level, and brought to the district platform for consideration. The reflect cycles play a key role in mobilizing local resources to meet local development needs, and in enabling farmers to proactively lobby for support from external actors. The participation of local government at various levels (village, sub-county, and district) allows needs to be matched to service providers and resources, and helps to legitimize local development efforts being undertaken.

The role of farmer learning cycles

Reflect cycles enhance farmer learning processes by providing opportunities for farmers to exchange views, and question different understandings through experience sharing and experimentation. This farmer learning process consists of a group of neighboring farmers, usually not more than 25 in number, meeting regularly for a period of time, as often as once per week, to study a certain subject or theme or take part in a practical activity. For agricultural and NRM issues, meetings are generally held in farmers' fields. The reflect cycles are characterized by democratic values and responsibility toward one's own situation. These reflect groups are often led by farmer innovators who have expertise in a topic of mutual interest. In this regard, reflect group leaders vary depending on the group's interests. Where expertise is lacking, the group seeks assistance externally. In the reflect process, farmers plan based on their needs and interests, exchange ideas, and acquire knowledge based on the collective wisdom of the group.

Sustaining the process

The main challenge is ensuring responsiveness of district-level actors and service providers to the needs articulated by farmers. This requires district-level commitment to demand-driven development and an organizational mechanism to leverage existing human and financial resources (from within and outside government) in response to demand. As illustrated in the sections on demand-driven service provision and multi-institutional processes, below, this mechanism takes a concerted effort and time to develop, requires frequent monitoring and adjustment (particularly initially, until the process is proven to be effective), and carries significant costs in the short run.

Lessons learned

The following lessons were learned from AHI's experience in fostering democratic process and vertical stakeholder collaboration:

- Ensuring effective representation in district development processes requires both bottom-up efforts to mobilize latent potential at community level and articulate demand, and efforts by district-level institutions to respond to this demand. Such an approach is therefore highly complementary to approaches highlighted in the upcoming section on multi-institutional partnerships at district level.
- Ad hoc approaches to vertical integration may be effective in addressing specific problems with limited transaction costs (e.g., without needing to invest in social structures and processes for demand-driven development at local and district levels). However, solutions will remain isolated in the absence of more systematic approaches. Therefore, while such ad hoc approaches may be effective in fostering vertical stakeholder collaboration, they are less effective in ensuring effective representation.
- A more systematic approach to farmer representation in district-level development planning and decision making requires significant up-front investment in institution building, which in turn requires a source of financing. However, costs are minimized and sustainability enhanced when building on the spirit of volunteerism and mobilizing local facilitators from communities or government agencies.

Responsive governance: The district role in participatory by-law reforms[4]

With recent decentralization efforts and the mainstreaming of participatory approaches in policy and development, considerable attention is now given to devolving decision making to the lowest level, and to refining participatory techniques by creating more inclusive spaces for hearing the voices of all (James et al., 2001; Ribot, 2002a,b; Scoones and Thompson, 2003). However, there is concern that decentralization has not resulted in improvements in NRM, nor has it affected the capacities and decision-making powers of local communities. It is only to a limited extent that policy makers seek the participation of local stakeholders in designing and formulating policies or by-laws. Yet, it is recognized that rural communities and local stakeholders would be more likely to see by-laws as addressing their own needs and constraints and more likely to implement them, if they had participated in their formulation (Nkonya et al., 2005).

Previous chapters of this book indicate that local NRM practices are shaped by a range of both formal and informal institutions. These chapters illustrate how AHI attempts to build "adaptive manager communities" (Fabricius et al., 2007)—communities empowered to formulate their own by-laws, and develop, adapt, and mobilize collective action and local innovations—have helped them better manage their landscape resources and even increase land productivity. The emphasis on local institutions and local innovations is based on literature suggesting that communities are more efficient than state structures in the management of natural resources (Agrawal and Gibson, 1999; Ostrom, 2000). While these and other studies have focused attention on the role of local institutions (formal and informal), little has been done to illustrate the benefits of linking these local institutions to higher, local government structures, nor how to go about it in practice. The role of local government in shaping, formalizing, and legitimizing these local by-laws has often been neglected. A critical component of INRM and of scaling INRM innovations is building capacity in the "middle," and particularly strengthening the institutions in local government that translate policy into action. There is therefore need for a better understanding of approaches and techniques for integrating local institutions and aspirations into formal policy objectives and processes, and for making government policies more responsive to, and representative of, local people's concerns and experiences.

In the natural resource management arena, by-laws are negotiated rules, social norms, and agreed behaviors that exist within communities to manage natural resources, and prevent and manage conflicts. They are a tool for managing natural resources in a way that places community interests above those of individuals (Bowles and Gintis, 2002; Coleman, 1988). They also give individuals confidence to invest in community activities, knowing that others will do so too (Pretty, 2003). In legal and policy terms, by-laws are a body of local laws and customs of a village, town, or city, or rules made by lower local government councils which provide the local guidelines to be followed in implementing sectoral policies and preventing agricultural and NRM practices that could be detrimental to the common good. This form of by-law is formulated at lower levels of decentralized government (villages, wards, districts) and often help to operationalize national policies.

This section draws from experience with participatory by-law reform in Uganda that involved moving beyond local communities and linking with higher levels of local government. Participatory by-law reforms are described in a logical and structured way to enable development practitioners and NRM researchers to consider how insights gained from AHI's work on participatory by-law reform, linking local communities with local government, might be relevant to their own practices and situations. The next section provides a simple description of legislated processes for formulating and enacting by-laws.

Historical and institutional frameworks for by-law reform in Uganda[5]

In Uganda and East Africa more broadly, many existing by-laws were inherited from the colonial administration and are thus seen as repressive and top-down (Okoth-Ogendo and Tumushabe, 1999). The majority were formulated before independence by British colonial administrators without local participation, with strict enforcement mechanisms—including force and coercion. In the colonial period, local chiefs and administrators strictly enforced by-laws as this was used as an indicator of their performance. There were also clear enforcement structures and coercive penalties for non-compliance. Enforcement of by-laws faded after independence, as most by-laws were regarded as instruments of colonial repression. Such top-down and centralized policies often resulted in disempowerment of local communities, the weakening of customary forms of governance, power imbalances, and the exclusion of vulnerable groups, and failed to provide appropriate incentives for community-based NRM (Means et al., 2002). Often locally unacceptable, many of the by-laws were left unenforced.

The Ugandan Constitution of 1995 and the 1997 Local Government Act sought to change this system and guarantee a process of consultation and participation at village, ward, community, and district levels for environmental plans and policies. Bottom-up involvement in policy formulation occurs within an overall context of administrative and political decentralization, which has the structure of a five-tier system of local councils and local government structures (Table 5.1). It includes the devolution of powers for development planning and the development and implementation of by-laws for land use, environmental management and agricultural production.

Within this structure, the village or LC 1 is the basic level of decentralization and of community participation in by-law formulation and implementation (see Box 5.3). The sub-county also has important political and administrative powers to develop by-laws and implement development plans, and is the lowest unit where policy reform can be effectively initiated. The district (LC 5) is the highest level of local government and therefore has important political and administrative powers to enact by-laws, consolidate development plans, and allocate budgets. It is also the most effective level for linking with the central government.

TABLE 5.1 Decentralized structures in Uganda: Levels and main functions

Local Council level	*Composition*	*Functions*
Local Council 1: Village (around 50 to 100 households)	9 members, at least 4 women	• Assist in maintaining law, order, and security • Initiate, support, and participate in self-help projects • Recommend persons for local defense units • Serve as communication channels with government services • Monitor the administration of projects • Impose service fees • Collect taxes • Resolve problems and disputes • Make by-laws
LC 2: Parish (composed of 3–10 villages)	Depends on the number of villages electing representatives, but must include at least 4 women	• Assist in maintaining law, order, and security • Serves as communication channels with government services • Initiate, support, and participate in self help projects • Monitor the administration of projects • Resolve problems and disputes
LC 3: Sub-county (Composed of 2–10 parishes)	Depends on the number of parishes, but must include at least: • 1/3 women • 2 youth • persons with disabilities • elected councilors from parishes	• Enact by-laws • Approve sub-county budget • Levy, charge, and collect fees and taxes • Monitor performance of government employees • Formulate, approve, and execute sub-county budgets • Resolve problems and disputes
LC 4: County (composed of 3–5 sub-counties)	5 members, including chairpersons or vice-chairperson from each sub-county	• Advise district officers and area Members of Parliament • Resolve problems and disputes • Monitor delivery of services
LC 5: District (composed of 3–5 counties)	36 members, including: • 12 women councilors • 2 youth • 2 people with disabilities • 19 elected councilors	• Exercise all political and executive powers • Provide services • Ensure implementation of and compliance with government policies • Plan for the District • Enact district laws and ordinances • Monitor performance of government policies • Levy, charge, and collect fees and taxes • Formulate, approve, and execute district budgets

BOX 5.3 FORMAL BY-LAW FORMULATION PROCESS IN UGANDA

The formal process of formulating and enacting by-laws consists of the following steps:

1. Any community can initiate the process of formulating a by-law or their councilor can draft a bill seeking to formulate a by-law;
2. The draft bill is introduced to the council by one councilor;
3. The bill is then published and distributed to all councilors by the Council Clerk;
4. The bill is debated and approved within 14 days after publication;
5. If passed, the bill is forwarded to the relevant higher council for certification of consistency with the constitution, ordinances, and other laws, after which it is returned;
6. The bill is then forwarded through the line Minister to Attorney General for certification of consistency with parliamentary laws and the Constitution, after which it is returned;
7. The certified bill is then signed by the District Chairperson to become an ordinance (for a district-level bill) or by-law (for lower council bills).
8. The ordinance or by-law is then published in the gazette, in local media, or posted in a conspicuous place.

Despite such clear guidelines, there are few available records on the formulation or revision of by-laws. Where there have been some attempts, the process has been far from participatory and has tended to be restricted to small editorials to existing by-laws and updated penalties. AHI's action research efforts were aimed at developing mechanisms for strengthening local participation in by-law reforms as a means to support improved NRM and more equitable development.

Approach development

This section reports on a single approach tested in south-western Uganda to strengthen the linkages between local-level by-law reform processes and higher levels of government.

Multilevel policy reforms emanating from the grassroots

The participatory by-law reform process tested in Kabale District, Uganda, consisted of the following iterative and complementary steps:

- participatory diagnostics
- district buy-in and goal setting
- bottom-up community learning and experimentation
- horizontal and vertical policy dialogue at the sub-county level
- district policy dialogues
- participatory monitoring and evaluation, feedback and reflection.

Implemented as a participatory action research process (Reason and Bradbury, 2001), the process of developing and testing these steps consisted of iterative series of action learning loops at diverse levels of policy innovation (Figure 5.1). The process was supported by a skilled action research team that motivated and facilitated people to participate in the process of action learning, while supporting platforms for policy dialogue and negotiation. The team created the conditions in which local people were able to participate, analyze and review existing by-laws, formulate appropriate by-laws, and monitor their implementation.

The above steps, each of which generated lessons on processes to be utilized within that step and on subsequent steps required to help achieve overall objectives, merit additional attention. This is done below, with particular emphasis on the linkages between levels and the role of local government in the process.

Step 1: Participatory diagnostics

Similar to participatory by-law processes profiled in Chapter 3, the first step of the participatory by-law reform processes carried out in Kabale was an intensive and iterative process of participatory diagnosis and community visioning (Sanginga and Chitsike, 2005). This was fundamental in stimulating collective

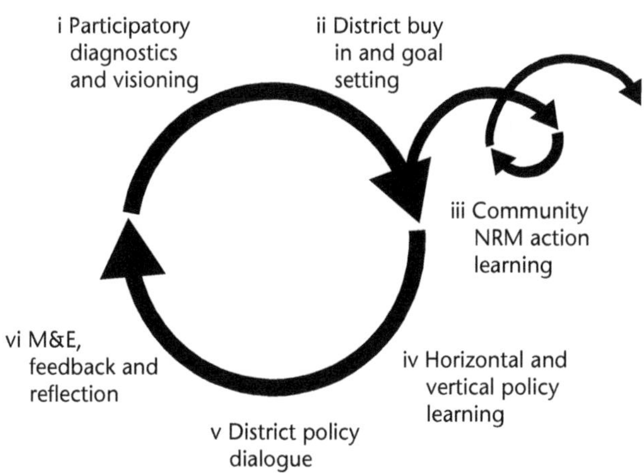

FIGURE 5.1 Operational framework for participatory policy action research

learning and the articulation of desired future conditions. Communities identified governance and institutional failures as critical aspects of community-based natural resources management. The lack of strong enforcement mechanisms of existing by-laws was seen as the main reason for the ineffectiveness of most prior NRM interventions. The community visioning and planning process encouraged farmers to think creatively about potential means to enhance compliance with and equitable implementation of by-laws. It was recognized that actions at the community level would not yield results unless they were linked with, supported, and legitimized by higher levels of governance and government institutions, given the latter's power to enact by-laws and impose sanctions for their enforcement. This required buy-in and support from district authorities.

Step 2: District buy-in and goal setting

The second step in participatory by-law reforms involved bringing together a number of stakeholders at the district level to begin to analyze the problem. A series of policy stakeholder workshops and learning events (seminars, field visits) were organized to catalyze local political support for sustainable NRM. These workshops revealed that the majority of policy makers and local leaders have a limited understanding of the policy and by-law formulation process, the existing policies and by-laws they are charged with implementing, and the implementation process itself. There was no systematic guidance on the processes and mechanisms for formulating and enforcing by-laws. In the first district NRM policy stakeholder workshop, three major recommendations were made: (i) to conduct an empirical study to provide evidence of people's awareness and level of compliance with existing by-laws, and constraints to their enforcement; (ii) pilot a participatory by-law reform project in selected communities; and (iii) establish a district Policy Task Force to provide oversight to the process of enhancing NRM governance.

An empirical study was then conducted to assess people's awareness, the effectiveness of existing by-laws, implementation constraints and strategies for improving their enforcement. Survey results showed that farmers often have high levels of awareness of existing regulations. For example, over 75 percent of farmers interviewed were aware of the regulation that requires farmers to construct soil bunds and other soil conservation structures along the contour. Over 60 percent of farmers were also familiar with the regulation requiring farmers to plant appropriate vegetation on these structures. Similarly, the majority of farmers (68 percent) knew about the tree planting by-law, which requires that any person who cuts a tree plants two and ensures they are protected. However, despite these high levels of awareness, by-law enforcement and implementation was weak. It was noted that the decentralization process had introduced multiple overlapping systems of governance and regulations (legal pluralism), as well as increased political interference, nepotism, confusion, and conflicts between different levels and structures of government. By introducing local councils at village level where local political and administrative powers are now concentrated, the decentralization process had weakened existing authorities and institutions for managing and regulating the use of natural resources. A combination of social, economic, and political factors had undermined the ability of local mechanisms, clan elders, and community organizations to manage conflicts (Means et al., 2002). This led to factionalism, with the more educated and wealthier farmers often not willing to accept decisions made by local communities and clan elders and preferring to take their cases to government institutions at higher levels for arbitration. Clan leaders were also found to exhibit biases in by-law enforcement and engage in corrupt behaviors.

Step 3: Bottom-up action learning processes

The second recommendation of the district policy stakeholder workshop was to pilot a participatory by-law reform process in select villages (LC1) in one sub-county. Four villages were selected in Rubaya Sub-County, where AHI was already established. The entry point was through existing farmers groups involved in participatory NRM activities and with high levels of social capital. The project's strategy was to build on existing social capital and to strengthen it through facilitating participatory social learning and policy dialogue processes. This approach contrasts with approaches highlighted in Chapter 4, in which processes for landscape governance were grounded in village-level fora or stakeholder-based negotiations. The approach presented here has the benefit of initiating with local level institutions with strong social capital and thus possessing many of the skills required to take on new challenges and sustain their engagement. However, it may be less effective in ensuring widespread buy-in by ensuring widespread representation or explicitly addressing conflicting interests. An effort was made by facilitators to explore multiple perspectives of resources users, with the aim of gaining credibility and support of different categories of farmers through more inclusive and consultative processes. Building on participatory diagnostics and community visioning, the next step involved the collective analysis of NRM issues and existing by-laws and participatory community action planning. As a result of this process, pilot communities reviewed and reformulated a number of informal by-laws—namely, those that exist outside the formal legislative structure but are addressing specific problems in the communities. At this time, technologies that could be implemented in conjunction with certain by-laws were also identified. Over time, participatory by-law reform processes were progressively institutionalized, in order to strengthen the capacity of village members to effectively engage with higher levels of governance (Box 5.4).

BOX 5.4 INSTITUTIONALIZING GRASSROOTS POLICY FORMULATION AND IMPLEMENTATION

Recognizing that power relations set limits and social conditions to people's participation (see also Cooke and Kothari, 2001; Stringer et al., 2006), the facilitation team played a proactive role in strengthening the capacity of farmers' organizations to engage effectively in policy dialogue. This included a range of participatory techniques (visioning, role plays, and other adult learning methods) for coaching and mentoring farmers' representatives to better articulate their policy needs and NRM visions with confidence. In order for ad hoc village by-law committees to become part of the policymaking process, there was a need to develop mechanisms to institutionalize participatory processes for policy formulation and implementation. The project therefore facilitated the formation and functioning of policy task forces (PTFs), with the following functions: (1) to create and facilitate a platform for dialogue between communities, local government councils and R&D organizations on the analysis of NRM issues and local by-laws; (2) to initiate and monitor the review, formulation, and implementation of by-laws; (3) to link the village with sub-county and district PTFs, local government and external agencies; and (4) to disseminate NRM technologies. The formation of these committees followed a more inclusive and participatory process for electing committee members and defining their roles and responsibilities.

Through the PTFs, proposed by-laws were debated, harmonized, and formalized into a set of five by-laws focused on: controlling soil erosion, tree planting, regulating the grazing of livestock, controlling bush fire and wetland management, each with its specific regulations and enforcement mechanisms. PTFs proved to be critical in building support for by-law review and formulation; for mobilizing the political, social, human, and technical resources needed to sustain the participation of local communities in policy dialogue and action; and for the adoption of NRM innovations. They also supported the evolution of collective action and other forms of social capital such as information exchange, resource mobilization, collective management of resources, cooperation and networking, and community participation in research and development activities. They increasingly became a vehicle through which farmers were pursuing wider concerns, initiating new activities, organizing collective action and extending relations and linkages with external organizations.

Step 4: Horizontal and vertical linkages at the sub-county level

Despite progress made at the village level, it was recognized that the strengthening of community-level governance processes would be insufficient in the absence of higher level reforms. Linkages to local government structures are a critical element to any policy process, particularly under decentralization—where the sub-county and district have important political and administrative powers in by-law formulation, the preparation of development plans, and budgeting. As the basic political and administrative unit of local government and with by-law formulation and dispute resolution functions, the sub-county was seen as an important nexus for stimulating democratic processes for the deliberation and influence of policies from the bottom up. A key component of participatory by-law reforms was therefore facilitating policy learning and dialogue between villages and the sub-county government.

It was particularly useful to sequence PTF meetings with farmer exposure visits and horizontal linkages between different communities, where farmers had the opportunity to harmonize their demands, share experiences, and rehearse the presentations they would make at sub-county level. These visits and deliberations centered on analyzing existing by-laws and identifying opportunities and needs for reviewing and reformulating existing by-laws or formulating new ones (Box 5.5). In addition to fostering experience sharing, these dialogues were a first step in by-law formalization. The different by-laws initiated at village level were then presented and debated at the sub-county level for harmonization and better coordination before they were enacted as formal by-laws, to be applied in all villages and parishes of the sub-county.

BOX 5.5 THE FOCUS OF DELIBERATIVE PROCESSES WITHIN PARTICIPATORY BY-LAW REFORMS

Deliberative processes at the local level focused on the following key issues:

1. *Content*: What is the by-law about? What is behind by-law formulation? What is the role of different types of resources (technology, information, social capital, labor, credit) in creating positive synergies between by-laws and development/conservation?
2. *Process of by-law formulation, implementation, and refinement*: What are effective approaches for crafting local institutions where they are deficient? How can by-laws be equitably assessed and formulated?
3. *Functions of by-laws*: What functions do by-laws currently play in diverse areas (community-based NRM, decentralization, landscape management, and technology adoption/dissemination)? What additional functions could by-laws effectively play?
4. *By-law enforcement:* What is the effective balance between formal and informal enforcement mechanisms in different contexts? What processes and conditions enhance compliance and minimize the need for strict enforcement?

5. *Legal and social foundations of by-laws*: How effective are customary and statutory laws in supporting by-laws under different land-use systems and conditions, and how and why does this effectiveness vary? For whom is legal pluralism beneficial/detrimental? What opportunities exist for building upon remnants of traditional governance systems and improving synergies through vertical policy linkages? To what extent can by-laws be used to operationalize statutory law in ways beneficial to local land users?
6. *Particular vs. general*: How can the need to adapt by-laws to the local context be balanced with standardization for legislation and enforcement? Can law enforcement agencies manage a high level of complexity in "adaptive" governance?
7. *Outcomes and impacts*: What are the impacts of improved (participatory) governance on poverty, equity, and environment in different contexts? How do processes and content affect outcomes?
8. *Vertical linkages and scale*: How can the scale of participatory by-law reforms be expanded without compromising quality in participatory processes? How can participatory by-law reforms be effectively reconciled with national policy formulation processes? What are effective processes for "going to scale"?

Step 5: Facilitating district-level policy dialogue

As noted earlier, the district is the highest level of local government that has powers to enact and formalize by-laws, and establish linkages with other sub-counties and the central government. In addition to the focused work at village and sub-county level, policy dialogues were facilitated at the district level to ensure coherence between policies at all three levels and to reach a wider consensus on by-law reform processes and outcomes. District-level policy workshops were usually high profile events aimed at re-focusing the policy dialogue and building a network of actors who could influence the policy process. Five policy stakeholder workshops were held over the course of three years, bringing together a large number of participants (80–100)—from district leaders and councilors to members of parliament, sub-county councilors, and representatives of local government technical services, research and development organizations, and farmers' organizations.

One strategy was to organize and facilitate field visits to showcase examples of successful village level by-law reforms. These visits had a profound effect in convincing policy makers, local leaders, and farmers alike of the benefits of participatory policy reforms, allowing them to see things with their own eyes and to share experiences with innovative farmers. Another important tool to stimulate learning at district level was the use of policy narratives and NRM scenarios—which help to simplify complex problems and enable more informed decision making (Keeley, 2001).

Recognizing that power relations are pervasive and always affect the quality and process of participation (Chambers, 2005), targeted efforts were necessary to empower the weakest stakeholders (farmers) and at the same time to enable policy makers and local leaders to acknowledge their own power, be aware of how they may habitually disempower others, learn to use power to empower those with less power, and avoid being inhibited by the learning process. A range of participatory techniques and other adult learning methods were used for engaging and empowering local communities directly in the articulation of their policy needs, and in the analysis, design, and implementation of policies and NRM innovations. This involved coaching and mentoring farmers' representatives to increase their assertiveness and confidence in articulating their policy needs and collective NRM visions. As a result, some of the most interesting moments during the stakeholder workshops were when farmers articulated their own visions and experiences with the participatory by-law review, formulation, and implementation process.

Out of the multilevel sharing processes emerged a genuine interest and willingness among stakeholders in Rubaya Sub-County to disseminate the approach to other villages and sub-counties and to the district at large. At the same time, other villages, sub-counties and districts (Kisoro, Kanungu, and Rukungiri) expressed interest in the process. NGOs such as CARE, Africare, and Landcare and government agencies such as NAADS took an interest in the process and began supporting selected communities. A series of sensitization meetings was held for farmer groups and development organizations in pilot communities to disseminate the participatory process of formulating and implementing local by-laws and NRM practices.

Step 6: Participatory monitoring and evaluation, feedback and reflection

At the end of each policy learning event and policy dialogue workshop, the research team facilitated a process of structured reflection using a tool called "After Action Review (AAR)" to help communities to reflect, analyze, and learn by talking, thinking, sharing, and capturing the lessons learned about the dialogues and workshops before these are forgotten (CIDA, 2003). AAR is usually facilitated using the following six questions: (i) What was supposed to happen and why? (ii) What actually happened and why? (iii) What accounts for the observed differences? (iv) What went well and why? (v) What could have gone better and why? and (vi) What lessons can we learn?

An important aspect of the participatory by-law reform process was to facilitate community-based participatory monitoring and evaluation (PM&E) to monitor progress, track outcomes, and enhance learning through critical reflection and feedback. To complement the PM&E system, systematic studies and process documentation were carried out to understand the outcomes of project interventions for equity, NRM, and sustainability. Boxes 5.6 and 5.7 illustrate some of the positive outcomes of the participatory by-law reform process. Box 5.8 illustrates what might go wrong with the process, and the importance of active monitoring to identify and address negative outcomes such as inequities in the flow of benefits and costs.

BOX 5.6 GENDERED OUTCOMES OF BY-LAWS

The number of collective action events and the level of participation of different stakeholders were two of the indicators used to track local buy-in to by-law reform processes. Results confirmed that women's participation in pilot communities was sustained over time (Figure 5.2). A linear trend line of women's participation shows a steady increase in the number of women participating over time (R2 = 0.83), from less than 20 to more than 60 women attending the different community meetings.

The relatively high participation of women is consistent with earlier analysis of the patterns and dynamics of participation in farmers' organizations in Africa (Sanginga et al., 2006). However, it is interesting to note that contrary to earlier findings on group dynamics which show decreasing participation of men in group activities, the findings of this study show that men's participation was also sustained over time. The process has increased women's confidence and changed perceptions of their status within communities. The vast majority of male and female farmers interviewed (95.6 percent) indicated that women's participation in decision-making and community leadership positions had improved in the three years since by-law reforms were initiated.

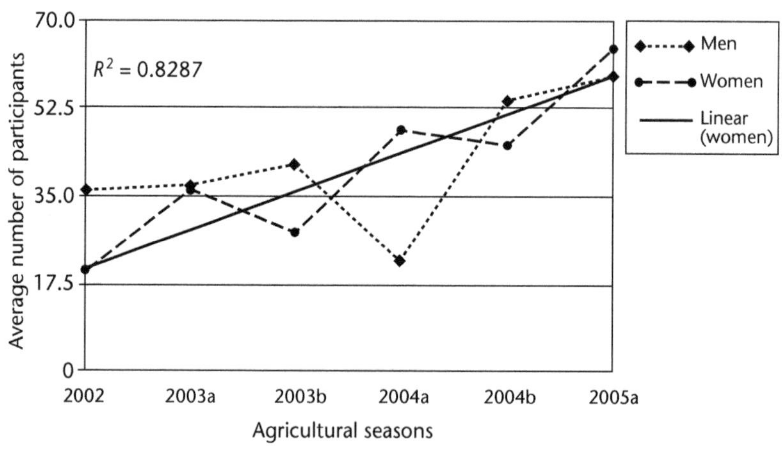

FIGURE 5.2 Gendered patterns of participation in by-law meetings over time in pilot communities

BOX 5.7 TECHNOLOGY ADOPTION AS AN INDICATION OF BY-LAW EFFECTIVENESS

NRM outcomes of by-law reforms were also tracked to assess how the process had influenced adoption rates and farmers' willingness to invest both labor and cash (e.g., for purchase of tree seedlings) in technology adoption. Results showed significant increases in adoption levels (Table 5.2).

TABLE 5.2 New soil conservation measures established in 2005 (% of farmers)

Soil conservation measure	*Female-headed households*	*Male-headed households*	*All households*
Construction of new terraces	38.6	45.3	42.1
Digging of trenches	32.9	38.7	35.9
Stabilizing soil conservation structures with agroforestry technologies	25.7	30.7	28.3
Planting grass strips	8.6	9.3	9.0
Use of trash lines	5.7	6.7	6.2

BOX 5.8 THE "DARK SIDE" AND LIMITS OF BY-LAW REFORMS

Although the previous results show that the outcomes of by-law reforms have been largely positive, the study also revealed some important downsides. We found that certain categories of farmers had difficulty in complying with some of the by-laws. These included older men and women, widows and orphans with limited family labor, or who lack money to hire labor or to buy farm implements needed to establish conservation structures. There had been instances of conflict among livestock owners and cultivators, which in some cases led to divisions and hatred within communities. It was also found that owners of small livestock, especially women with small farm sizes, had problems with the by-law to control free grazing. Strict enforcement of this by-law forced the poor to sell their livestock, thereby perpetuating the poverty trap. A focus group discussion in one of the villages revealed that two factions had emerged as a result of the controlled grazing by-law. One group (*Nkund'obutungi*, the wealthier farmers) disliked the system of free grazing and did not allow other farmers to graze in their plots, because they have large farms in which they graze their livestock. It is this group that was pushing for strict enforcement of the controlled grazing by-law. The second faction (*Nkund'obutungi*, the poorer farmers) had smaller plots where livestock could not be grazed and limited labor for controlled grazing. This group was forced to confine their animals or be fined for non-compliance. The *Nkund'obutungi* passed a by-law against grazing on their plots, thus negatively affecting the *Nkund'obutungi.* In turn, the *Nkund'obutungi* organized themselves into a group and agreed to allow grazing in each other's land. This conflict led to the failure of the controlled grazing by-law, with implementation left to the wealthier households who would benefit from it. Clearly, viable feeding alternatives were required by poorer households to enable them to restrict their grazing activities and avoid experiencing negative livelihood impacts from by-law reforms. The stakeholder-based planning processes highlighted in Chapter 4 would also have been useful in reconciling divergent views on the problem and solutions.

Lessons learned

The main thrust of action research was to support and facilitate the integration of participatory approaches in policy decision-making at district level, and to strengthen local-level processes and capacity for developing, implementing, and enforcing by-laws to improve natural resource management. Some of the lessons learned from the participatory by-law reform process are summarized below:

1. The understanding and analysis of existing by-laws and policy processes is an important first step in participatory by-law reforms, as it enables innovations to target key gaps in both the content and process of these reforms.
2. While by-laws can be effectively formulated at village level, their enforcement may require involvement of a higher level authority with the power to sanction the by-laws and enforce their implementation, such as the local government.
3. Participatory by-law reforms must involve capacity building for both local communities and decentralized local government structures. The inadequacy of human capital at different levels of local government is a key constraint to by-law formulation and implementation. Building capacity in local government structures linking communities to higher level authorities is critical for effective by-law reforms.
4. Linking local communities with local government requires an "honest broker" from the research or NGO community, or from the community itself, with the capacity and skills to provide evidence-based analysis and to facilitate policy dialogue.
5. As observed in Chapter 4, participatory by-law reforms involve both institutional and technical innovations. Not only are technologies important

for by-law implementation by providing alternative livelihood options for activities curtailed by by-laws. By-laws are also important for technology adoption, by enhancing their uptake and/or effectiveness.

6. Mature social capital can help in the establishment of local institutions for environmental governance, and in ensuring the effectiveness and continued participation in such institutions. Participatory by-law reforms require the ability of farmer groups and local communities to self-organize and to engage with and influence NRM governance processes. With an appropriate catalyst (external facilitator, strongly felt need), rural communities have the capacity to develop their own institutions, skills and networks for improved NRM governance. External agents can play a critical role in building social capital for by-law reforms and for the pursuit of other long-term development efforts.
7. Piloting is important. Many policies and by-laws have failed because they tried to do too much too soon, with little time to learn by doing and build upon these successes in taking on new challenges. Piloting the by-law reform process and particular by-laws in selected communities offers policy makers, development agents and other stakeholders the opportunity to test an approach and its effectiveness in addressing NRM challenges before expanding to other areas.
8. There are some "dark sides" of participatory by-law reforms. Enforcement of by-laws does not always ensure fairness, especially to women, the elderly, and others endowed with fewer human, financial, social, and political resources. Caution must be used to ensure that participatory processes do not reproduce existing patterns of social exclusion by ignoring those who are less able to negotiate their rights and shape social relationships to their advantage (see also Cleaver, 2005; German and Stroud, 2007).
9. In order for participatory by-law reforms to become part of the formal policymaking process, mechanisms are needed to institutionalize the approach. Decentralization policies now prevalent in many eastern African countries offer an opportunity for achieving this, as districts and other decentralized local government bodies have received legislative and executive powers to formulate and implement their own policies and by-laws in NRM. However, support from research and development organizations is required to ensure reforms are accompanied by effective means of engaging local communities in by-law formulation and implementation.
10. Given the policy resistance, implementation failures, and defensive routines (Sterman, 2006) common in local government structures, R&D professionals may need to stay close to the policy process and exploit opportunities that come along to get political buy-in to participatory processes. This may require opportunism in diagnosing the policy environment, identifying points of leverage, and recognizing short-term opportunities associated

with legislative calendars, planning, and budgeting activities, and changes in political leadership and government personnel.

Systems for demand-driven information provision

With the support of IDRC's ACACIA initiative,[6] AHI embarked on an action research experiment to develop and field test a system for demand-driven information provision at district level. The experience was piloted in Kabale District, Uganda, with the aim of learning lessons that could be scaled up to other districts and countries. While information is needed for all realms of human well-being, the pilot experience focused on the areas of agricultural production, marketing, and natural resource management. One of the first activities was to assess the challenges associated with current patterns of information access and sharing (Box 5.9)—which suggests the strong need for a coordinated approach to information and communication.

BOX 5.9 CHALLENGES IN INFORMATION SOURCING AND DISSEMINATION IDENTIFIED IN KABALE DISTRICT

The consulted stakeholders expressed facing challenges in sourcing information as well as disseminating it. The following were identified as constraints to information sourcing:

- Information is scattered (diversity of sources).
- Some information is inaccurate.
- Information access requires having personal contacts in institutions that are information sources.
- They lack awareness of what information is available.
- Information available is most likely to be in English rather than the local language (*Rukiga*).
- There is a general culture of waiting to be informed or told rather than being proactive information seekers.

Meanwhile, the following were identified as constraints to information dissemination:

- There is limited capacity and resources to package information in a suitable form.
- Because stakeholders involved in information dissemination would prefer using the least-cost dissemination form, the adequacy of information and quality of information delivery may be compromised.

- Stakeholders involved in information dissemination would prefer that recipients pay for the service, so the tendency is to provide information to those who are willing and able to pay for it.
- There is inconsistency in the information delivered to farmers from diverse sources.
- There is a repetition of efforts, with different organizations disseminating the same information to the same population without coordination.

Approach development

Approach 1—Demand-driven information provision at district level

The approach for demand-driven information provision required both effective articulation of information needs from farmers, and the development of a system for information gathering, packaging, and delivery. Figure 5.3 illustrates the key steps in this process and how information needs to flow in order to link farmer needs with information sources and their ultimate dissemination and application among target groups.

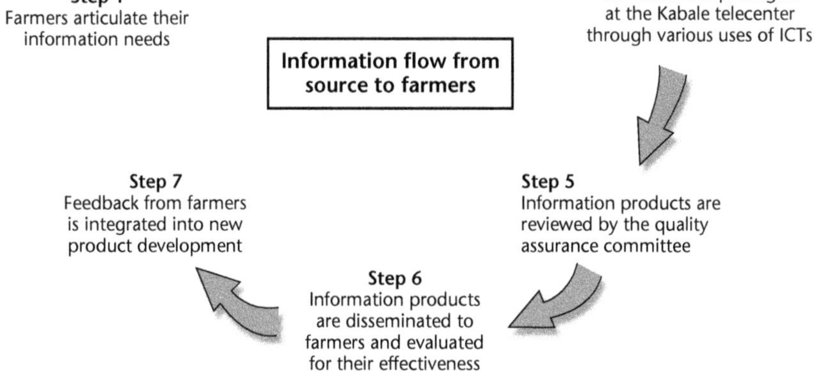

FIGURE 5.3 Information flow in demand-driven information provision

These steps may be summarized as follows:

1. *Farmers articulate their information needs.* Farmer groups at the village level meet with community-based facilitators (CBFs) to identify issues of concern in their farming, marketing, or natural resource management practices. The CBFs may use tools such as "needs trees" to generate an open-ended discussion on current information needs. A more formal Information Needs Protocol (Box 5.10) is then applied to categorize identified needs into three main subject matters (agriculture, natural resources management, and markets), to improve gender equity in information needs articulation and to identify preferred information sources and channels.

BOX 5.10 BASIC COMPONENTS OF THE INFORMATION NEEDS PROTOCOL

- Challenges and related information needs in:
  - Agricultural production
  - Marketing
  - Natural resource management
- Whether there are any gender-specific information needs that have been missed
- Preferred information sources (to determine whether there is a specific source they know of where the information can be sourced and which is considered reputable)
- Preferred communication channels (e.g. radio, pamphlets, posters, SMS, and demonstration)

2. *Farmer information needs are collated at parish and sub-county levels.* Following the articulation of information needs at village level, CBFs deliver the results to parish level committees or village information centers (VICE), who then compile the information to distil priority needs throughout the parish. Priorities at the parish level are then submitted to the sub-county telecenter, where priorities at sub-county are distilled (see Table 5.3 for an example). The telecenter collates all the information received from the six parishes and responds by: (i) distributing existing information available at the telecenter, and/or (ii) sending information needs to the district level telecenter for identification and packaging.
3. *Information is gathered from selected sources.* While the original idea was to source information through the sub-county telecenters, language barriers and problems with internet connectivity made this difficult. Therefore, an evaluation is made on whether the information sources preferred by farmers is feasible, based on information availability. Decisions on information sources are then

TABLE 5.3 Categories of information needs articulated by groups in different parishes of Rubaya Sub-County (N= 55 groups)

Topic	*Number of groups in each Parish*						*Total*
	Mugandu	*Karujanga*	*Buramba*	*Rwanyena*	*Kitooma*	*Kibuga*	
Soil selection using local indicators	6	1	2	8	9	7	33
Clean seed management	*6*	*10*	*9*	*10*	*10*	*6*	*51*
Making organic fertilizer using local resources	2	4	5	6	1	8	26
Post-harvest handling	*6*	*10*	*9*	*10*	*10*	*9*	*54*
Market information	*6*	*10*	*9*	*10*	*10*	*10*	*55*
Control of crop and animal pests and disease	6	1	6	6	8	7	34
Work plan development	0	1	0	4	0	6	11
Credit schemes	1	1	2	6	6	5	21
NRM technologies and by-laws	3	3	8	10	6	9	39
Fertilizer sourcing and application (quality, quantity)	3	2	3	8	9	5	30
Apiculture	0	4	2	1	2	4	13

made and the information is gathered. This often includes the sourcing of information on the internet through the Kabale telecenter. A checklist was developed to guide the service provider in gathering information from various sources (Box 5.11). Information may also be gathered by community members working on behalf of the larger community, as illustrated by the efforts made by parish-level marketing committees to source weekly market information at various local markets within the district.

4. *Information is packaged at the Kabale telecenter through various uses of ICTs.* The district telecenter serves as a hub where information acquired from diverse sources is organized and scrutinized for its clarity, quality, and relevance. Based on the communication medium preferred by farmers, the budget and the nature of the information itself, decisions are also made at this stage on the means of dissemination—as it influences how information is packaged for end users. This evolved from a heavy reliance on paper-based products to posters and radio broadcasts, and eventually, to the piloting of collectively managed mobile phones as parallel means to enhance information access at parish level (Box 5.12). The information is then prepared for the identified dissemination medium and translated into the local language for dissemination.

BOX 5.11 SAMPLE CHECKLIST TO AID SERVICE PROVIDERS IN SEEKING INFORMATION FROM DIVERSE SOURCES

Internet

1. What is the source? Is it reputable? Is it relevant to your context?
2. What do farmers need to know to be able to apply the information in their farms/lives?
3. Can you find all of this information on the internet? If not, can the experience of other knowledgeable local actors help to fill the gaps?

NGOs

1. General description about the knowledge or innovation that farmers have demanded, from the perspective of the NGO (Why is it so popular? How does it differ from other options?)
2. What are the key steps in implementation? [*Please put yourselves in the farmer's shoes and find out enough detail so that you could apply the innovation yourself if you needed to, as this will enable you to describe it in sufficient detail for others.*]
3. What are the main challenges to its implementation, and how can farmers overcome them?
4. Do you have any written material on the innovation that we could use to develop an information product for farmers (final products, grey literature, field reports)?
5. Would you like to co-author the publication and help us in the writing?
6. Who can farmers or NGOs contact to find out more information?

Expert farmers

1. General description about the knowledge or innovation that farmers have demanded for, from the perspective of the model farmer (Why is it so popular? How does it differ from other options available to you?)
2. How did you acquire the experience? What lessons can it offer to other farmers wishing to learn from you?
3. What are the key steps in implementation? [*Please find out enough detail so that you could apply the innovation yourself if you needed to, as this will enable you to describe it in sufficient detail for others*]
4. What are the main challenges in its implementation, and how can they be overcome by other farmers wishing to repeat the experience?
5. Can we use your name in the publication, to publicize to other farmers the good work you have done?
6. Can other farmers or NGOs contact you to find out more information?

BOX 5.12 USE OF WIRELESS PHONES TO ENHANCE FARMER INFORMATION ACCESS

Wireless telephones powered by solar energy were distributed to each of the six parishes where the ACACIA project was piloted. Users (members of parish committees and other local farmer groups) pay a small fee for the service, which is standardized across parishes and is designed to cover the cost of airtime and general maintenance. To monitor the effectiveness of these phones, log books were distributed so that records for each call being made through the phone could be maintained (including characteristics of the user and the use). While the original emphasis was placed on phone use specifically for the project's focus on information related to NRM, agriculture, and markets, actual usage was monitored to observe the extent to which these phones are useful for the intended purpose—and the extent to which unanticipated usage can also contribute to improved livelihoods. Results indicated that calls focused on personal and social issues were by far predominant, while marketing and agricultural production information was also actively sought out (Figure 5.4). Findings also showed a higher proportion of women using the phones to request information on NRM and agriculture, and men for personal reasons and to search for market information.

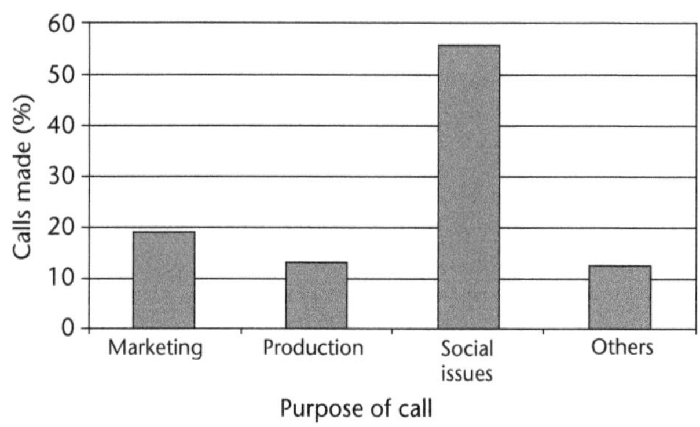

FIGURE 5.4 Use of VICE phones in 2008, Rubaya Sub-County, Kabale District, Uganda

5. *Information products are reviewed by a Quality Assurance Committee.* The Quality Assurance Committee (QAC) was established to oversee district-level efforts to respond to farmer information demands, and to ensure that products are effectively disseminated and utilized. Members of the Committee were selected by farmer representatives at sub-county and district levels based on jointly agreed selection criteria. Members of the QAC consisted of representatives from sub-county and district-level

farmer organizations, NGOs serving as active information providers, representatives of district line ministries (District Veterinary, Agriculture and Fisheries Officers, District Secretary of Production), and agricultural research. Formal terms of reference were drafted to help orient the QAC in its responsibilities related to the production and dissemination of information products and to ensure adequate representation of the views and needs of communities.

6. *Information products are disseminated to farmers and evaluated for their effectiveness.* After adjusting information products based on feedback from the QAC, they were either aired on the radio or printed and taken to the subcounty telecenter and VICE for dissemination. Participatory monitoring and evaluation was then carried out with farmer groups in six parishes, with a focus on product content (relevance, intelligibility), means of dissemination, and usefulness in decision making.
7. *Feedback from farmers is integrated into new product development.* Over time, farmers' feedback is a means through which general qualities of effective information products and delivery are distilled. Lessons learned through this feedback can then become mainstreamed within future approaches to information sourcing, packaging and dissemination.

Lessons learned

The AHI–ACACIA project generated a number of lessons that may be of more widespread interest, which include the following:

- Developing a system for demand-driven information provision utilizing ICTs is a challenge in contexts where the ability to pay for services is limited. It requires the concerted efforts of multiple actors (government, civil society, farmers, and research), close attention to mid-term outcomes (to enable the introduction of corrective changes), and ability to identify and capture opportunities.
- The technical challenges associated with effective systems of ICT-for-development are not just related to hardware and connectivity. They have to do with the development of human skills in the areas of information needs assessment, information capture, information processing and packaging, monitoring and evaluation, and adaptive learning.
- Use of ICTs for development is not simply a technical matter; a host of institutional and governance challenges must also be addressed to get it to work effectively. Institutional challenges include developing and sustaining farmer institutions capable of and motivated to work in the collective interest; a transition in the role of ender users from receivers of advice to active seekers of information; and multi-institutional collaboration at district level to achieve synergies and economies of scale. Governance challenges may be identified in the equitable articulation of information

needs; the management of resources owned or managed collectively (e.g., services, ICT infrastructure); and clear mechanisms (decision processes, terms of reference, incentives) to govern interactions among stakeholders.
- Developing user friendly and cost-effective ways of linking rural communities to information sources is an ongoing challenge, requiring additional commitments to action-based learning and experimentation.

Fostering multi-institutional partnerships at district level

Each of the above district-level approaches, and indeed many of the methodological innovations in this entire volume, requires some form of collaboration among organizations with complementary mandates, skills, and resources to be effectively implemented. This raises significant challenges, given the tendency for development and conservation initiatives to be conceived and often implemented by specific government agencies or non-governmental organizations. These organizations tend to specialize in production or conservation, research or development, livelihoods or governance, with minimal collaboration among institutions with complementary mandates. At the district level, rural development and natural resource management initiatives have not lived up to their potential as a result of lost opportunities for joint planning and resource sharing. Poor structural and functional linkages among different organizations and poorly coordinated planning have led to inefficiencies and opportunities lost in fostering synergies in resources and mandates. These constraints have clearly hindered innovation, undermined impact, and reduced opportunities for fostering more integrated, "win–win" solutions.

There is a need for a holistic approach that facilitates decision making at landscape and district levels as a substitute for isolated efforts. This approach to NRM necessitates a functional and well organized partnership. To achieve this goal, the spirit of collective action endemic in many societies in eastern Africa needs to be drawn upon in development and conservation activities. At the district level, partnerships among research, development, and conservation agencies can play a crucial role in ensuring more inclusive decision making at all levels and in exploiting synergies that enable multiple goals to be met simultaneously (e.g., livelihood improvements and conservation).

Approach development

AHI has experimented with two approaches to district-level institutional partnerships: multi-stakeholder platforms and informal partnerships. The latter largely emerged as a natural step in the implementation of other NRM innovations, whereas the former was intentionally designed as a district-level institutional innovation to be tested and improved upon through action research. We present both, owing to the lessons that may be learned through drawing comparisons between them.

Approach 1—Multi-stakeholder platforms

The approach is based on the development of an alliance of institutions with a shared vision and coordinated actions, in this case a vision for integrated natural resource management. That vision should encompass multiple objectives (e.g., development and conservation), as well as a set of core values that help to sustain the partnership and enhance its relevance (for example, local ownership, flexibility, shared credit, and a spirit of voluntarism). As the platform derives its legitimacy from a demand-driven approach to development, its members must be diverse so as to enhance the ability to respond effectively to articulated needs. Ideally, partners should include local government, NGOs, CBOs, and farmer groups, research and conservation institutions, and individual community members (Box 5.13). Given the relationship between good governance and good environmental practices, involvement of government agencies responsible for by-law formulation and enforcement may also be useful. The private sector may also be called in to explore opportunities to link local livelihood needs to market opportunities. Such a platform provides a mechanism for negotiation and decision-making in the articulation of strategic development plans and in the sharing of responsibility for their implementation.

The formation of such a platform is likely to involve the following steps:

1. Hold individual consultations (person to person and organization to organization) to identify the weaknesses of the current way of doing business and bolster commitment for a new approach.
2. Conduct consultations with farmers, farmer groups, and other intended beneficiaries of development efforts on their concerns related to livelihoods, natural resources, and the quality of governance and service provision.
3. Host a workshop with potential platform members to develop a joint understanding of the deficiencies in current (disconnected) development and conservation initiatives, explore goals and desired functionalities of the platform, articulate the core strengths of different partner organizations in supporting the effort, solicit commitments from partner organizations, and agree on next steps in a collaborative planning process.
4. Initiate a bottom-up diagnostic, visioning, and planning process starting at the farmer group level to ensure adequate coverage of diverse sub-counties, parishes, and villages in the district. Collate plans at parish, sub-county, and district levels, distil the forms of support requested from outside actors and discuss how to effectively support these plans at each level.
5. Hold a meeting of the platform to agree how to support the action plans and reflect on what needs to be done by the platform to steward the initiative into action.

6. Hold a facilitators' training with volunteer facilitators from different levels, to impart the necessary skills for facilitating participatory and deliberative reflection and planning processes.[7]
7. Formulate a constitution and strategic plan to guide the operations of the platform.
8. Establish a Secretariat to guide the implementation of the platform's business plan.
9. Carry out periodic evaluations of the effectiveness of the platform through consultations with partners and beneficiaries, and replanning to improve the platform's effectiveness and responsiveness to feedback.

BOX 5.13 DEVELOPMENT OF A DISTRICT MULTI-STAKEHOLDER PLATFORM IN KAPCHORWA DISTRICT, UGANDA

In Kapchorwa District, an alliance of institutions was formed with a shared vision for integrated natural resource management and inspired by the Landcare approach. Members of this platform (the Kapchorwa District Landcare Chapter, or KADLACC) include NGOs, CBOs, farmer groups, local government, research and conservation organizations an individual community members. The platform objectives are:

- To create a forum for government, civil society, research organizations, and other stakeholders involved in land and natural resource management to harmonize their activities and work collaboratively.
- To build the capacity of member organizations in planning, influencing policy and resource mobilization to enhance performance at district level.
- To advocate for democratic processes for NRM and land-use policies.
- To conduct action research on ways to support integrated approaches to land use and livelihoods.

Key focal areas and activities of KADLACC are summarized in Table 5.4.

So what is the added value of enhancing district-level collaboration? For KADLACC, the benefits may be summarized through a before/after comparison of development practice (Table 5.5).

TABLE 5.4 Focal areas and activities of KADLACC

Focal areas	*Activities*
Protected area collaborative management	Facilitating negotiations among communities and protected area managers Working with displaced and indigenous peoples to enhance access to customary resources
Watershed management	Filling knowledge gaps through training and action research Negotiation support for socially optimal solutions
Farmer institutional development and learning	Conduct farmer skills needs assessments Supporting institutional capacity building of farmer groups Matching innovations and technologies to farmer needs
Marketing and enterprise development	Seek and develop market niches and opportunities for income generation Awareness creation on ecosystem health-based product branding
Partnership and networking	Affiliation and participation in the African Landcare Network and Landcare International

TABLE 5.5 Comparison of development practice before and after the establishment of the multi-stakeholder platform

Before	*After*
NRM not mainstreamed in development initiatives but carried out through "lone ranger" approaches	Integrated development and NRM planning at multiple levels, with the involvement and support of local government
Limited access to development and extension services for a large number of households	Farmer groups linked to trained facilitators from various member institutions, enabling more widespread access to services
Role of local government in pro-poor, ecologically friendly policy support process undefined or unclear	Strengthened role of local government structures in integrated NRM planning; involvement of community members in policy reform
Conservation efforts delinked from rural development and marketing; livelihood needs seen as contradictory to conservation objectives	Strategic approach for linking livelihood goals to conservation objectives and supporting the marketing of ecologically friendly products in place

Approach 2—Informal partnerships in INRM

Informal approaches to collaboration at district level are the norm, and generally emerge on an as-needed basis. Such a need may arise from community- or project-level needs that cannot be met through the community's efforts or a single support institution, the desire to exploit an opportunity that is conditional upon partnership (for example, funding streams), or commitments to donors. Such partnerships are largely ad hoc in nature, ephemeral (lasting for as long as the specific activity or need lasts), and carry limited transaction costs given the limited investments in partnership building (relative to actual implementation). Common steps in the development of informal partnerships in AHI have involved the following:

1. A challenge or opportunity arises that calls for linkages to new organizations with the required skill base, mandate, or resources.
2. Constituent-building to seek buy-in, often from individuals who come to represent the wider organization—but at times through a formal agreement with the partner organization.
3. Planning workshop to agree on the division of roles and responsibilities and budgets for supporting partner activities.
4. Implementation (including any number of steps associated with engaging the beneficiaries in planning and/or implementing activities for which partners have assumed co-responsibility).
5. Joint monitoring and evaluation (largely focused on the work plan, but at times including a reflection on the partnership itself), and adjustment as needed to address challenges that have emerged through implementation.

Lessons learned

The following lessons were derived from a comparative analysis of the two approaches to multi-institutional partnerships:

- The transaction costs of more formal partnerships are higher than informal approaches, and the benefits gained from the former must be worth the effort. Achieving such benefits will often require the development of rather ambitious goals supported by significant buy-in from partner organizations.
- Success, particularly with the more formal institutional platform approach to partnership, is more likely where there is a strong spirit and practice of voluntarism. This generally comes from the establishment of trust and rapport among group members, and from a sense of accomplishment that goes beyond what individual member institutions have achieved in isolation. Success of the multi-institutional platform also rests on building the capacity of volunteer facilitators, for whom a set of "soft skills" can go a long way in sustaining community engagement.

- To support truly bottom-up approaches to development at district level with wide geographical coverage, more formal approaches to district level partnership are likely to be required—given the limitations in the skill base, mandate, and resources of any given organization, and the need to match local expectations with a firm commitment by district-level service providers.
- Strong complementarities exist between local government and civil society, owing to their unique skill sets and institutional mandates. One key function of local government is lending legitimacy to the platform as a mainstream (rather than marginal) mechanism for the coordination of development activities in the district. Efforts should therefore be made to ensure district-level partnerships include these two sets of actors.
- Start-up activities often require an external source of funding, to sustain activities until partners have bought into the idea and begin contributing their own resources (often in the form of staff time and operations) to ensure the platform's financial viability.
- The effectiveness of district-level institutional partnerships is constrained by staff turnover or shifts in the focus of partner organizations, a problem which is likely to be more acute in informal partnerships than in established platforms where continuity is more likely due to institutional level rather than individualized commitments.

Missing links

While significant progress has been made in understanding the elements of effective approaches to district institutional and policy innovations, a number of methodological gaps remain. These gaps suggest a number of priorities for future research and methodological innovation on the topic:

1. *Sustainability of district-level institutional and policy innovations.* The experiences shared in this chapter derived from project-based experiences lasting a number of years (3 to 6) and supported by external funding. Lessons are needed on how to sustain such innovations with existing financial resources once human resources and institutions are strengthened. Lessons on how to sustain such innovations have begun to emerge with the ACACIA experience (through efforts to institutionalize demand-driven information provision within NAADS, to be discussed in Chapter 6) and with the Landcare experience (through efforts to build self-sustaining district platforms and farmer reflect cycles). However, exit strategies require time to implement and financial resources—both to ensure the sustainability of initiatives and to learn lessons on how approaches change as they are institutionalized.[8] It is these lessons that are perhaps most useful when scaling out district-level innovations to new districts.
2. *Linking methods for farm- and landscape-level innovation to district institutional innovations.* The host of approaches described in earlier chapters needs a

home if they are to be applied on a wider scale. That institutional home could be within district-led innovations, as suggested in this chapter, or within national-level institutions, as suggested in Chapter 6. With the exception of participatory by-law reforms, AHI has yet to make a systematic effort to scale up specific proven methodological innovations (e.g., watershed management) through district-level institutions or initiatives. Efforts have instead largely focused on piloting novel innovations at this level.

3. *Scaling out.* With the exception of our work with ACACIA/NAADS and Landcare, little effort has been made to scale out proven district-level innovations to new districts (where a set of institutional and political conditions similar to the pilot district is more likely to prevail) or countries (where a unique set of contextual factors is likely to affect an approach's feasibility). Even where these efforts have been made, they have in some cases been ephemeral owing to limited funding horizons. Both experiences are urgently needed if we are to capitalize upon the investments made to date in pilot experiences in AHI.

Conclusions

This chapter illustrates a set of methodological innovations designed to enhance the potential of districts as engines of rural development and sustainable natural resource management. Our experiences point to the fundamental role of *institutional* innovations at multiple levels (particularly village, sub-county, and district) to enable cross-scale communication, exploit synergies in the human and financial resources found at diverse levels, and tap the latent potential that exists at each level of socio-political organization. It also points to the fundamental role of institutional innovations in getting technological innovations to work and in supporting improved natural resource management at a meaningful scale. Thus, findings also suggest that it is high time that meaningful investments be made in the "soft skills" (such as facilitation and institutional strengthening) required to revitalize public institutions and the *modus operandi* of the agricultural and NRM sector.

Notes

1 More critical reviews suggest that decentralization is simply a means for central governments to transfer their fiscal and administrative burdens to decentralized actors (Nsibambi, 1998).

2 The Tanzanian Local Government (District) Authorities Act of 1982 empowered district councils to pass by-laws and the 1997 Local Government Act of Uganda provides the legal framework for the participation of local communities in policymaking (Sanginga, 2003; see also www.leat.or.tz/publications/decentralization/4.3.district.authorities.php).

3 Please note that at this stage, the team may not know which stakeholders are relevant to the kind of problems that may emerge later on. Thus, if the research and development

team has little familiarity with actors in the district, it is best that this activity be more comprehensive than what is thought to be needed.

4 This section draws heavily from Sanginga et al. (2010a).

5 For details see Sanginga et al. (2010b).

6 ACACIA works with African partners to help countries in Africa apply information and communication technologies (ICTs) to social and economic development. ACACIA's mission is to support research on ICTs that improve livelihood opportunities, enhance social service delivery, and empower citizens while building the capacity of African researchers and research networks. For more information, visit: www.idrc.ca/acacia/.

7 Please note that this step was introduced here owing to feedback received from participants. However, it may be useful to have this step come earlier, prior to Step 4.

8 In recognition of the fact that changes must occur as an approach moves from an independent initiative with external funding to its institutionalization within government structures and programs, as illustrated by the NAADS experience.

References

ACACIA (2006) ACACIA Prospectus 2006–2011. www.idrc.ca/acacia/ev-113431-201-1-DO_TOPIC.html

Agrawal, A. and C. Gibson (1999) Enchantment and disenchantment: The role of community in natural resource conservation. *World Development* 27(4): 629–649.

Bowles, S. and H. Gintis (2002) Social capital and community governance. *Economic Journal* 112: 412–426.

Brinkerhoff, D. (2001) Taking account of accountability: A conceptual overview and strategic options. Draft report for the Implementing Policy Change Project, Phase 2, Center for Democracy and Governance, USAID. Washington, D.C.P: ABT Associates, Inc.

Brockington, D. (2007) Forests, community conservation, and local government performance: The village forest reserves of Tanzania. *Society and Natural Resources* 20: 835–848.

Chambers, R. (2005) *Ideas for Development.* London: Institute of Development Studies and Earthscan.

CIDA (Canadian International Development Agency) (2003) *Knowledge Sharing. Methods, Meetings and Tools.* Ottawa: Canadian International Development Agency.

Cleaver, F. (2005) The inequality of social capital and the reproduction of chronic poverty. *World Development* 33(66): 893–906.

Coleman, J. (1988) Norms as social capital. In: E. Ostrom and T.K. Ahn (eds.), *Foundations of Social Capital.* Cheltenham: Edward Elgar, pp. 136–158.

Cooke, B. and Kothari, U. (eds.) (2001) Participation: The new tyranny?, London: Zed Books.

Davids, I. (2003) Developmental local government: The rural context and challenges. *Development Update* 4(1): 31–54.

Fabricius, C., C. Folke, G. Cundill, and L. Schultz (2007) Powerless spectators, coping actors, and adaptive co-managers: A synthesis of the role of communities in ecosystem management. *Ecology and Society* 12(1): 29. [online] www.ecologyandsociety.org/vol12/iss1/art29/ (accessed July 9, 2007).

Flor, A.G. (2001) ICT and poverty: The indisputable link. Paper presented at the Third Asia Development Forum on 'Regional Economic Cooperation in Asia and the Pacific,' 11–14 June, 2001, Bangkok, Thailand. Available at: www.fsp.usp.br/acessibilidade/ICTandPoverty-TheIndisputableLink2001.pdf (accessed October 14, 2009).

Galvin, M. (1999) The impact of local government on rural development in South Africa. *Transformation* 40 (1999): 87–111. Available at: http://www.transformation.ukzn.ac.za/archive/ tran040/tran040005.pdf (accessed Nov. 5, 2009).

German, L. and A. Stroud (2007) A framework for the integration of diverse learning approaches: Operationalizing Agricultural Research and Development (R&D) Linkages in Eastern Africa. *World Development* 35(5): 792–814.

German, L., W. Mazengia, H. Taye, M. Tsegaye, S. Ayele, S. Charamila, J. Wickama (2010) Minimizing the livelihood trade-offs of natural resource management in the Eastern African Highlands: Policy Implications of a Project in "creative governance". *Human Ecology* 38, 31–47.

German, L., W. Mazengia, W. Tirwomwe, S. Ayele, J. Tanui, S. Nyangas, L. Begashaw, H. Taye, Z. Adimassu, M. Tsegaye, F. Alinyo, A. Mekonnen, K. Aberra, A. Chemangeni, W. Cheptegei, T. Tolera, Z. Jotte, and K. Bedane (2008) Enabling equitable collective action and policy change for poverty reduction and improved natural resource management in the eastern African highlands. In: E. Mwangi, H. Markelova, and R. Meinzen-Dick (eds.), *Collective Action and Property Rights for Poverty Reduction: Lessons from a Global Research Project*. Washington, D.C.: CAPRi, pp. 11–12.

Government of Uganda (2004) Poverty Eradication Action Plan (2004/5—2007/8). Kampala: Ministry of Finance, Planning and Economic Development.

Government of Uganda (2005) National Strategy for Growth and Reduction of Poverty of Uganda. Kampala: Government of Uganda.

IULA (1993) IULA world-wide declaration of local self-government. Available at: www.bunken.nga.gr.jp/siryousitu/eturansitu/charter/iula_decl_txt.html (accessed November 6, 2009).

James, R., P. Francis and G. Ahabwe (2001) The institutional context of rural poverty reduction in Uganda: Decentralisation's dual nature. *LADDER Working Paper* No. 6. Available at: www.uea.ac.uk/dev/odg/ladder/ (accessed March 22, 2005).

Keeley, J. (2001) Understanding and influencing policy processes for soil and water conservation. In C. Reij and A. Waters-Bayer (eds.), *Farmer Innovation in Africa: A Source of Inspiration for Agricultural Development*. London: Earthscan Publications, pp. 281–291.

Khan, S. (2006) Local government and participatory rural development: The case study of district government in north western Pakistan. PhD thesis, Gomal University. Available at http://eprints.hec.gov.pk/1011/1/742.html.htm

Mamdani, M. (1996) *Citizen and Subject: Contemporary Africa and the Legacy of Late Colonialism*. Princeton: Princeton University Press.

Means, K., C. Josayma, E. Neilsen, and V. Viriyasakultorn (2002) *Community-based Forest Resource Conflict Management: A Training Package*. Vols. 1 and 2. Rome: Food and Agriculture Organisation of the United Nations.

Nahdy, S. (2004) The Ugandan National Agricultural Advisory Services. In: W. Rivera and G. Alex (eds.), *Volume 1: Decentralized Systems, Case Studies of International Initiatives*, pp. 46–52. *Agricultural and Rural Development Discussion Paper* No. 8.

Nkonya, E.M., J. Pender, E. Kato, S. Mugarura, and J. Muwonge (2005) Who knows, who cares? Determinants of enactment, awareness and compliance with community natural resource management. *CAPRi Working Paper* No. 41. Washington, D.C.: International Food Policy Research Institute (IFPRI).

Norris, P. (2001) *The Digital Divide: Civic Engagement, Information Poverty and the Internet Worldwide*. Cambridge: Cambridge University Press.

Nsibambi, A. (ed.) (1998) *Decentralization and Civil Society in Uganda: The Quest for Good Governance*. Kampala: Fountain Publishers.

Okoth-Ogendo, H.W.O and G. Tumushabe (1999) *Governing the Environment: Political Change and Natural Resource Management in East and Southern Africa*. Nairobi: ACTS Press.

Ostrom E. (2000) Collective action and the evolution of social norms. *Journal of Economic Perspectives* 14(3): 137–158.

Oyono, P.R. (2005) The foundations of the *Conflit de langage* over land and forests in southern Cameroon. *African Study Monographs* 26(3): 115–144.

Perret, S. (2004) Matching policies on rural development and local governance in South Africa: Recent history, principles and current challenges. Paper presented at the workshop on *Local Governance and Rural Development*, organized by GTZ and the University of Pretoria Post Graduate School for Agricultural and Rural Development. Available at: www.up.ac.za/academic/ecoagric/fulltext/2004-06.pdf (accessed November 5, 2009).

Pretty, J. (2003) Social capital and the collective management of resources. *Science* 32: 1912–1914.

Reason, P. and H. Bradbury (2001) *Handbook of Action Research: Participative Inquiry and Practice*. London, UK: Sage.

Ribot, J. (2002a) African decentralization: Local actors, powers and accountability. *Democracy, Governance and Human Rights Working Paper* No. 8. Washington, D.C.: WRI.

Ribot, J. (2002b) *Democratic Decentralization of Natural Resources: Institutionalizing Popular Participation*. Washington, D.C.: World Resources Institute.

Ribot, J.C. (2003) African decentralization: Actors, powers and accountability. *Democracy, Governance and Human Rights Paper* No. 8. Geneva: UNRISD.

Rivera, W. and G. Alex (eds.) Volume 1: Decentralized systems, case studies of international initiatives. *Agricultural and Rural Development Discussion Paper* No. 8.

Sanginga, P. (2003) Strengthening social capital for improving policies and decision-making in natural resources management. Natural Resources Systems Programme Project Report. London: DfID.

Sanginga, P. (2004) Facilitating participatory processes for policy change in natural resource management: Lessons from the highlands of southwestern Uganda. *Uganda Journal of Agricultural Sciences* 9: 958–970.

Sanginga, P. and C. Chitsike (2005) The power of visioning. A handbook for facilitating the development of community action plan. *Enabling Rural Innovation Guide* #1. Kampala, Uganda: International Centre for Topical Agriculture (CIAT).

Sanginga, P., R. Kamugisha, and A.M. Martin (2010a) Strengthening social capital for adaptive governance of natural resources: A participatory action research for by-law reforms in Uganda. *Society and Natural Resources* 23: 695–710.

Sanginga, P., A. Abenakyo, R. Kamugisha, A. Martin, and R. Muzira (2010b). Tracking outcomes of social capital and institutional innovations in natural resources management: Methodological issues and empirical evidence from participatory by-law reform in Uganda. *Society and Natural Resources* 23: 711–725.

Sanginga, P., J. Tumwine, and Nina Lilja (2006) Patterns of participation in farmers research groups. *Agricultural and Human Values* 23(4): 501–512.

Scoones, I. and I. Thompson (2003) Participatory processes for policy change. *PLA Notes* 6(10): 51–57.

Sheridan, M. (1994) Environmental consequences of independence and socialism in North Pare, Tanzania, 1961–88. *Journal of African History* 45(2004): 81–102.

Smoke, P. (1993) Local government fiscal reforms in developing countries: Lessons from Kenya. *World Development* 21(6): 901–923.

Sterman, J. D. (2006) Learning from evidence in a complex world. *American Journal of Public Health* 96: 505–514.

Stringer, L.C., A.J. Dougill, E. Fraser, K. Hubacek, C. Prell, M.S. Reed (2006) Unpacking 'participation' in the adaptive management of social-ecological systems: A critical review. *Ecology and Society* 11.

Tacconi, L. (2007) Decentralization, forests and livelihoods: Theory and narrative. *Global Environmental Change* 17(2007): 338–348.

Veit, P. with G.Z. Banda, A. Brownell, S. Mtisi, P. Galega, G. Mpundu Kanja, R. Nshala, B. Owuor Ochieng, A. Salomao, and G. Tumushabe (2008) *On Whose Behalf? Legislative Representation and the Environment in Africa.* Washington, D.C.: WRI.

Wily, L.A. and P.A. Dewees (2008) From users to custodians: Changing relations between people and the state in forest management in Tanzania. *World Bank Policy Research Working Paper* No. 2569. Washington, D.C.: World Bank.

Wunsch, J.S. and D. Olowu (1990) *The Failure of the Centralized State: Institutions and Self-Governance in Africa.* San Francisco: Westview Press.

6

INSTITUTIONAL CHANGE AND SCALING UP

Chris Opondo, Jeremias Mowo, Francis Byekwaso, Laura German, Kenneth Masuki, Juma Wickama, Waga Mazengia, Charles Lyamchai, Diana Akullo, Mulugeta Diro, and Rick Kamugisha

Context and rationale

> Cases of participatory watershed management … managed by NGOs, are becoming increasingly abundant. Yet, almost without exception, they are very small in scale and can be expanded only by repeating the same slow, costly, in-depth techniques in successive villages. Many government-sponsored approaches have expanded rapidly, but often lack the local ownership and group coherence necessary for sustainable management of the common pool components of watersheds. If approaches … are to be participatory and rapidly replicable, then the preconditions for scaling up have to be identified and introduced into the design of projects and programmes.
>
> *Farrington and Lobo, 1997: 1*

Over the last decade, there has been a growing concern among donors and development agencies about the limited impact that natural resource management (NRM) technologies and practices have had on the lives of poor people and their environment. Interventions have often failed to reach the poor at a scale beyond the target research sites (Ashby et al., 1999; Briggs et al., 1998; Bunch, 1999). Acknowledgment of this fact has resulted in a recent surge of interest in the concept and practicalities of "scaling up." Yet organizations accustomed to work at a certain scale struggle with the organizational, methodological, and financial challenges of "going to scale" (Snapp and Heong, 2003). Technologies that are relatively easy to assimilate into farming systems and bring rapid returns to farmers can often spread of their own accord (Chapter 2, this volume). Yet moving beyond socio-cultural and institutional barriers to access, and disseminating more complex NRM technologies at larger scales, pose more complex challenges (Middleton et al., 2002). If technologies and novel approaches to research and development are to be rapidly replicable, then the preconditions for scaling up have to be identified and introduced into the design of projects and programmes (see, for example, Farrington and Lobo, 1997). This chapter explores AHI experiences with "scaling out" from benchmark sites and facilitating institutional reforms for more widespread impact.

Scaling out and institutional change defined

The proliferation of terminology around efforts to "go to scale" has created a lot of confusion, with the terms scaling out, scaling up, horizontal scaling up and vertical scaling up, among others, often used interchangeably. For example, for the World Bank (2003) the term scaling up is used in reference to the replication, spread, or adaptation of techniques, ideas, approaches, and concepts (the means), as well as to increased scale of impact (the ends), while for Lockwood (2004) scaling up implies expanded coverage rates to rapidly meet the needs of diverse groups or to ensure that "islands of success" are maintained at expanded scale.

The use of different terminology to say the same thing requires that one's definitions be clarified up front. AHI adopts definitions similar to those proposed by Gündel et al. (2001), which clearly differentiate between the horizontal and vertical dimensions of "going to scale" as a question of geographical expansion vs. changes in structures, policy, and institutions. To be more precise, AHI defines "scaling out" as a process of reaching larger numbers of a target audience through expansion of activities *at the same level of socio-political organization*. In short, it implies doing the same things but over a larger area. "Scaling up," on the other hand, involves innovations at a new level of socio-political organization—namely, support to institutional changes which enable tested innovations or the *process of innovation itself* to be supported over a larger area (Millar and Connell, 2010). In short, it involves doing new things at a level where it will make a bigger difference. It often involves taking the lessons and experiences from pilot projects to decisions that are made at the upper levels of management, such as what kind of approaches to support and where. Institutionalization is the process through which new ideas and practices become acceptable as valuable and become incorporated into normal routines and ongoing activities in society (Norman, 1991). It is a more permanent form of scaling up, as it involves assimilation of the innovation into the everyday structures, procedures or practices, or organizations. According to Jacobs (2002: 178), institutionalization is a change that has "relative endurance" or "staying power over a length of time," or "has become part of the ongoing, everyday activities of the organization." Figure 6.1 helps to visualize how scaling out and scaling up are conceived of within AHI.

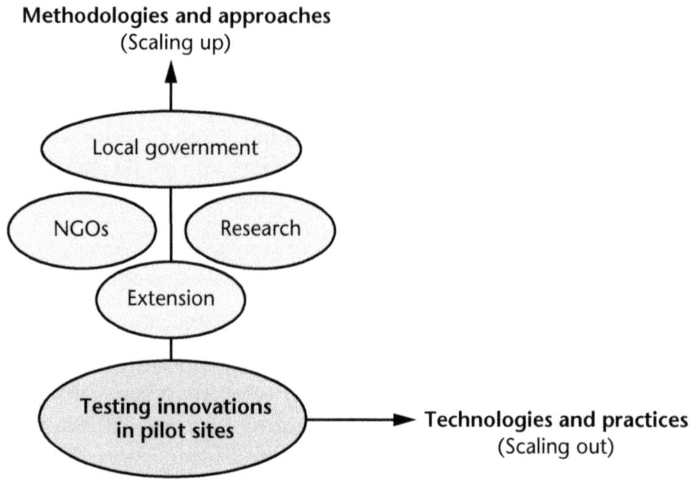

FIGURE 6.1 Scaling out and scaling up in AHI

In the context of AHI, many activities—whether at farm, landscape, or district scale—have been designed as pilot or demonstration projects enabling the design and testing of innovations to explore "what works, where and why." This has enabled the program to test what does or does not work well and identify needed adjustments, before engaging in costly (and risky) innovations at a broader scale. This "piloting" strategy is essential for enhancing innovation while ensuring efficient use of resources.[1] It helps to avoid the traps of sticking to the status quo (which may or may not be working well) for fear of making costly mistakes and supporting costly innovations before they are proven to work. Yet it also leaves the innovation process incomplete, as the process of taking pilot experiences to a larger scale remains—and is also likely to involve further refinements for the approach to become more widely applicable to new contexts.

Methodological innovation in AHI has been structured around a set of "learning loops"—key analytical thrusts that have been the subject of action research-based learning (Figure 6.2). It is important to note that while the innermost loops are largely focused on developing novel methodological innovations, the outermost loop focuses on "going to scale." This encompasses the dissemination of lower-level social and biophysical innovations as well as new types of institutional innovations to enable the former to be applied as part of everyday institutional practice, and to support the institutionalization of the overall approach to action research.

Before closing the section on definitions, it is important to acknowledge the partnership dimension of scaling up. To some authors (Uvin and Miller,

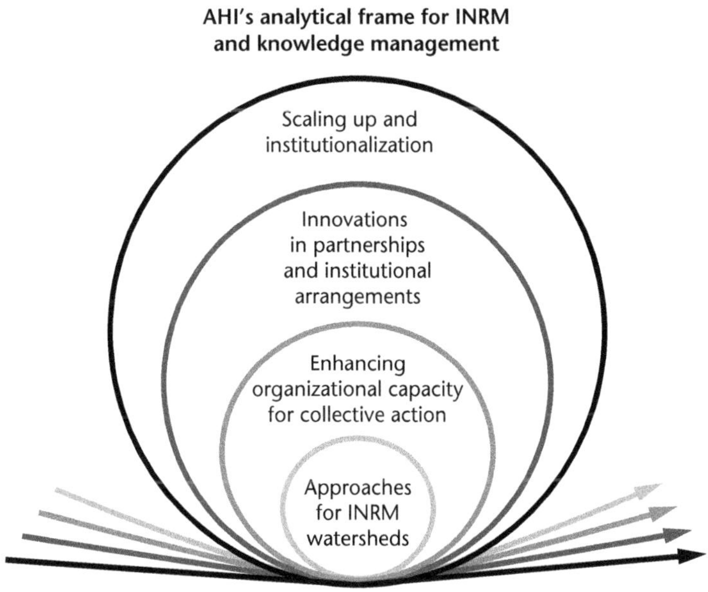

FIGURE 6.2 AHI "Learning loops"

1994; 1996), scaling up implies increased interaction with diverse stakeholders. FARA (2006) has also emphasized the need for widening the scope of participants in agricultural research beyond traditional actors to bring impact to rural communities. This view is in tandem with proponents of the innovation systems approach (Hall, 2005; Sumberg, 2005) and beyond farmer participation (Scoones et al., 2007). While this is not an explicit feature in AHI definitions, it is explicit in the innovations tested by the program.

Elements of "scalable" innovations

An objective of AHI is to spread successful innovations (whether new approaches to development and NRM or tangible technologies) from pilot benchmarks to new environments, be they communities or research and development organizations. Scalability may be defined as the ability to adapt an innovation to effective usage in a wide variety of contexts (Clarke et al., 2006). From both empirical work and interactions with farmers, researchers, and managers, AHI has harvested some of the elements regarded as key ingredients for innovations to be scalable in a given context (Box 6.1).

BOX 6.1 CHARACTERISTICS THAT DETERMINE THE POTENTIAL OF AN INNOVATION TO GO TO SCALE

Valued outcomes: The ability of the innovation to generate income and enhance well-being at community level or to achieve policy objectives at the institutional level. An example of the former is a high-value cash crop with a ready market in urban centers. An example of the latter is a methodological innovation within a research or development organization that promotes institutional objectives (e.g., farm-level value capture, market-oriented research), or demand-driven service provision.

Effectiveness: The ability of the innovation to meet the goals and aspirations of beneficiaries. For example, an approach or process that emphasizes equitable technology distribution or sharing among individuals and among villages will appeal to the majority of farmers, especially those with meager resources and formerly excluded by research and development programs.

Efficiency: What is being piloted is cost-effective, thus enhancing its potential for scaling up and out. Production of a unit of good or service is termed economically efficient when that unit of good or service is produced at the lowest possible cost, relative to the value it generates. With limited financial resources, this consideration is particularly important in an organization's decisions to invest in particular research or development activities.

Sustainability: The potential for the benefits from the innovation process to be enjoyed over prolonged periods by the recipients, even after those supporting its dissemination are no longer involved. This is a characteristic of a process or state that can be maintained at a certain level indefinitely. Although the term is used more in environmental circles, it is relevant to social processes (e.g., participation, collective decision-making, institutional collaboration) that work in tandem with technologies.

Yet it is not just the characteristics of the innovation that matter, but the nature of the scaling up/out process itself, that will determine its success. For *scaling out* to be effective, the following conditions must be met:

- The participatory process of problem identification and prioritization, and the matching of innovations to these priorities, must be effective and sustained over time.
- Efforts must be made to overcome the social and institutional constraints to spontaneous and mediated forms of technology dissemination, such as the tendency for gender-based patterns of technology access or the bias exhibited by extension agents in some countries or locations toward wealthy male farmers (who can more easily innovate).

- Adequate attention is given to developing the financial, human, and social capital required to apply the technology successfully (Adato and Meinzen-Dick, 2002; Knox McCulloch et al., 1998).
- Adequate attention is given to adapting the technology to local conditions (Chambers et al., 1989).

For *scaling up* to be effective, the following conditions must be met:

- Committed leadership to identify and support new strategic directions.
- Budgetary reallocations and the provision of sufficient financial resources to support proposed institutional changes.
- Behavioral and attitudinal changes that exhibit a willingness to make reforms (e.g., the decentralization of authority and resources) (Gillespie, 2004).
- Realignment of institutional incentive mechanisms such as staff performance appraisals to new policy objectives.
- Strategic networks to build upon complementary skill sets, institutional mandates, and resources.
- Scaling up fast-track interventions needs to be well aligned with government policies and procedures so as to ensure sustainability (Buse et al., 2008).
- Scaling up requires that the host organization has the capacity to interest people and enable them to adopt new ideas or diffuse the intended innovations (Senge et al., 1999).
- Ability to cope with and adapt to a diversity of contexts and dimensions that are political, institutional, financial, technical, spatial, and temporal (Gonsalves and Armonia, 2000).

Learning organizations

As part of an introduction to institutionalization, it is important to consider what is known about characteristics that make organizations effective in meeting new challenges and adapting to change. One highly relevant body of literature in this regard is that which explores the nature of "learning organizations." Just what constitutes a learning organization is a matter of ongoing debate (Argyris and Schon, 1996; Senge, 1990). In this sub-section, we explore some of the themes that have emerged in the literature and among key thinkers on the subject.

The concept of a learning organization emerged in response to an increasingly unpredictable and dynamic business environment. Organizational learning involves individual learning, and those who make the shift from traditional thinking to the culture of a learning organization develop the ability to think critically and creatively. According to Meinzen-Dick et al.,

> [this] can be fostered by a spirit of critical self-awareness among professionals and an open culture of reflective learning within organizations. In such an environment, errors and dead ends are recognized as opportunities for both individual and institutional learning that can lead to improved performance.
>
> *Meinzen-Dick et al., 2004*

The term "learning organization" was coined in the 1980s to describe organizations that experimented with new ways of conducting business in order to survive in turbulent, highly competitive markets (see Argyris and Schon, 1996; Senge, 1990). The aim in such organizations is to become effective problem solvers, to experiment with new ideas and to learn from internal experiences and the best practices of others. In the learning process, positive results accrue to individuals and the organization or to the organizational culture as a whole. However, concrete cognitive (mental) and behavioral traits, as well as specific types of social interaction and the structural conditions to enhance the likelihood that the necessary organizational qualities are achieved and sustained over time, need to be in place. Some of these key qualities are communication and openness; a shared vision, open inquiry and feedback; adequate time allocation for piloting new ideas; and mutual respect and support in the event of failure. Senge (1990) notes that for learning to be effective, the personal goals of staff in such organizations must be in line with the mission of the organization.

The process of evolving into a learning organization therefore involves behavioral change, and changes in the ways of thinking and information processing (Garvin, 1993). It may take as long as five to ten years for institutional change to become part of the corporate culture, given the fragility of change and resistance that often accompanies it (Kotter, 1995). This is because most people practice defensive reasoning; because people make up organizations, those organizations also tend to exhibit this culture (Argyris, 1991). So at the same time that an individual or organization is avoiding embarrassment or the threat of failure, it is also avoiding learning. Senge et al. (1999) point out that there is also a need to focus on understanding the factors limiting change, such as lack of systems thinking, fear and anxiety in the face of change, and the danger of innovations acquiring "cult status" and thus becoming isolated from the organization. Kotter (1995) has suggested that the failure to "anchor" cultural change is a key challenge to learning organizations. Thus, until new behaviors are rooted in social norms and shared values of the organization, they are subject to resistance.

Research and development organizations also operate in a dynamic world and must learn to adjust to changes in their context in order to be effective and, often, in order to survive. The institutional learning and change (ILAC) initiative of the Consultative Group on International Agricultural Research represents a formalized response of the agricultural sector to embrace the concept of institutional learning.[2] The ILAC mission is to strengthen the capacity of collaborative programs to promote pro-poor agricultural innovation, and to ensure that research and development activities are managed more effectively vis-à-vis contributions to poverty reduction. Institutional learning and change is a process that can change behavior and improve performance by drawing lessons from the research process and using them to improve future work. The ILAC framework encompasses a set of emerging interventions that will strengthen performance by encouraging new modes of professional behavior associated with continuous learning and change (ILAC, 2005). Research in the ILAC model involves multiple stakeholders in a process that is more participatory, iterative, interactive, reflective, and adaptive.

According to ISNAR (2004), research activities that would allow for institutional learning and change include new modes of working, such as: (i) public–private partnerships as research organizations embrace a market-led research agenda; (ii) new research approaches oriented towards innovations in the commodity value chain; and (iii) new paradigms that link research, extension, universities, and farmers' organizations in participatory knowledge quadrangles. While R&D actors engage in these activities, institutional change and learning take place, leading to the generation of lessons that further inform thinking and organizational practice.

Institutional change in agricultural research systems

As much of AHI's work on institutional change involved national agricultural research systems (NARS), it is worth taking some time to summarize what is known to date about institutional change and innovation in these organizations.

Historical evolution of research approaches

In the past, most agricultural research and extension organizations have carried out research and extension in a top-down or linear manner. Technologies have in large part been generated on-station, with minimum inputs of end users to define desirable characteristics of the technology, and then transferred to the end users using the unidirectional "transfer-of-technology" model (research to extension to farmers) (Hagmann, 1999). This approach tends to be commodity based and employs a unidirectional communication model, undermining the extent to which the socio-economic circumstances of the end users are considered. With such an approach, R&D institutions are unable to adequately respond to the demands of end users.

Historically, the focus of research in the CGIAR and in NARS has been on food crops and on high yielding varieties. This has led to some undeniable successes, with nearly 71 percent of production growth since 1961 occurring owing to yield increases (Hall et al., 2001). However, recognition of the huge gap between what scientists do and can do on-station, and what farmers do and can do on-farm, led to a conviction that major changes were necessary in the way in which technologies were designed and evaluated (Collinson, 1999).

There is evidence from adoption studies and direct feedback from farmers that technologies developed by research were not always relevant to farmers' needs because the socio-economic and agro-ecological circumstances of the end users were seldom considered (Baur and Kradi, 2001). Probst et al. (2003) show that the complexity and magnitude of farmers' problems have increased considerably and "new" approaches, concepts, and theoretical perspectives are needed. They argue that research should shift from a focus on the production of scientific "goods" to support more integrated and complex livelihood options. Collinson (2001) lays the blame for low research impact among smallholder farmers in developing countries on conventional approaches and paradigms undergirding applied research institutions. He argues that in this paradigm, scientists and managers give their allegiance to commodities, disciplines, and institutions rather than to the intended beneficiaries as the appropriate drivers of research programming and organization.

In response to these critiques, research approaches have gone through a series of transformations in an attempt to enhance the effectiveness of research in achieving impact. One of the first shifts was from on-station research to on-farm research, notably through the farming systems research (FSR) approach. Even FSR has gone through a series of conceptual and methodological transformations, with its guiding conceptual framework expanding from an initial emphasis on cropping systems in the 1970s to an emphasis on farming systems in the 1980s and on watershed level work in the 1990s (Hart, 1999). Part of this transformation involved an increasing emphasis on farmer participatory research (FPR) (Ashby and Lilja, 2004; Collinson, 1999), where farmers' circumstances and criteria become central to problem definition and research design. The move from FSR to FPR was necessitated by the tendency of researchers to lead the research process in FSR, with limited involvement of farmers. Other reasons were that: (i) smallholder farmers, particularly in marginal areas, were not benefiting from the yield increases achieved through FSR; and (ii) the commodity orientation, which places emphasis on finding the best germplasm and the best husbandry to maximize yields, isolates results from the everyday realities of farmers (Collinson, 1999). The shift to FPR, where effectively applied, led to the more active participation of farmers in decision-making at all stages of research, from problem identification to experimentation and implementation, and even the dissemination of research results.

While these approaches led to important changes in research methodology and involvement of smallholders, they failed to catalyze large-scale impacts from agricultural research and extension or widespread changes in institutional practice. The demand for demonstrable impacts of research and development efforts is now high on the agenda of governments, donors, and civil society (FARA, 2005; IAC, 2004; NEPAD, 2001; Williamson, 2000). In this regard, national agricultural research organizations in developing countries are seeking ways of improving the involvement of stakeholders in research and development (R&D) processes to achieve greater impact and more efficient research systems in times of shrinking budgets (AHI, 2001, 2002; ASARECA, 1997). This has led to a drive for institutional change throughout the system.

New drive for institutional change

The demand for demonstrable impacts on poverty has led to a call for institutional learning and change (ILAC) within the agricultural profession and institutions (Ashby, 2003; Meinzen-Dick et al., 2004; Okali et al., 1994). A number of initiatives have supported institutional change in the agricultural sector in Africa and beyond. One of the most prominent is the World Bank's efforts to work with developing countries to improve the ability of their national agricultural research organizations (NAROs) to generate technology that increases agricultural productivity while alleviating poverty (World Bank, 1998). In so doing, diverse reforms have been carried out in an effort to make research systems more effective.[3] These have included efforts to: (i) ensure greater administrative flexibility to enable the pursuit of financing from diverse sources, guarantee timely disbursement of funds and provide a system of open and merit-based recruitment, pay, and promotion; and (ii) deepen the involvement of different stakeholders (farmers and others) to help focus research on client needs. Interdisciplinary and multi-stakeholder approaches have been promoted as part of the latter effort (and adopted with varying degrees of success) and as a means to accommodate the diversity of farmers' needs, integrate production with the management of natural resources that sustain agricultural productivity, and ensure contextual factors (constraints and opportunities beyond local control) are considered.

More recent efforts such as Integrated Agricultural Research for Development (IAR4D) and innovation systems approaches recognize the need to extend beyond technological innovation to cover a wider set of institutional and policy innovations essential for rural development. Such developments must be multi-directional, expanding the institutional knowledge base of research institutions to enhance the contribution of research to agricultural innovation, while also enhancing the capacity of individuals, organizations, and innovation systems to catalyze this innovation and better articulate the contribution of research to it (ISNAR, 2004). ICRA-NATURA (2003) further argues that capacity building in key components of IAR4D is essential for bringing about the desired impacts from research investments. This is because the shift from the traditional commodity-based and disciplinary approach to more integrated approaches will expand the complexity of challenges faced, and thus the knowledge systems and skill sets required to confront these. It is also important to recognize that change will only be meaningful and sustained when there is buy-in from within organizations.

When involving institutional leaders themselves in institutional learning and change, change objectives can emerge from within—shaping the nature of institutional aims and the strategies seen as most likely to support these.

In support of this call for institutional reforms in agricultural research, AHI has worked with NARS to support institutional learning and change using an action research/action learning approach. The approach aimed to ensure that institutional change objectives and processes within national research systems respond to the concerns of managers and national policy priorities, as well as the needs and interests of the intended beneficiaries. The challenges facing AHI and its NARS partners included: (i) how to build in-house capacity for critical reflection and experiential learning at institutional level; and (ii) how to form strategic partnerships with organizations beyond agricultural extension.

The sections which follow highlight approaches taken to scale out proven innovations from benchmark sites and for supporting institutional change, and lessons learned in the process.

Scaling out proven innovations from benchmark sites

As illustrated in Chapters 1 through 5, AHI places an emphasis on action research for the purpose of developing and testing innovative approaches to INRM. This requires sustained investment by donors and site teams in specific locations (plots, farms, villages, micro-watersheds, and even districts) in a process of experiential learning and trial and error. Thus, activities are implemented as pilot or demonstration projects to test what is and is not working and undertake any needed adjustments, before translating lessons from benchmark or pilot sites to a broader scale. Once specific solutions or approaches to generating these are validated in benchmark sites, R&D teams face the challenge of scaling these out to wider areas. The overall goal of scaling out is to reach more people with technologies, approaches, and tools that have been validated in pilot learning sites and thus expand impacts on livelihoods and landscapes.

Lessons learned from the original testing of methods and approaches in benchmark sites are key ingredients to scaling out to new farms, villages, watersheds, or districts. However, scaling out is also in and of itself a learning process, owing both to the process of discriminating among past interventions to highlight those that worked the best and to the surprises that come from implementing a similar set of actions in new locations. The process also requires some methodological innovations, as scaling out implies using new techniques or approaches to spread the same innovations but over a wider area.

Approach development

Two primary approaches were tested for scaling out successful innovations from AHI benchmark sites.

Approach 1—Implement a high-profile activity and advertise it well

One approach that was tested involved few steps, as follows:

1. Replicate the successful activity in a highly visible location that is accessible to a large number of people;
2. Publicize the activity, approach used, and impacts obtained using mass media; and
3. Find means to effectively harvest expressions of interest in adopting the approach, and support endogenous efforts to learn from project experiences.

A notable example of this approach is the water source rehabilitation activities carried out in Lushoto, Tanzania. It is worth noting that AHI research teams in the benchmark sites not only considered and dealt with agricultural technologies and approaches, but also ecosystem services that were seen as central to community livelihoods. The idea was not only to build rapport with the community based on the program's responsiveness to NRM priorities highlighted in the participatory diagnostic exercises (thus hoping to catalyze interest in a wider set of activities), but also to link water source improvement to broader processes of INRM at landscape level (e.g., reduced erosion, labor saving for investment in other NRM activities, etc.).

Approach 2—Document successes and demonstrate them to the target audience

A second approach has been to monitor implementation of the innovation and gather proof of its effectiveness, and to host targeted dissemination activities. The following generic steps were taken to scale out integrated solutions to new watersheds or districts:

1. Ensure the development and documentation of successful innovations in pilot sites by following a minimum set of necessary steps:
   a) Facilitate a participatory process of problem identification and prioritization, planning, and implementation for the most pressing concerns related to agriculture and NRM, as described in Chapters 2 through 5;
   b) Provide close follow-up to implementation through periodic monitoring and reflection meetings with intended beneficiaries, to ensure any emerging problems are rapidly addressed. This enables adjustment

and replanning as needed, while identifying, addressing, and documenting success factors and challenges; and

c) Evaluate the effectiveness of the approach using participatory monitoring and evaluation or measurement of project-level indicators through a more formal impact assessment.

2. Host an event to showcase successful innovations to relevant stakeholders and decision-makers to stimulate demand and bring pride to communities hosting the innovation.
3. Provide follow-up to organizations interested in scaling out the approach to their respective areas of operation.

Field days with diverse stakeholders have been widely used by AHI as mechanisms for scaling out successful technologies and approaches. This is an opportunity for participating farmers to demonstrate and testify to the performance of the technology or approach to non-participating farmers, policy and decision-makers, input suppliers, NGOs, CBOs, and other relevant stakeholders. Box 6.2 illustrates how such a field day is carried out through a case study from Areka BMS in 2007.

BOX 6.2 FIELD DAY IN AREKA, SOUTHERN ETHIOPIA

Phase III of AHI at the Areka BMS started with a participatory exploration of problems of Gununo watershed, in 2002/2003. Problems were identified and prioritized by groups disaggregated by gender, wealth, and age. This was followed by the development of a formal action research proposal. Since then, a number of research activities have been launched to develop integrated solutions to prioritized problems related to the management of soil and water, vertebrate pests, trees, and springs. Two years after the commencement of activities, positive results were observed for most of the activities undertaken—many of which had not been attempted before by research institutions or NGOs (e.g., equitable dissemination of technologies, porcupine control). The team decided to undertake a field day to demonstrate and scale out the initial results to various stakeholders from within and outside the pilot site (see Plate 15).

In 2007, the Areka site team organized a field day to scale out experiences from Gununo watershed. Over 236 participants from various institutions participated, including research centers (Awasa and Holleta Agricultural Research Institutes), directors from national and regional agricultural research institutes (EIAR, SARI), institutes of higher learning, the Bureau of Agriculture at different levels (regional, zonal, district), Council members at different levels (zonal, district, and peasant association), NGOs and farmers in and outside of the watershed. News agencies (national and regional TV and radio) were invited to help document the event and related innovations and share them with a wider public. In the course of the day, participants visited sites where different technologies and approaches had been applied. Leaflets on different topics were prepared for participants, and over one hundred copies were distributed among them. News of the experiences received media coverage in the national language on Ethiopian Television and Southern Nation Television and Radio. The team also made an arrangement with the news agencies to host a special TV program for wider dissemination, and to have an additional radio program aired in the local language spoken by farmers in Gununo watershed and surrounding areas. The impact of the field day has included increased demand for technologies and approaches by farmers in neighboring villages, government agencies, and NGOs. It has also led to an expanded membership in village research committees, including farmers from neighboring villages who had not participated in pilot research activities, thus scaling out technologies and approaches beyond the watershed.

Lessons learned

The following lessons have been learned through AHI's early efforts to scale out proven innovations from benchmark sites:

1. The first approach to scaling out, in which a high profile activity is implemented and well advertised, is effective only for activities that carry a very high value among local communities, and can therefore muster political support for scaling out.
2. Farmers often need organizational training more than technical training to enable them to adopt or sustain an innovation. Key competencies include articulating their demands, developing institutional capacities (e.g., for accessing input or output markets or ensuring technologies are equitably multiplied and disseminated), improving natural resource governance (e.g., implementing local by-laws in support of technological or other innovations), and monitoring and evaluating innovations.
3. Practical field demonstrations and inter-community visits are vital elements of scaling out, because they enable farmers and other stakeholders to understand how the technology or practice works and observe the benefits in situ.
4. Institutional dependency needs to be overcome if scaling out is to be sustained. Success cases have indicated that in order to overcome any given problem, farmers need ready access to all the necessary elements that enable them to adopt, adapt, and disseminate new technologies and practices that they have found attractive. These include increased organizational capacity, access to appropriate materials for implementation and maintenance of the innovation, and technical support when problems arise.

5. The use of participatory monitoring and evaluation and process documentation tools is helpful not only in guiding the change process, but also in providing information on outcomes and impacts accruing from R&D efforts. These tools generate information that is complementary to that which is normally collected by researchers (which tends to focus on quantitative, and often biophysical indicators such as yield or soil fertility), such as the performance of indicators of importance to farmers. This information is essential to bolstering support for the innovations among actors seeing them for the first time.
6. Radio is a very effective tool for raising awareness on an innovation to a large audience; however, it is insufficient for imparting the necessary knowledge and skills to ensure effective implementation. Thus, effective scaling out requires both awareness creation and support to formal efforts to train others and help them trouble shoot during their efforts to implement complex methodological innovations. Fortunately, piloting an innovation develops human resource capacities within villages and institutions that can in turn be leveraged to support the spread of innovations, provided these individuals are empowered with the necessary training/facilitation skills and financial resources.
7. Responsiveness of research and development institutions to farmers' articulated needs is critical to ensuring the effectiveness of any scaling-out effort. In order to maintain the interest of the intended beneficiaries of any scaling-out effort, it is important that time is taken by researchers or development professionals to understand their problems, that attractive options be made available and that timely actions are taken to respond to farmer demand. This represents a challenge for most institutions or projects, particularly those constrained by inadequate human and financial resources. For instance in Tanzania, the ratio of extension agent to farmers is 1:1600. It is also constrained by institutional approaches that are supply- rather than demand-driven.
8. A comprehensive approach to scaling out that considers various requirements to successful adoption of the technology or approach is essential. This includes strategies for raising awareness, for building capacity, for availing the necessary inputs (technological or other), and for monitoring the spread and performance of the innovations. Capacity development efforts often involve more than a one-off training in the classroom; practical implementation is generally required for adequate assimilation of new technologies or practices.
9. For scaling out to be effective, partnerships with new actors beyond agricultural research are important. This is true for several reasons. In the absence of such partnerships, bottlenecks to participatory processes are quickly reached as communities express needs that go beyond the institutional mandate of research. Second, with limited communication and harmonization of efforts, different organizations may pursue conflicting goals or duplicate efforts in some locations while leaving other locations

with no services. There may also be a need for research institutions operating at different levels or in different locations to coordinate their activities for greater effectiveness. While partners may face challenges associated with divergent approaches, mandates, and resource levels, most important is that they share the same goals, philosophy, and eventual credit.

Self-led institutional change

In addition to scaling out technological and methodological innovations from benchmark sites to new watersheds and districts, AHI has supported processes of institutional change at national level. Institutional change is aimed at structural, procedural, and systems changes within R&D organizations to enhance the effectiveness of these organizations and their relevance to clients.

Most institutional change work in the region has been catalyzed by actors and factors outside of the organizations undergoing change (Chema et al., 2003). While this may also be true in the case of AHI, the institutional change processes carried out in partnership with NARS may be termed "self-led" because while AHI and partner organizations have provided the facilitation, the change process was largely self-propelled and self-managed by senior managers. In participatory reflection workshops, NARS managers and partners noted that for sustainable change to be realized, the change process had to start from within the organizations themselves (AHI, 2001). The managers indicated that self-led institutional change should start with development of better dialogue between researchers and managers, and then with external partners such as NGOs, extension departments, the private sector, and institutes of higher learning. AHI has engaged with NARS managers to facilitate their efforts to steer self-led institutional transformations, with the aim of ensuring that change is initiated and managed from within the organization. This has enabled changes to be aligned with organizational and national policies and priorities. This means that NARS stakeholders jointly reflect on the key aspects that they would like to change and then internally develop solutions and strategies for managing the change process.

While a general focus on poverty alleviation is clear in the emphasis on institutional change in eastern African research institutions, more specific change objectives needed to be clearly articulated. As a result, AHI held meetings with NARS managers and researchers to discuss some of the main objectives of agricultural research, and the changes that are needed in national agricultural research systems to achieve these objectives (AHI, 1998; 2001). Participants highlighted the following changes in organizational policies, structure, and function that are required to enhance their effectiveness in contributing to poverty alleviation:

1. *Enacting policy reforms in research and extension agencies to ensure that participatory approaches become part and parcel of researchers' daily routines.* It was observed that such reforms necessitate new incentive schemes so as to

motivate staff and reward them for their efforts. Stakeholders identified reluctance of managers to experiment with innovative reward mechanisms as a key barrier to institutionalization of participatory approaches. They stressed decentralization of authority from headquarters to research stations and accountability to stakeholders as the main strategies to ensure demand-driven approaches to R&D.

2. *Strengthening and targeting research activities through improved consultation with end users.* This includes two key elements: farmer participation in defining desired changes and the contributions of research to these, and the need to strengthen farmer organizations to enable them to participate effectively in defining research priorities and articulating demand for advisory services. The National Agricultural Research Organization (NARO) of Uganda, for example, has restructured its programs to shift from a commodity-based to a thematic focus to ensure that farmers' priorities and participation are at the center of research activities (see www.naads.org).
3. *Strengthening the interface between research and extension, especially at local, district, and regional levels.* The general trend is for research and extension to work independently of one another, creating unnecessary competition over resources and resulting in the dissemination of contradictory information to the public. In Uganda, stakeholders envisioned a stronger relationship between NAADS and NARO through zonal agricultural research institutes and NAADS district operations.
4. *Developing systems for experiential learning and action research in support of institutional learning and change.* Stakeholders identified the iterative process of planning, action, reflection, and feedback (among implementers, and with farmers and policy makers), replanning, and continuous improvement as key to meeting any challenge associated with impact-oriented research and extension. Part of this strategy includes the need for an effective and flexible system for data capture and analysis, so that information on the effectiveness of approaches under development is accessible to decision-makers.
5. *Creating and strengthening innovation platforms, networks, and systems for communication, documentation, and dissemination.* Strategic partnership arrangements were seen as a key ingredient to achieving impact, to capitalize upon synergies in institutional mandates, skills, and resources. For example, demands emerging out of participatory problem diagnosis and prioritization that do not fall within the mandate of one organization can be more easily accommodated if organizations with diverse mandates are working in partnership. Similarly, action research to develop methodological innovations in research and development requires contributions from both research and development agencies.
6. *Bolstering support from senior leadership.* As institutionalization of participatory and impact-oriented research approaches was required to accelerate

impacts, AHI and NARS partners were of the view that the process needed the support of the top leadership within the organization—both to reflect on their readiness to engage in change, and to develop the necessary mechanisms to support it.

AHI support to self-led institutional change agendas has gone through several phases. From 2002, an attempt was made to pilot the institutionalization of participatory research approaches together with researchers, managers, and their development partners so as to make them common practice in selected NARS. These efforts have evolved to encompass a wider array of approaches, such as integrated research and innovation systems approaches. Organizations that have been involved in these efforts included the Department of Research and Development of Tanzania, the Ethiopian Agricultural Research Organization,[4] and, more recently, the National Agricultural Research Organization of Uganda and the *Institut des Sciences Agronomiques du Rwanda*. AHI and its partners aimed to catalyze changes among the NARS partners so that approaches proven to be effective become institutionalized. This implied developing new ways of interacting and engaging with other stakeholders, among other internal changes in organizational structure and function.

Approach development

Two approaches may be highlighted based on where the impetus for change originated from, whether the external environment or the beneficiaries themselves.

Approach 1—Self-led institutional change catalyzed by external drivers

The first approach involves changes induced by external drivers, such as donor agencies, new government policies or global trends (e.g., newly acquired knowledge or development strategies, shifts in the global economy or climate change). Since change cannot be effective without local ownership, the approach involves collaboration with internal managers and leaders of R&D organizations so that the change process is driven and managed from within.

The following primary steps were taken in the AHI context:

1. An external push for change in the way research and development practices are undertaken occurs. One prominent example is the recent push by donors, politicians, and civil society for research and development organizations to show impact, and increase the rate and scale over which impact is achieved.
2. Managers and other stakeholders visualize the changes they would like to see, often with the help of an external facilitator. Visualizing change often

involves a search for evidence of what works in practice, so as to ground change in proven practices rather than in theory alone. Within public research organizations, for example, managers are demanding that evidence of impact from new approaches be gathered to enable them to make informed decisions about new investments (e.g., reallocation of budgets and staff time). Visualizing change may also involve developing a framework to highlight the scope of changes required, to plan for these, and to evaluate the effectiveness of the process as it is implemented (Box 6.3).

BOX 6.3 DEVELOPING PERFORMANCE CRITERIA FOR EVALUATING INSTITUTIONAL CHANGE

In Ethiopia, an assessment framework was developed by asking workshop participants, "If research were highly effective, what would stakeholder X be doing differently?" The question was asked of farmers, farmer organizations, researchers, and research organizations, to highlight behavioral changes that would occur at diverse levels if the envisioned outcomes were achieved. Answers to this question helped in the development of a set of performance criteria against which institutional changes were evaluated during implementation.

3. "Best bet" approaches are identified. Before getting started with the self-led change process, organizational leaders assess internally their own projects or those of other agencies to identify key approaches and lessons that can be institutionalized. In Ethiopia and Tanzania, for example, eight projects using participatory research were assessed to give managers insights on best practices and conditions for successful outcomes to be achieved. A standardized structured questionnaire was employed to collect data about the approaches employed by and results emanating from each of these projects. Analysis of all the case studies and discussions that followed provided a basis for defining critical success factors or "cornerstones" for effective research (Figure 6.3). These cornerstones were in turn utilized to design an institutional change strategy, and to monitor and evaluate the performance of this strategy during its implementation.
4. Piloting of the innovation in selected sites or research centers. Prior to engaging in organization-wide change processes, pilots are used to test whether the new ideas/approaches are feasible within the new institutional context. This helps to build capacity in applying the innovation in practice and also highlights key activities that must be undertaken to ensure the innovation is successfully internalized. Only once these pilot experiences have proven successful are efforts made to institutionalize them within the organization as a whole.

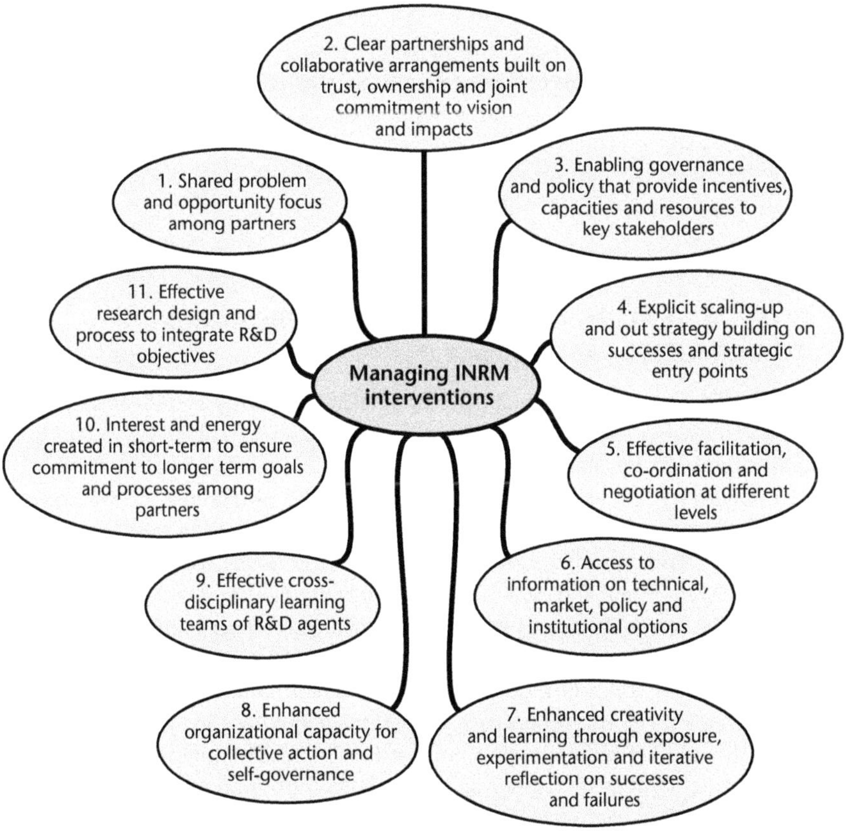

FIGURE 6.3 Cornerstones for effective research in Ethiopia and Tanzania

5. Synthesis of lessons learned. The lessons from the change process are documented continuously and synthesized for wider dissemination. This synthesis enables key lessons to be distilled on what did and did not work well, which then guides efforts to expand the approach within the wider organization.
6. Institutionalization of changes based on lessons learned. Once the lessons are synthesized, the managers of the research or development organization utilize them as ingredients for institutionalizing the approaches. Institutionalization is meant to ensure the approaches become routine and are applied in everyday activities of the organization. This requires allocating budgets and staff time to these activities in all research centers and/or among all staff, and mainstreaming the activities into annual planning and review processes. In some cases (only where needed for effective implementation), structural changes

may be required—for example, the creation of new units to enhance interactions among disciplines or with outside actors. Such institutional reforms begin during the piloting phase, but are often expanded at this time as they require commitments from senior management or the organization's headquarters.

An example from Rwanda is illustrated in Box 6.4.

BOX 6.4 SCALING UP AND OUT AHI APPROACHES TO INRM: THE CASE OF ISAR IN RWANDA

Limited adoption of NRM practices by smallholder farmers in Rwanda had led to increasing soil erosion and subsequent siltation of cultivated wetlands and valley bottoms. Research and development approaches were ineffective in catalyzing shifts toward more sustainable NRM. Limitations of the conventional approach included an emphasis on individual components rather than on component interactions or systems; a top-down approach to technology development and dissemination with limited involvement of intended beneficiaries (and developed technologies failing to reflect famers' priorities or realities); a focus on the plot and farm level (which left issues operating at other spatial scales or requiring collective action unaddressed); failure to link technological innovation with other complementary innovations (e.g., supportive market linkages or policies); and limited collaboration among relevant research and development partners. The composition of research teams (with few social scientists or researchers with skills in integrated approaches) and high staff turnover were further impediments to implementing desired institutional changes. This called for an approach for encompassing broader units of analysis and intervention and which takes into account both the biophysical and social dimensions of NRM. To accommodate these requirements, the *Institut des Sciences Agronomiques du Rwanda* (ISAR) adopted the AHI approach to INRM to address the diverse factors responsible for natural resource degradation in Rwanda.

Initiatives taken to address the problem

The development and implementation of watershed management plans for integrated NRM was identified as a major thrust over the next two decades in Rwanda's agricultural sector master plan. Subsequently, ISAR sent two scientists to India to explore the possibility of learning from this country's experience with participatory watershed management and explore the possibility of a south–south partnership to leverage benefits from past experiences in both countries. With World Bank support, ISAR also recruited a group of "experts" in different professional fields outside Rwanda to support capacity development following the genocide. In 2005, one of the senior scientists recruited by ISAR, a former AHI benchmark site coordinator, was hired to support institutionalization of the watershed management approach within ISAR. The following steps were followed:

1. A country-wide tour to different agro-ecological zones was organized for newly recruited scientists.
2. The integrated watershed approach was introduced to ISAR management and scientists.
3. The ISAR DG requested that capacity building on INRM be conducted for all ISAR scientists.
4. Two training workshops, sponsored by the Government of Rwanda, were conducted at ISAR headquarters.
5. The ISAR DG agreed to sponsor watershed-level INRM pilot activities in three pilot sites (and to subsequently increase this to four sites).
6. Watershed teams comprised of all disciplines were formed, and local development partners engaged.
7. Selection of pilot sites by research teams in collaboration with farmers and district partners.
8. Introduction of the approach to watershed communities.
9. Selection (by farmers) of representatives to work with the research team, taking into consideration hamlet representation and farmer categories (age, wealth, gender, landscape location of plots).
10. Implementation of participatory diagnostic surveys to identify constraints and opportunities for overcoming these constraints.
11. Prioritization of identified issues by farmers, with facilitation of the research team.
12. Feedback of results to watershed communities and other stakeholders.
13. Participatory preparation of community action plans (CAPs).
14. Implementation of CAPs, and periodic follow-up by the research and development team.
15. Lessons learning from pilot watersheds to explore the potential for institutionalizing the approach throughout the organization.

Although the above steps in self-led institutional change have taken place within the selected NARS, variations in the approaches used in different countries have been noted. Box 6.5 illustrates how self-led institutional change may be catalyzed by different drivers—whether national policy priorities or donors.

BOX 6.5 NATIONAL POLICY PRIORITIES AND DONORS AS DRIVERS OF INTERNAL CHANGE IN NARS

In Ethiopia, the director of research, managers, and researchers were under pressure by government ministers and members of parliament to provide evidence of impact from agricultural research in order to secure ongoing funding for their activities. The recurrent drought and food insecurity had created pressure on the government to deliver interventions to mitigate these challenges. In the case of Uganda, donors demanded evidence of impact from the work that had been funded. This led the Director General of NARO to bring in external consultants from ICRA and Makerere University to design a workshop on integrated agricultural research for development (IAR4D) as a new way of conducting research that would show rapid impacts among target beneficiaries. All 13 zonal agricultural research and development institutes (ZARDI) attended these workshops and developed action plans that they implemented when they returned to their respective research stations to practice what was learned in the workshops.

Box 6.6 illustrates how the key steps in institutional change may vary according to context and the priorities of stakeholders involved.

BOX 6.6 VARIATIONS IN INSTITUTIONAL CHANGE PROCESSES LED BY EIAR AND NARO MANAGERS

Steps in institutional change in the Ethiopian Institute of Agricultural Research (EIAR):

1. Inception workshop with managers on what needs to change
2. Learning and experience sharing workshops combined with training
3. Field-based implementation of action plans generated in workshops
4. Follow-up by AHI Regional Research Team and managers on implementation of action plans
5. Synthesis of lessons and insights from workshops and the field
6. Dissemination of lessons and insights to managers, researchers, and regional stakeholders.

Steps in institutional change in the National Agricultural Research Organization (NARO):

1. Institutional change facilitators' design workshop for senior managers

2. Workshop on institutionalized responses to research and development challenges for representatives from zonal agriculture research and development institutes (ZARDI)
3. Workshops for staff of research stations implementing pilot experiences
4. Piloting of innovations at station level
5. Mid-term evaluation
6. Adjustments in the approach formulated based on recommendations from the evaluation, and proposal developed and submitted for funding.

Approach 2—Self-led institutional change catalyzed by grassroots demand

A second approach to institutional change is catalyzed not by external actors, but by grassroots demand. While change is initiated from below, it also requires responsiveness to farmer demands among service organizations—and is thus often conditional on a favorable institutional and policy environment or organizational leadership.

The following steps are key in enabling institutional change based on grassroots demand:

1. Grassroots problems identification, facilitated by an independent party. In contexts where farmers are not adequately organized, the articulation of grassroots demand may require the involvement of an independent party. In cases where community-based organizations are strong and networked at higher levels, this demand may be expressed spontaneously.
2. Information sharing or advocacy with district and national policy makers. Once demands are articulated, identified changes must be advocated to leverage the necessary political will to support these changes among policy makers or institutional managers (depending on the nature of changes desired).
3. Gathering of evidence that identified changes are in fact able to leverage the purported benefits. As mentioned above, policy makers and managers will often require evidence that the proposed change works in order to justify changes in policies, institutional practices or budgets. Where such an innovation has already been implemented in practice, evidence can be gathered from these existing cases. Where such cases do not exist, evidence can only be gathered through the piloting of innovations and documentation of observed changes (for example, using an action research approach).

4. Lessons sharing at diverse levels (e.g., national fora, cross-district exchange visits) to influence other actors to invest in the innovation.
5. Scaling up or institutionalization of the innovation with monitoring, to enable mid-course adjustments to be made.

For an example of this approach, please see Box 6.7.

BOX 6.7 LINKING FARMERS TO POLICY MAKERS: THE ROLE OF ACTION RESEARCH IN FARMER INSTITUTIONAL DEVELOPMENT

A host of challenges have been experienced in the delivery of public sector services in developing countries: matching services to felt needs of beneficiaries, enabling effective stakeholder participation, ensuring equitable coverage and representation of diverse social groups, efficiency and effectiveness in service delivery, and overcoming barriers to information flow. The National Agricultural Advisory Services of Uganda (NAADS) is a program for demand-driven extension provision that relies on local institutions (farmer fora) to articulate farmers' demands from private service providers. While NAADS policies provided the institutional framework for effective demand-driven service delivery, a number of concerns were raised by intended beneficiaries about its effectiveness in practice.

Initiatives to address the problem

AHI, CARE, and other organizations operating at community level throughout the district had observed a host of complaints leveraged by farmers about the implementation of NAADS. One key concern was the limited effectiveness of farmer fora in representing all of the villages in their jurisdiction and in ensuring downward accountability in the management of financial resources and services. In response to these concerns, AHI and CARE formed the Coalition for Effective Extension Delivery (CEED) with other concerned organizations to support stakeholders in addressing these concerns. Key steps in the process included the following:

1. Identification of critical bottlenecks to the effective functioning of NAADS. This was done by consulting diverse stakeholders involved in implementing the program or intended to benefit from it (e.g., male and female farmers). The limited effectiveness of farmer institutions was identified as one key barrier to effective program implementation.
2. A participatory diagnostic activity was carried out in one parish where CARE was working to assess the problem more deeply from the perspective

of the intended beneficiaries. The primary concern related to farmer representation was poor coverage of services for parishes geographically distant and politically disconnected from the farmer fora.

3. A participatory action research process was initiated with CEED facilitation, to support parish residents to find their own ways to overcome the problems of representation and accountability. This involved the piloting of institutional innovations recommended by farmers—in this case, the formulation of parish-level farmer fora to help represent and advocate on behalf of parish residents at sub-county level.
4. The performance of these pilot experiences was evaluated in order to formulate recommendations.
5. Feedback was provided to NAADS at national and district level, and CEED lobbied for institutionalizing the approach in other NAADS parishes/districts.
6. NAADS commissioned a national-level study by CEED on farmer institutional development to explore whether the same problems exist in other NAADS districts, as a key step in leveraging institutional commitment for reforms. Findings suggested the problems were very similar to those experienced in other districts.
7. Parish-level farmer representative bodies were adopted by NAADS and implemented in other districts under the name of parish coordination committees (PCCs).

Outcomes

NAADS implementation at parish level (i.e., planning, monitoring, and quality assurance of service delivery) has been strengthened as a result of PCCs. This has increased awareness and interest of farmers in the NAADS program, leading to improved outputs from program activities. PCC chairpersons have also been integrated into the sub-county farmer fora for improved information flow and overall coordination. The primary challenge is sustaining and maintaining the spirit of volunteerism within farmer institutions, as the members of NAADS farmer fora and PCC perform their roles without a wage.

Approach 3—Self-led institutional change catalyzed from within

The final approach involved institutional changes for which the impetus largely comes from within the organization itself. The external environment may be instrumental in either the creation of the institution or in providing inspiration to reforms, but the management is self-motivated to innovate in the development and/or reform of key organizational processes as a means to meet the core objectives of the organization. The following are key steps involved in such reforms:

1. Development of an institutional mandate, policies or guidelines that structure learning within the organization.
2. Exposure to innovations related to the overall mandate of the organization. This may occur through partnerships, literature review, field visits or internal monitoring systems that enable the identification of best practices or 'nodes of innovation' within the organization itself. This step may be particularly instrumental or time consuming cases involving a new organization for which all operational systems must be generated from scratch.
3. Direct involvement in an innovation and lessons learning process, either within partner organizations or from isolated cases.
4. Active improvement on the approach as the innovation process unfolds through periodic monitoring, reflection, synthesis of lessons learned and documentation. This has multiple functions, from improved performance of the innovation to better alignment of the approach with the broader institutional mandate and procedures.
5. Development of a strategy for institutionalizing the approach, including the formulation of action plans and formalization of partnership agreements required to apply the innovation at a larger scale (e.g., nationally, or institution-wide).

An example of this approach is summarized in Box 6.8, which profiles efforts to institutionalize the system for demand-driven information provision within NAADS operations in Kabale District. An example in which AHI played a more minor role in exposing the lead institution to innovations of possible relevance to the organizational mandate (as highlighted in Step 2, above) is presented in Box 6.9.

BOX 6.8 EFFORTS TO INSTITUTIONALIZE DEMAND-DRIVEN INFORMATION PROVISION IN NAADS

Once the system for demand-driven information provision developed under the ACACIA project and described in Chapter 5 was running on a pilot basis, AHI faced the challenge of how to institutionalize it before the project came to an end. Several options were considered. The most feasible of these was to institutionalize the initiative within NAADS, the Ugandan system for demand-driven extension delivery described in Box 6.7. NAADS had both the vision and the institutional infrastructure to accommodate demand-driven information provision. Parish coordination committees (PCCs), sub-county farmer fora and district farmer fora under NAADS provided a hierarchy of farmer institutions through which information needs could be articulated and delivered. PCCs were already responsible for articulating agricultural service delivery needs within NAADS, and their role could easily be expanded to encompass information needs. NAADS also had a district monitoring and evaluation team drawn from various government departments (production, planning, information) and civil society that could assume the functions of the Quality Assurance Committee set up under AHI–ACACIA. In an exploratory meeting, we discovered that not only was NAADS an opportunity for AHI, AHI was also an opportunity for NAADS. NAADS had faced a series of challenges in their efforts to nationalize a system for demand-driven service provision in agriculture, and saw the model as a potential means to address the following concerns:

- Ensuring service providers have quality and up-to-date information.
- The proliferation of service providers in NAADS districts had raised challenges for quality control, with some service providers less informed than farmers. At other times, contradictory information was provided by different service providers. NAADS saw the AHI–ACACIA model as a means to access quality information and to deliver it to service providers.
- Ensuring cross-fertilization among farmers and communities. Even within farmer groups at village level, farmers were unaware of what happens on the plots of other farmers. As one moves to the district level, such lost opportunities are magnified. Farmers also have indigenous technologies and knowledge that may be of relevance to other farmers and villages. Thus, a centralized information capture and delivery mechanism was seen as an excellent opportunity to achieve economies of scale in knowledge management at district level.

The final year of the AHI–ACACIA project was spent piloting the management of demand-driven information provision within NAADS with NAADS leadership, based on the following steps:

1. Document how the current system works for the NAADS Secretariat and district, to bolster commitment among a wider array of NAADS stakeholders. This included past activities, the value added, lessons learned, and implementation guidelines derived from the pilot phase.
2. Develop and pilot test a mechanism for sustainable demand-driven information provision by the district telecenter. This included: (i) the handover of ownership of the telecenter to the District Farmer Forum and development of mechanisms for its effective management; (ii) developing the terms of reference for contracting a private sector service provider to manage the telecenter; and (iii) contracting a service provider on trial basis under NAADS technical procedures and procurement system, with close follow-up monitoring by AHI and NAADS.
3. Bolster commitment and buy-in from the NAADS Secretariat. This was done by sharing preliminary experiences at the annual NAADS planning

meeting, featuring the initiative in the Kabale District semi-annual review report and hosting a site visit for the Secretariat.
4. Harmonize NAADS and AHI procedures for articulating service delivery needs. This included: (i) developing an integrated protocol for articulating farmer needs for advisory services and information and for synthesis of information at the district level; (ii) pilot testing the protocol; (iii) conducting a training on the use of the modified protocol; (iv) mainstreaming the process into standard NAADS information needs assessments (twice yearly); and (v) raising awareness among farmers on the pathways through which they may request information on a regular basis.
5. Pilot test a mechanism for information needs articulation and delivery at sub-county level, and for linking farmers within the sub-county to the district service provider.

In practice, a number of challenges were encountered, among these:

- Frequency of information delivery. Information needs were articulated on a bi-annual basis within NAADS, limiting the agility of information feedback to farmers. An entire production season could come and go within such a period. Establishment of new service contracts at sub-county level and encouraging more proactive articulation of information needs by farmers on a regular basis were two ways envisioned to overcome this problem.
- Articulation of information needs. Under NAADS, sub-counties must prioritize few enterprises for service delivery in order to enhance the efficiency and effectiveness of service delivery. Under AHI–ACACIA, the focus was much broader—encompassing just about any information need in the area of agriculture, marketing, and NRM. These challenges were addressed by adjusting information needs assessments under NAADS to accommodate a wider set of components (production, marketing, and NRM interests specific to the chosen enterprises). However, information on other enterprises, and on natural resource management concerns that go beyond specific crops or enterprises, was effectively excluded.
- Effective governance of the telecenters. Mechanisms for maintenance and upkeep of computers and other equipment in the telecenters raised a major challenge. NAADS was not in a position to pay salaries; the only means to embed telecenter operations in NAADS was by means of service contracts. The district telecenter would be treated as a priority enterprise for the district, with the same holding true at sub-county level. To ensure effective ownership and upkeep, ownership was to be given to farmers rather than local government—and managed through the district and sub-county farmer fora. The challenge, then, became how to ensure facilities owned by farmers but operated by private service providers would be well maintained.

This helps to illustrate the complexity of trying to mainstream a complex approach within existing institutional structures and mandates—and provides a clear case for embedding new institutional innovations within existing institutional structures that can potentially ensure their sustainability should they prove to be effective. While the learning process was ongoing at the time of writing, efforts to institutionalize the demand-driven information provision model under the NAADS framework had received acceptance by stakeholders at different levels. The NAADS Secretariat was also keen to scale it up to other districts.

BOX 6.9 THE IMPORTANCE OF "OWNERSHIP" OF THE CHANGE INITIATIVE BY KEY DECISION-MAKERS

Experiences of AHI and partner organizations in supporting institutional change suggest that senior level decision-makers are crucial to supporting any change process. These individuals play an essential role in aligning institutional policies and incentives in support of the envisioned change, so that staff are encouraged and enabled to participate in new kinds of activities. They also play an important role in availing the necessary resources for competence building at station and national level and for monitoring and evaluation processes to track learning and outcomes. In Ethiopia, high levels of political and financial commitment have now taken institutional change processes in new directions, with a focus on partnerships with development actors and the private sector, as well as clear impacts on agricultural practices and technology adoption. The pressure and will for change is filtering down to the level of researchers, who are keen to also see impact from their work. At the time of writing, pilot learning was ongoing in certain research stations in Uganda, but greater financial support from government and donors was required to support learning and scaling up.

Lessons learned

A host of lessons were learned through AHI's efforts to support institutional change processes within partner organizations. These include the following:

1. The essential role of political will and ownership of institutional reforms by key decision-makers. AHI experiences suggest that for changes to be successful, key stakeholders—particularly senior management—must be convinced that the new changes are needed and that the changes are both effective in achieving a desired outcome and feasible. This ensures internal

ownership of the change process and enables the alignment of organizational resources with change objectives and processes (see Boxes 6.8 and 6.9 above). Building consensus for change requires time and resources, and should be carefully planned.

2. The importance of identifying and supporting "champions" of institutional change. The change process depended a great deal on local "champions" within the respective organizations to motivate others and catalyze change. In Ethiopia, the Director of EIAR and his deputies were the champions in steering the process. In Uganda, the Director General and Deputy Director in charge of outreach managed the process, providing leadership in the process of developing a strategic vision and building local commitment.
3. The critical importance of grounding institutional change on a clear vision, supported by evidence of what works in practice. In each of the cases profiled above, one of the first steps in improving the effectiveness of institutional practice involved a thorough analysis of current performance of priority innovations in order to assess their effectiveness. This is important both for leveraging the necessary political will for reforms, and for learning lessons on what works that can be built upon in efforts to institutionalize new approaches. Acquiring evidence will involve either comprehensive evaluations of approaches already being implemented (building upon rigorous impact assessment or evaluation methods), or the piloting of new concepts using rigorous action research. In the second case, approaches used as they are tested and adjusted over time, and the outcomes achieved at these different stages of innovation, are documented. It is also important to keep visions and expected outputs and outcomes realistic so that visible impacts can be demonstrated and the motivation for reforms can therefore be sustained. This can often be done by reflecting on challenges likely to be faced in realizing the vision and planning accordingly (e.g., to reduce the scope of ambitions or to implement measures to overcome these challenges).
4. The need to support synergies between different levels of organization when supporting institutional change. AHI worked very closely with NAADS staff at district level in implementing the NAADS program and piloting initiatives at community level, while also carrying out strategic nation-wide studies on behalf of the Secretariat. Feedback to the district NAADS coordinator and the Kampala-based Secretariat on lessons learned from these different levels of intervention and knowledge generation helped to inform the NAADS Secretariat on issues concerning institutional policy and strategies.
5. The fundamental importance of systematic lessons learning and support to transitional phases of institutional change, given the complexity of the challenge. The application of action–reflection processes within AHI and partner organizations was catalytic in stimulating researchers and managers to reflect internally on their strengths and weaknesses and further

develop strategies to overcome identified weaknesses. This was the case in Ethiopia, Rwanda, Tanzania, and Uganda, where change consultants worked with AHI regional team members and NARS partners to introduce and apply action–reflection processes with a view of developing an effective national agricultural research system. In order to apply what was learned in workshops, participants were asked to develop action plans and tasked with their implementation following the workshop. Through this process, they were able to accumulate experience which would then serve as an input to subsequent workshops (and thus the design of institutional change processes) and to future capacity development efforts within the organization. Effective documentation of institutional learning and change processes, and the outcomes induced by these changes, is also essential to learning lessons that will then orient the change process in ways beneficial to the ultimate aims.

6. The importance of "pilots" in supporting organizational learning prior to scaling up. The piloting of institutional changes in select locations or departments within an organization is essential to the institutional learning process. In the absence of such pilot experiences, significant institutional resources will be allocated to implementing changes throughout the organization before they are tested or proven to work. While new approaches may be tested in other contexts either outside the organization or in selected branches of the organization, institutionalizing the approach requires new skills to be built and adjustments to be made for the approach to be effective and affordable in its new institutional context. Piloting is therefore essential in enabling key decision-makers to understand what kind of reforms will be required, and at what cost, for the reforms to be effective. This enables effective prioritization of investments with high returns while avoiding costly and politically risky institutional investments in changes with uncertain outcomes.
7. The need to reward innovation and change within organizational cultures and procedures such as performance appraisals and monitoring systems. Rewarding innovation in these ways will encourage managers and staff to engage in innovative approaches. In cases where innovations are effectively identified, documented, and built upon, it should also ultimately yield improvements in institutional performance and impact.

Missing links

While AHI engaged in a host of experiences aimed at scaling out proven innovations and enabling institutional change, a number of key gaps remain. These include the following:

1. *Approach for fostering synergy between local opinion leaders and technical organizations.* The support of political leaders is key to mainstreaming and

institutionalizing participatory and integrated approaches, or to ensuring the success of any institutional transformation. Local political, religious, and opinion leaders also have a role to play in bolstering support and mobilizing communities to take up new development initiatives in their localities, but approaches for enhancing their involvement have not been adequately explored in AHI.

2. *Strategies for ensuring institutional buy-in and commitment.* While AHI has experimented with institutional change in a number of agricultural research and development organizations, we have yet to distill the minimum set of commitments required to sustain change. Some pre-conditions that seem to matter include: co-financing arrangements for the change initiative, integration of the initiative into standard planning and evaluation procedures (so that it is part and parcel of everyday business), and regular monitoring visits by senior management. However, new efforts are needed to distill the basic conditions for effective institutional change and the processes through which these conditions can be developed or met.

   Efforts are also needed to involve human resource and institutional policy experts in institutional change efforts. When capacity development is done in isolation from human resource departments and policy makers, these efforts may be ephemeral owing to lack of institutional commitment and follow-up. Each of the NARS in the eastern African region has human resource departments that are responsible for organizing capacity-building events. These need to link with new projects that come into their organizations that have capacity-building components, to minimize duplication of efforts and explore ways in which skills developed in the context of projects can be internalized in the everyday practices of organizations.
3. *A strategy for broadening the base of institutional support.* A critical mass of champions is needed in all organizations, particularly those undergoing a process of institutional change, in order to minimize the effects of turnover on organizational memory and performance. This applies in all public research and development organizations. AHI has not developed the necessary lessons on how to create such a critical mass and thus cement commitment to reforms into the future.
4. *Approaches to institutionalize methods developed by AHI in the everyday practices of development organizations.* With the exception of demand-driven information provision, few of the institutional change initiatives carried out by AHI partner organizations involved institutionalizing methods tested in AHI benchmark sites—representing a disconnect between the piloting of innovations and institutional change efforts. While AHI conducted a series of regional trainings to share our methods with stakeholders throughout the ASARECA region, with the exception of Rwanda, no follow-up was given in-country to ensure the effective assimilation of these methods

within national research and development organizations. Thus, it is possible that these methods were largely lost to the organizations that the participants represented. Such trainings should be linked to organizational change processes, to enable active evaluation of methods by national development agencies and support in-field application and institutionalization of those methods deemed to be most desirable.

5. *Mainstreaming action research.* In addition to mainstreaming the most promising methods and approaches presented in this volume, institutionalizing the action research process itself within research and development organizations is a "must." This will help to ensure that methodological innovations are not only produced by pilot projects of short duration, but continue to be developed as everyday practice of these agencies in partnership with local communities. Research organizations largely lack the vision and necessary skill sets to conduct research on "process" (observing the outcomes associated with different approaches to development support), and lack the funding base required to sustain heavy facilitation tasks. This undermines their commitment to action-based learning and their ability to effectively bridge research and development. Development actors and NGOs, on the other hand, seldom give adequate attention to assessing the effectiveness of the approaches they employ to support rural communities and how these approaches advance or undermine livelihood goals, equitable benefits capture, sustainability, and innovation. They also tend to undervalue the contribution that research might make to help them learn more proactively from practice. Both sides have much work to do to bridge the gap between systematic learning and practice. New research should look into the appropriate institutional models for action research based on new forms of research–development partnership.
6. *Performance assessment and incentive-based systems that recognize and reward achievement in the application of new approaches.* If research organizations are to take on the complex challenge of institutionalizing process-based action research and other forms of innovation, individuals in these organizations need to receive adequate support in the form of both training and reward systems. This would create an enabling institutional environment for scientists and other R&D professionals to overcome their fear of change, re-evaluate their roles in new approaches, change their mindsets, and enhance the "soft skills" that are so fundamental to institutional innovation and change.
7. *Partnership model to sustain research–policy and research–practice dialogue and synergy.* National and district level institutional arrangements are needed to strengthen the linkage between research and policy on the one hand (for evidence-based policy development, as seen in the NAADS case) and research and development on the other. Multi-institutional efforts are identified in the literature as a way to harness the relative strengths of extension, NGOs, universities, and the private sector in testing innovations or "going

to scale" (Snapp et al., 2003). Models for such arrangements need to be developed and pilot tested in an action research mode to distill relevant lessons, particularly in cases where an action-based approach to research is envisioned—given the need for high-quality research *and* facilitation skills to operationalize such an approach.

Conclusions

Scaling out and institutional change are essential for leveraging more widespread and sustained impacts from development interventions. All too often, such interventions are transient, owing to the frequent failure to move beyond pilots to exit strategies that build the necessary institutional competencies to sustain change and/or to "go to scale." Scaling out from benchmark sites and institutionalizing innovations in the everyday practices of organizations are processes that need to be treated as organizational and research objectives in their own right by research and development organizations.

This chapter distilled AHI's experiences with each of these processes, while also highlighting critical gaps that remain. While significant advances have been made in understanding the critical ingredients to success, the cost of these initiatives and time taken to understand their ramifications over the long run mean that much more must be done to understand the processes through which these success factors may become manifest at reasonable cost. It is important to recognize that "going to scale"—whether through the horizontal spread of innovations to new farms, communities, and watersheds or the changes in institutional policies and practices—requires its own set of professional skills and competencies. R&D organizations and donors alike must recognize the need to invest in such "soft skills" if past and future development investments are to yield more meaningful returns.

Notes

1 Such a piloting approach was also integral to the "learning process" approach to participatory development adopted by the Ford Foundation and USAID in their programs in Southeast Asia from the late 1970s to the early 1990s (Korten, 1981).
2 For more information, see: www.cgiar-ilac.org/
3 Similar reforms have been carried out among agricultural extension agencies in eastern Africa. Programs such ATIRI and NALEP in Kenya, PADEP in Tanzania, NAADS in Uganda and Research Extension Advisory Councils (REAC) in Ethiopia are among the initiatives aiming at strengthening the role of end users in the design, implementation, and evaluation of research and extension and aligning these programs to the needs of end users.
4 The Ethiopian Agricultural Research Organization (EARO) has since changed its name to the Ethiopian Institute of Agricultural Research (EIAR) to reflect the broad partnerships under which it now works.

References

Adato, M. and R. Meinzen-Dick (2002) Assessing the impact of agricultural research on poverty using the sustainable livelihoods framework. *FCND Discussion Paper* No. 128 and *EPTD Discussion Paper* No. 89. Washington, D.C.: IFPRI.

AHI (1998) *African Highlands Initiative: An Eco-regional program in Eastern Africa*. AHI Phase 2 Proposal January 1998 to December 2003.

AHI (2001) *African Highlands Initiative Annual Report 2001*. Kampala, Uganda: AHI.

AHI (2002) *African Highlands Initiative Annual Report 2002*. Kampala, Uganda: AHI.

Argyris, C. (1991) Teaching smart people how to learn. *Harvard Business Review* 4(2): 4–5.

ASARECA (1997) *Association for Strengthening Agriculture Research in East and Central Africa Annual Report 1997*. Entebbe, Uganda: ASARECA.

Ashby J.A., J.I. Sanz, E.B. Knapp and A. Imbach (1999) CIAT's research on hillside environments in Central America. *Mountain Research and Development* 19(3): 241–250.

Ashby, J. (2003) Introduction: Uniting science and participation in the process of innovation – Research for development. In: B. Pound, S. Snapp, C. McDougall and A. Braun (eds.), *Managing Natural Resources for Sustainable Livelihoods: Uniting Science and Participation*. London: Earthscan/IDRC.

Ashby, J. and N. Lilla (2004) Participatory research: Does it work? Evidence from participatory plant breeding. *Proceedings of the 4th International Crop Science Congress*, September 26–October 1. Brisbane Australia.

Argyris, C. and D. Schön (1996) *Organisational Learning II: Theory, Method and Practice*. Reading, MA: Addison Wesley.

Buse, K., E. Ludi, and M. Vigneri (2008) Sustaining and scaling the millennium villages: Moving from rural investments to national development plans to reach the MDGs. Formative Review of MVP Synthesis Report, September 2008.

Baur, H. and C. Kradi (2001) Integrating participatory research methods in public agricultural research organizations: A partially successful experience in Morocco. *AgREN Network Paper* No. 109. London: ODI.

Briggs, S.R., M. Tenywa, and B.R. Nalileza (1998) A review of past and present agricultural and environmental research in the highland areas of Uganda. Report for NARO, DFID-NRSP. UK and Kampala: SRI and Makerere University.

Bunch, R. (1999) Reasons for non-adoption of soil conservation technologies and how to overcome them. *Mountain Research and Development* 19(3): 223–219.

Chambers, R., A. Pacey and L.-A. Thrupp (1989) *Farmer First: Farmer Innovation and Agricultural Research*. London: Intermediate Technology Publications.

Chema, S., E. Gilbert and J. Roseboom (2003) A review of key issues and recent experiences in reforming agricultural research in Africa. *ISNAR Research Report* No. 24.

Clarke, J., C. Dede, D. Jass Ketelhut, B. Nelson, and M. Collinson (2006) A design-based research strategy to promote scalability for educational innovations. Available at: http://muve.gse.harvard.edu/rivercityproject/documents (accessed January 4, 2011).

Collinson, M. (eds.) (1999) *A History of Farming Systems Research*. New York: CAB International.

FARA (2005) Promoting inclusiveness of civil society organizations (CSOs) in African agricultural research agenda: A FARA/GFAR collaborative initiative. *Meeting Report* 26 p.

FARA (2006) Promoting inclusiveness of civil society organizations (CSOs) in agriculture in African ARD. *Concept Note*. 4 p.

Farrington, J. and C. Lobo (1997). Scaling up participatory watershed development in India: Lessons from the Indo-German watershed development programme. *Natural Resource Perspectives* No. 17. London: ODI.

Garvin, D.A. (1993) Building a learning organization. *Harvard Business Review* 71(4): 78.

Gillespie, S. (2004) Scaling up community-driven development: A synthesis of experience. *FCND Discussion Paper* No. 181. Washington, D.C.: International Food Policy Research Institute.

Gonsalves J. and R. Armonia (eds.) (2000) *Scaling Up: Can We Bring More Benefits to More People More Quickly?* Rome: Global Forum on Agricultural Research (GFAR).

Gündel S., J. Hancock, and S. Anderson (2001) *Scaling-up Strategies for Research in Natural Resource Management: A Comparative Review*. Chatham, UK: NRI.

Hagmann, J. (1999) *Learning Together for Change: Facilitating Innovation in Natural Resource Management through Learning Process Approaches in Rural Livelihoods in Zimbabwe.* Weikersheim: Margraf Verlag.

Hall, A. (2005) Embedding agricultural research in a system of innovation. Paper presented at the CGIAR *Science Forum on Strengthening Research-for-Development Capacity*, December 6, 2005, Marrakech, Morocco.

Hall, M., J. Dixon, A. Gulliver, and D. Gibbon (eds.) (2001) *Farming Systems and Poverty: Improving Farmer's Livelihoods in a Changing World.* Rome and Washington: FAO and The World Bank.

Hart, R. (1999) Understanding farming systems: FSR's expanding conceptual framework. In: M. Collinson (ed.), *History of Farming Systems Research*. Rome: FAO and CABI, pp. 41–57.

Inter Academy Council (IAC) (2004) *Realizing the Promise and Potential of African Agriculture: Science and Technology Strategies for Improving Agricultural Productivity and Food Security in Africa.* Amsterdam: InterAcademy Council.

ILAC (2005) The Institutional Learning and Change Initiative: An Introduction. *ILAC Brief* No. 1.

ISNAR (2004) Inception Report, ISNAR Program. Washington, D.C.: IFPRI.

ICRA-NATURA (2003) Mobilizing partnerships for capacity building in integrated agricultural research for development (IAR4D). Paper presented at the a workshop held at the International Agricultural Centre, Wageningen, Netherlands, 27–29 November, 2003.

Jacobs, R.L. (2002) Institutionalizing organizational change through cascade training. *Journal of European Industrial Training* 26: 177–182.

Knox McCulloch, A., R. Meinzen-Dick, and P. Hazell (1998) Property rights, collective action, and technologies for natural resource management: A conceptual framework. *SP-PRCA Working Paper* No. 1. Washington, D.C.: CAPRi.

Korten, D.C. (1981) The management of social transformation. *Public Administration Review* 41(6): 609–618.

Kotter, J.P. (1995) Leading change: Why transformation efforts fail. *Harvard Business Review* 73: 59–67.

Lockwood, H. (2004) *Scaling Up Community Management of Rural Water Supply*. The Hague: IRC International Water and Sanitation Centre.

Meinzen-Dick, R., M. DiGregorio, and N. McCarthy (2004) *Methods for Studying Collective Action in Rural Development*. Washington, D.C.: IFPRI.

Middleton, T., M.A. Roman, J. Ellis-Jones, C. Garforth, and P. Goldey (2002) Lessons learned on scaling-up from case studies in Bolivia, Nepal and Uganda. Report No. IDG/02/21. Bedford, UK: Silsoe Research Institute.

Millar, J. and J. Connell (2010) Strategies for scaling out impacts from agricultural systems change: The case of forages and livestock production in Laos. *Agric Hum Values* 27: 213–225.
NEPAD (2001) *Annual Report 2001.* Johannesburg: NEPAD.
Norman D.A. (1991) Cognitive Artefacts. In: J.M. Carroll (ed.), *Designing Interaction – Psychology at the Human-Computer Interface.* Cambridge: Cambridge University Press.
Okali, C., J. Sumberg and J. Farrington (1994) *Farmer Participatory Research: Rhetoric and Reality.* London: Intermediate Technology Publications.
Probst, K., J. Hagmann, M. Fernandez, and J. Ashby (2003) Understanding participatory research in the context of natural resource management: Paradigms, approaches and typologies. *AgREN Network Paper* No. 130. London: ODI.
Scoones, I., J. Thompson, and R. Chambers (2007) *Farmer First Revisited: Farmer Innovation and Agricultural Research and Development Twenty Years On.* Brighton: University of Sussex, Institute of Development Studies.
Senge, P. (1990) *The Fifth Discipline: the Art and Practice of the Learning Organization.* New York: Doubleday.
Senge, P., A. Kleiner, C. Roberts, R. Ross, G. Roth, and B. Smith (1999) *The Dance of Change: The Challenges of Sustaining Momentum in Learning Organizations.* London: Nicholas Brealey.
Snapp, S.S., M.J. Blackie, and C. Donovan (2003) Realigning research and extension to focus on farmers' constraints and opportunities. *Food Policy* 28: 349–363.
Snapp, S.S. and K.L. Heong (2003) Scaling up: Participatory research and extension to reach more farmers. In: B. Pound, S.S. Snapp, C. McDougal, and A. Braun (eds.), *Uniting Science and Participation for Adaptive Natural Resource Management.* London: Earthscan, pp. 67–87.
Sumberg, J. (2005) Systems of innovation theory and the changing architecture of agricultural research in Africa. *Food Policy* 30: 21–41.
Uvin, P. and D. Miller (1994) *Scaling Up: Thinking Through the Issues.* City: World Hunger Program. Available online at: www.globalpolicy.org/component/content/article/177/31630.html (accessed June 10, 2003).
Uvin, P. and D. Miller (1996) Paths to scaling up: Alternative strategies for local nongovernmental organizations. *Human Organization* 55(3): 344–354.
World Bank (1998) Reforming agricultural research organizations: Good practice for creating autonomous bodies and managing change. *AKIS Good Practice Note* No.01/99.
Williamson, J. (2000) What should the World Bank think about the Washington Consensus? *The World Bank Research Observer* 15 (2) August 2000: 251–264
World Bank (2003) *Scaling-Up the Impact of Good Practices in Rural Development.* A working paper to support implementation of the World Bank's Rural Development Strategy. Report Number 26031. Washington, D.C.: The World Bank.

ANNEX I

Program management and governance

Introduction

AHI's complex structure and accountability channels demanded that due attention be given to the program's governance. As stated earlier, AHI was, for most of its history, both an ASARECA network[1] and a CGIAR ecoregional program convened by ICRAF. Program management is shared by the AHI regional office, AHI implementing partners, AHI Steering Committee and AHI stakeholders.

AHI site teams and national focal points

AHI site teams are composed of a site coordinator and a group of between seven and ten research and development specialists from different disciplines and organizations. The site coordinator and site team members are employees of national research and extension organizations in each partner country. They are responsible for coordinating and facilitating all activities in AHI benchmark sites, in partnership with partner organizations and the regional office. National focal points within National Agricultural Research Institutes are responsible for providing coordination between AHI and national research programs, and for monitoring team performance.

Regional research team

Everyday management of the program falls under the Regional Coordinator, administrative staff and a small regional research team. The regional research team (RRT) collaborates with and supports site teams directly in the field through structured exchanges (for joint learning) and technical follow-up. Each RRT member is responsible for an AHI research theme around which his/her interactions with site teams and partners is structured. They are also responsible for fostering methodological innovation and lessons learning regionally around their assigned theme.

A technical support group (TSG) comprised of the RRT and site coordinators assists the Regional Coordinator in ensuring high-quality implementation and technical outputs, agrees on annual work plans and budgets, and guides the overall technical direction of AHI.

AHI steering committee

The major task of the Regional Steering Committee (RSC) is to provide strategic direction and technical support to the network. The AHI RSC is the highest body governing the program and comprises the Director General of host country research institutes (or his/her designate), AHI site coordinators, representatives from partner International Agricultural Research Centres (IARCs), selected donor representatives, and the Executive Secretary of ASARECA. The RSC is chaired on a rotational basis by host NARI representatives.

AHI partners and stakeholders

AHI has worked with a broad range of partner institutions at different stages of program evolution, depending on the nature of work being done. These include: NARS and extension organizations in ASARECA member countries; International Agricultural Research Centres (ICRAF, CIAT, CIP, IFPRI); NGOs (IUCN, Eco-agriculture Partners, Lishe Trust, Africare, Action-Aid, Care International); local government; networks (African Mountain Forum, the International Mountain Forum, Landcare International, Mountain Research Initiative); and universities inside and outside the ASARECA region (Makerere University, Wageningen University, Uppsala University).

A broader set of stakeholders is also involved in the governance, management, and implementation of AHI. These include farmers in AHI benchmark sites; district institutions (NGOs, CBOs, and local government); national, regional, and international institutions; and donor organizations. These stakeholders are either beneficiaries of the different AHI products, provide strategic direction and financing, or provide in-kind support to project implementation. Local communities provide labour and land for testing of approaches and methods in INRM, while district institutions participate in scaling out and up approaches, backstopping farmers, and providing in-kind support. Other AHI stakeholders provide funding or participate in scaling up INRM.

Note

1 This was true until September, 2007 when ASARECA reorganized from 17 networks to 7 programs. Some of the activities that were undertaken by AHI were relocated to the new NRM program of ASARECA. However, AHI has maintained its mandate as a CGIAR ecoregional program convened by ICRAF.

ANNEX II

Key phases in AHI's evolution

Phase I (1995–1997)

Phase I of AHI employed a competitive grant system to foster multidisciplinary teamwork among diverse areas of agronomic expertise and to foster partnerships among national agricultural research institutes, agricultural extension institutions, and local communities. As could be expected, multidisciplinarity was relatively new to those involved and a great deal of effort was spent during this phase to raise awareness on the merits of collaborative research among scientists accustomed to deriving professional recognition from achievements within narrow scientific disciplines. Four countries were involved in this phase, namely Ethiopia, Kenya, Madagascar, and Uganda. Two themes served to structure research cooperation: Integrated Pest Management (IPM), focusing on pests and diseases caused by soil nutrient depletion and agricultural intensification, and the Maintenance and Improvement of Soil Productivity (MISP). Regional research fellows (RRFs) with expertise in these areas were also hired to provide technical support to teams of grantees in the implementation of research programs. In an evaluation commissioned by ICRAF in May 1996, changes in direction were proposed to strengthen multidisciplinary collaboration. This led to a shift in operational modalities from a competitive grant system in Phase I to the use of benchmark sites to foster research cooperation and innovation in Phase II.

Phase II (1998–2000)

During this phase the country coverage was expanded to five to include Tanzania, and benchmark sites were selected in areas with high population density, evidence of natural resource degradation, and representative of broader eco-regions. This resulted in eight benchmark sites initially, two in Madagascar, two in Ethiopia, two in Kenya, one in Uganda, and one in Tanzania. Site teams composed of NARI scientists representing diverse disciplines (soils, plant breeding, livestock husbandry, entomology, and "socio economics") were constituted to carry out the work in each benchmark site. Site coordinators were identified and seconded to AHI either partially or fully (depending on the distance of sites to research stations) to coordinate the work in benchmark sites. The number and disciplinary diversity of RRFs were also expanded to include systems and participatory research perspectives. The second phase of AHI also saw the formation of a technical support group (TSC) consisting of site coordinators and the regional research team (regional research fellows and Regional Coordinator) to provide technical oversight to the work. The small grants were replaced with larger projects designed to enable multidisciplinary and multi-institutional teams from research and development (agricultural extension and/or civil society) organizations to work more holistically. An internal Planning, Monitoring, and Evaluation (PM&E) framework outlining the program purpose, specific goals, outputs, and strategy for implementation was also developed.

This phase led to the generation of a host of technological and methodological innovations for integrated natural resource management at farm level, including the integration of high-value crop varieties with soil nutrient management practices; "linked technologies" enhancing adoption through synergies among crop, soil, and livestock technologies; and innovations for pest and disease control.

An external review conducted in 2000 led to a number of further recommendations to orient future work, including:

- bolstering the commitment of research managers from partner NARIs;
- improving communication and documentation;
- focusing and phasing of activities to concentrate on process and partnerships;
- reducing the number of benchmark sites to enable more focused attention to research and dissemination, and to enable an expansion in scope to include innovations beyond the farm level;
- introducing structural and procedural changes to reduce transaction costs;
- using zonation for the purpose of dissemination and marketing of products;
- strengthening the incorporation of socio-economic aspects into the research program; and
- increasing the participation of farmers.

These recommendations clarified AHI's purpose in developing and testing methodological innovations for integrated natural resource management and supporting their institutionalization within the NARIs. It also encouraged an expanded focus of INRM to address issues that manifest themselves or require interventions beyond the farm level. The program was given a one-year (2001) transition period to refocus its activities in response to the demands of reviewers and key AHI stakeholders. The transition period was marked by the involvement of resource persons more experienced in "process-oriented" research to assist in charting out major areas of concentration in the next phase of AHI.

Phase III (2002–2004)

With experience gained through focused efforts in benchmark sites, and the greater confidence evident among participating institutions, farmers, and other stakeholders, the AHI entered its third phase. Defining features of the new mandate included:

- stronger integration across disciplines, including social science;
- expansion of the scale and scope of activities to include INRM at watershed level;
- integration of technological and other forms of innovation at watershed and district levels; and
- emphasis on decision-making processes as a foundation for the selection and application of technologies to suit specific socio-economic and biophysical situations expected to help in setting priorities and improving the focus of AHI activities.

During Phase III, a host of methodological innovations to harmonize interactions among local interest groups at landscape level were developed and tested. These innovations included efforts to enhance the equitability of technology dissemination and access, to foster collective action in pest control and soil and water management, to enhance the compatibility of trees with different landscape niches, and to integrate livelihood improvements with improved management of resources. They also included efforts to adapt the Landcare approach to the eastern Africa region.

Parallel initiatives were also undertaken during this phase with NARI managers, to support them in facilitating processes of self-led institutional change as a means of enhancing the impact orientation of research. This initiated a regional assessment of participatory research initiatives to derive "best practices," and a series of national level workshops and pilot experiences at local research stations to design and test institutional innovations to mainstream such practices within national agricultural research institutes.

Phase IV (2005–2007)

The major focus of the fourth and final phase reported on in this volume was to scale out lessons and approaches through district institutional innovations, and to institutionalize the INRM approach within partner NARIs and other institutions in the region. Key targets for Phase IV included:

- further development of the watershed approach, focusing on enabling collective action and integrating biophysical, social, and economic and market dimensions of farm and landscape management;
- use of experiences and methods from the pilot sites combined with information from wider syntheses to derive good practices and methods for development agencies;
- research to understand linkages and dynamics between vulnerability, poverty, livelihood strategies, economic growth, and NRM;
- provision of relevant and timely information to district and national development actors and decision-makers; and
- dissemination of "how to" information for INRM, with follow-up mentoring for institutional change in select research institutions in the ASARECA region.

Emphasis was also placed in Phase IV on knowledge management, including the documentation of past lessons and active learning, and on institutional innovation for broader uptake of lessons learnt.

AHI at present (2011)

While this book does not present work carried out by AHI after 2007, the program has a number of ongoing activities under the rubric of the Eastern Africa Programme of ICRAF – with the Regional Coordinator of AHI also coordinating ICRAF's work in the region. Ongoing projects which have either been developed under the AHI umbrella or include AHI as a major component include:

- An initiative funded by IDRC, covering Ethiopia and Uganda, on "Going to Scale." The project aims to further enhance the adaptive management capacities of rural communities for sustainable land management and devolve AHI approaches to national agricultural research institutes.
- A project funded by IFAD, covering Kenya, Uganda, and Tanzania, on "Enabling Rural Transformation and Grassroots Institution Building for Sustainable Land Management (SLM), Increased Incomes and Food Security." The project aims to strengthen local institutions as key structures in implementing effective SLM by smallholder farmers.
- An EU/IFAD funded project on "Evergreen Agriculture" in eastern and southern Africa. The initiative aims to scale up agroforestry-based conservation agriculture for improved nutrition, income, and environmental resilience in the region.

ANNEX III

AHI benchmark sites

Areka site

The Areka site is located in the south-central highlands of Ethiopia, the home of sedentary Wolaita farmers. The area is a mixed crop–livestock system with a high diversity of staple and cash crops (enset, wheat, maize, barley, sorghum, sweet potato, Irish potato, faba bean, field pea, and horticultural crops). Livestock are grazed in a large communal grazing area or in semi-communal fenced plots. Despite the diversity of enterprises characterizing the system, landholdings are extremely small (0.74 and 0.26 hectares on average for high and low wealth categories, respectively) and the area is subject to chronic food deficits. Unique to this site are a large number of landless families who earn a living as sharecroppers or through petty trade.

Key NRM challenges in this site included: a) enhancing the productivity and returns from crop, livestock, and tree components without further exacerbating system nutrient decline; b) arresting water resource degradation and resource conflicts through more optimal land management practices and improved governance; and c) increasing the viability of agriculture (through intensification and value addition) as a pathway to food security.

Ginchi site

The Ginchi benchmark site is located in the Western Shewa Zone, Ethiopia, home to the Oromo ethnic group. It is a mixed crop–livestock system that is more extensively managed than other sites. The system is very limited in biomass. Indiscriminate cutting of remnant trees and contiguous forest stemming largely from prior land reforms and from regime change, and the resulting ambiguity in tenure systems (Bekele, 2003), as well as failure to invest in NRM practices with delayed returns due to perceived tenure insecurity, have contributed to large areas of landscape devoid of vegetation and with very low nutrient stocks. This has placed increased burden on women and children who must walk long distances to gather firewood, and negative impacts on soil nutrients due to the sharp increase in the use of dung for fuel in recent decades (Omiti et al., 1999). Loss of tree cover and cultivation of Eucalyptus around springs have led to the degradation of springs, the sole source of water for both humans and livestock. Yet the tendency for humans and livestock to share common watering points has made water quality more of a concern than water quantity in the minds of local residents.

High-value crops such as Irish potatoes and garlic are grown on fenced homestead plots, while extensive outfield areas are used almost exclusively for barley production and livestock grazing. Valley bottoms are used exclusively for livestock grazing. While all land is officially owned by the government, individuals have de facto ownership over all land in the watershed. Yet management is collective in certain spatial and temporal niches. Households own outfield areas on both sides of the catchment, cultivating one side and leaving the other for grazing during the rainy season. The side of the catchment that is left for grazing is done so by all households with contiguous plots, enabling free movement of livestock by those households owning land in the area. Valley bottoms are grazed year-round, with access during the cropping season restricted to those households owning plots of land in these areas. During the dry season, outfields and valley bottoms are open access resources. This scenario makes systems innovation very challenging, requiring collective action not only among households living within the watershed but involving others who graze their livestock in the area.

The key challenges for integrated NRM included: a) intensifying production (of crops, livestock, and trees) while ensuring sustainable nutrient management in the system; and b) reversing water resource degradation by fostering positive synergies between trees, soil conservation structures and water in micro-catchments. Furthermore, seasonal open-access grazing makes investments in afforestation and soil conservation structures in the outfields challenging, as cattle can easily destroy such investments. Site teams and local leaders have highlighted this as a key challenge for this site, and targeted local negotiations and integrated policy and technological innovations as avenues for innovation.

Kabale site

The Kabale benchmark site is located in Kigezi highlands of southwestern Uganda, home to the Bakiga ethnic group. The area is characterized by high population density, steep cultivated slopes, fragmented landholdings, land shortages, and adequate rainfall. This site is also a mixed crop–livestock system with a relatively small livestock component. Communal grazing areas are negligible, making zero grazing a necessity, and free grazing – where it does occur – a source of conflict due to crop damage. In addition to limited numbers of livestock, enterprises include Irish potatoes and vegetable crops in the valley bottoms, and cereals (sorghum, maize, wheat, finger millet), pulses, and bananas on the hillsides. Trees are few and declining in number, a trend which has been exacerbated in recent years as a result of the high demand for wood from a nearby gin distillery.

Key NRM challenges in this site have included: a) integrating technological innovation with improved natural resource governance to minimize the incidence of conflict emanating from small landholdings, limited economic opportunities, and gender inequalities; b) improving incomes from small and fragmented landholdings through soil fertility management, diversification, and value addition; and c) managing the dependency syndrome, acute in this site due to a high density of non-governmental organizations (NGOs) and community-based organizations (CBOs) with short-sighted support strategies.

Kapchorwa site

Kapchorwa District is located on the slopes of Mt. Elgon in eastern Uganda. The district has a total population of 193,510 as per the 2002 population and housing census. The district population growth rate is at 4.33 percent which is high compared to the national average of 3.3 percent. The district has three ecological zones: lowlands (33 percent), which are almost deserted due to insecurity caused by cattle rustling; highlands (34 percent), which are heavily settled and cultivated; and forest (33 percent), which is a protected area. Agriculture is the main economic activity, engaging over 82.1 percent of the working population. The primary crops are maize, bananas, coffee, beans, wheat, barley, sunflower, and vegetable crops, with 82.1 percent of households living from farming.

The district is also home to the Mt. Elgon National Park, established as a Crown Forest in 1930. Management of the area within and surrounding the park has been subject to the whims of shifting government policies on forest management, changes which have affected most severely the native Benet ethnic group who have occupied the moorlands inside the park for the past 200 years. These changes have also negatively affected conservation in the area, as park officials and local residents alike have exploited the loosely guarded protected area under the current land tenure arrangement and ambiguity of rights of adjacent communities.

Key challenges include equitable resource access given histories of ethnic conflict (cattle raiding); managing resources sustainably within the buffer zone of the national park given the history of displacement and conflict; and limited quality of and access to support services due to a sparse NGO presence, limited coordination among sectors, and weak civil society.

Lushoto site

Lushoto District is located in the West Usambara Mountains of northeastern Tanzania. The district is home to the Wasambaa and small numbers of Wapare ethnic groups and migrants from other areas of Tanzania. The pilot watershed covers an area of 6,006 hectares and spans the Baga and Bumbuli Wards, six villages and a population of 13,163 (Meliyo et al., 2004). The land-use system is relatively intensified and involves the cultivation of cash crops in the valley bottoms, staple crops and tea on the hillsides, and small livestock holdings. From the 1950s onward, a number of afforestation programs designed to reduce pressure on State forest while contributing to conservation and livelihood goals were initiated, resulting in a dramatic increase in tree cover within farmland.

Challenges to INRM in this site have included: a) intensifying production of crops, livestock, and trees while ensuring sustainable nutrient management in the system; b) reversing water resource degradation by fostering positive synergies between trees, soil conservation structures, and water in micro-catchments; and c) managing environmental degradation stemming from cultivation from steep hillsides and mountain tops, and damage caused by rapid movement of water across the landscape (e.g., burial of fertile valley-bottom soils).

References

Bekele, M. (2003) Forest property rights, the role of the state and institutional exigency: The Ethiopian experience. Ph.D. Thesis, Swedish University of Agricultural Sciences, Uppsala. pp.220.

Meliyo, J.L., A. Mansoor, K.F.G. Masuki, J.G. Mowo, L. German, and R.S. Shemdoe (2004) Socio-economic and biophysical characteristics of Baga Watershed in Lushoto District, Tanzania. *AHI Site Report* No. 1; Lushoto Benchmark Site, Tanzania.

Omiti, J.M., K.A. Parton, J.A. Sinden, and S.K. Ehui (1999) Monitoring changes in landuse practices following agrarian de-collectivisation in Ethiopia. *Agriculture, Ecosystems and Environment* 72: 111–118.

ANNEX IV

Key AHI publications

Peer-reviewed journal articles

Amede, T., A. Stroud, and J. Aune (2004) Advancing Human Nutrition Without Degrading Land Resources through Modeling Cropping Systems in the Ethiopian Highlands. *Food and Nutrition Bulletin* 25(4): 344–353. United Nations University, Tokyo.

Barrios, E., R.J. Delve, M. Bekunda, J.G. Mowo, J. Agunda, J. Ramisch, M.T. Trejo, and R.J. Thomas (2006) Soil Quality Indicators: A South–South Development of a Methodological Guide to Integrate Local and Scientific Knowledge. *Geoderma* 135: 248–259.

Campbell, B., J. Hagmann, A. Stroud, R. Thomas, and E. Wollenberg (2006) What Kind of Research and Development is Needed for Natural Resource Management? *Water International* 31(3): 343–360. International Water Resources Association, Montpellier.

German, L. (2006) Moving Beyond Component Research in Mountain Regions: Operationalizing Systems Integration at Farm and Landscape Scales. *Journal of Mountain Science* 3(4): 287–304.

German, L., S. Ayele, and Z. Adimassu (2008) Managing Linkages Between Communal Rangelands and Private Cropland in the Highlands of Eastern Africa: Contributions to Participatory Integrated Watershed Management. *Society & Natural Resources* 21(2): 134–151.

German, L.A. and A. Keeler (2010) "Hybrid institutions": Applications of common property theory beyond discrete property regimes. *International Journal of the Commons* 4(1). Online at: http://www.thecommonsjournal.org/index.php/ijc/article/view/108/95.

German, L., B. Kidane, and R. Shemdoe (2006) Social and Environmental Trade-Offs in Tree Species Selection: A Methodology for Identifying Niche Incompatibilities in Agroforestry. *Environment, Development and Sustainability* 8: 535–552.

German, L., H. Mansoor, G. Alemu, W. Mazengia, T. Amede, and A. Stroud (2007) Participatory Integrated Watershed Management: Evolution of Concepts and Methods in an Ecoregional Program of the Eastern African Highlands. *Agricultural Systems* 94(2): 189–204.

German, L., W. Mazengia, H. Taye, M. Tsegaye, S. Charamila, and J. Wickama (2009) Minimizing the Livelihood Trade-Offs of Natural Resource Management in the Eastern African Highlands: Policy Implications of a Project in "Creative Governance." *Human Ecology* 38(1): 31–47.

German, L., J.G. Mowo, and M. Kingamkono (2006) A Methodology for Tracking the "Fate" of Technological Innovations in Agriculture. *Agriculture and Human Values* 23: 353–369.

German, L. and H. Taye (2008) A framework for evaluating effectiveness and inclusiveness of collective action in watershed Management. *Journal of International Development* 20: 99–116.

German, L., H. Taye, S. Ayele, W. Mazengia, T. Tolera, M. Tsegaye, K. Abere, K. Bedane, and E. Geta (2008) Institutional Foundations of Agricultural Development in Ethiopia: Drawing Lessons from Current Practice for Agricultural R&D. *Quarterly Journal of International Agriculture* 47(3): 191–216.

German, L., G. Villamor, S. Velarde, E. Twine, and B. Kidane (2009) Environmental Services and the Precautionary Principle: Using Future Scenarios to Reconcile Conservation and Livelihood Objectives in Upper Catchments. *Journal of Sustainable Forestry* 28(3): 368–394.

German, L. and A. Stroud (2007) A Framework for the Integration of Diverse Learning Approaches: Operationalizing Agricultural Research and Development (R&D) Linkages in Eastern Africa. *World Development* 35(5): 792–814.

Kaluski, D.N., E. Ophir, and T. Amede (2001) Food Security and Nutrition: The Ethiopian Case for Action. *Public Health Nutrition* 5(3): 373–381.

Masuki, K.F.G., J.G. Mowo, T.E. Mbaga, J.K. Tanui, J.M. Wickama, and C.J. Lyamchai (2010) Using Strategic "Entry Points" and "Linked Technologies" for Enhanced Uptake of Improved Banana Germplasm in the Humid Highlands of East Africa. *Acta Horticulturae* 879(2): 797–804.

Mazengia, W., D. Gamiyo, T. Amede, M. Daka, and J. Mowo (2007) Challenges of Collective Action in Soil and Water Conservation: The Case of Gununo Watershed, Southern Ethiopia. *African Crop Science Conference Proceedings* 8: 1541–1545. African Crop Science Society, El-Minia, Egypt.

Mowo, J.G., L.A. German, M.N. Kingamkono, and K.F. Masuki (2010) Tracking the Spillover of Introduced Technologies: The Case of Improved Banana Germplasm in North-eastern Tanzania. *Acta Horticulturae* 879(2): 695–704.

Mowo, J.G., B.H. Janssen, O. Oenema, L.A. German, J.P. Mrema, and R.S. Shemdoe (2006) Soil Fertility Evaluation and Management by Smallholder Farmer Communities in Northern Tanzania. *Agriculture, Ecosystems and Environment* 116(1–2): 47–59.

Mowo, J., C. Opondo, A. Nyaki, and Z. Adimassu (2010) Addressing the Research–Development Disconnect: Lessons from East and Central African Highlands. *Development in Practice* 20(8): 1000–1012.

Sanginga, P. C., R.N. Kamugisha, and A.M. Martin (2007) The Dynamics of Social Capital and Conflict Management in Multiple Resource Regimes: A Case of the South-western Highlands of Uganda. *Ecology and Society* 12(1): 6. Online at: http://www.ecologyandsociety.org/vol12/iss1/art6.

Sanginga, P.C., R. Kamugisha, A. Martin, A. Kakuru, and A. Stroud (2004) Facilitating Participatory Processes for Policy Change in Natural Resource Management: Lessons from the Highlands of Southwestern Uganda. *Uganda Journal of Agricultural Sciences* 9: 958–970. National Agricultural Research Organization, Kampala.

Books

Adimassu, Z., K. Mekonnen, and Y. Gojjam (2008) *Working with Communities on Integrated Natural Resources Management.* Ethiopian Institute of Agricultural Research (EIAR), Addis Ababa. 134p. ISBN: 978-99944-53-23-8. Online at: http://www.kef-online.at/en/r4d-news/thiopien-working-with-communities-on-integrated-natural-resources-management.html. Volume includes 13 separate chapters not listed here.

Amede, T., L. German, S. Rao, C. Opondo, and A. Stroud (eds) (2006) *Integrated Natural Resource Management in Practice: Enabling Communities to Improve Mountain Livelihoods and Landscapes.* Kampala, Uganda: African Highlands Initiative. Volume includes 63 separate chapters not listed here.

Hurni, H. and J. Ramamonjisoa (eds) (1999) *African Mountain Development in a Changing World.* Antananarivo, Tokyo and Nairobi: African Mountain Association, the United National University and the African Highlands Initiative. 332p. ISBN: 3-906151-33-6. Volume includes 19 separate chapters not listed here.

Book chapters[1]

Amede, T. (2003) Pathways for Fitting Legumes into the Farming Systems of East African Highlands: A Dual Approach. In: Waddington, S. (ed.), *Grain Legumes and Green Manures for Soil Fertility in Southern Africa: Taking Stock of Progress*, pp. 21–30. Soil Fert Net and CIMMYT-Zimbabwe, Harare.

Amede, T. and R. Kirkby (2004) Guidelines for Integration of Legumes into the Farming Systems of East African Highlands. In: Bationo, A. (ed.), *Managing Nutrient Cycles to Sustain Soil Fertility in Sub-Saharan Africa*, pp. 43–64. Academic Science Publishers, Nairobi.

Amede, T. (2003) Opportunities and Challenges in Reversing Land Degradation: The Regional Experience. In: Amede, T. (ed.), *Natural Resource Degradation and Environmental Concerns in the Amhara National Regional State: Impact on Food Security*, pp. 173–183. Ethiopian Soils Science Society, Addis Ababa.

German, L., S. Charamila, and T. Tolera (in press) Managing Trade-Offs in Agroforestry: From Conflict to Collaboration in Natural Resource Management. In: Klappa, S. and D. Russell (eds), *Transformations in Agroforestry Systems*. Berghahn Books.

German, L.A., B. Kidane, and K. Mekonnen (2008) Watershed Management to Counter Farming System Decline: Towards a Demand-Driven, System-Oriented Research Agenda. In: Menon, S.S.V. and P.A. Pillai (eds), *Watershed Management: Concepts and Experiences*, pp. 71–86. ICFAI University Press, India.

German, L., W. Mazengia, W. Tirwomwe, S. Ayele, J. Tanui, S. Nyangas, L. Begashaw, H. Taye, Z. Adimassu, M. Tsegaye, F. Alinyo, A. Mekonnen, K. Aberra, A. Chemangeni, W. Cheptegei, T. Tolera, Z. Jotte, and K. Bedane (2008) Enabling Equitable Collective Action and Policy Change for Poverty Reduction and Improved Natural Resource Management in the Eastern African Highlands. In: Mwangi, E., H. Markelova, and R. Meinzen-Dick (eds), *Collective Action and Property Rights for Poverty Reduction: Lessons from a Global Research Project*, pp. 11–12. CAPRi, Washington, D.C.

German, L., W. Mazengia, W. Tirwomwe, S. Ayele, J. Tanui, S. Nyangas, L. Begashaw, H. Taye, Z. Adimassu, M. Tsegaye, S. Charamila, F. Alinyo, A. Mekonnen, K. Aberra, A. Chemangeni, W. Cheptegei, T. Tolera, Z. Jotte, and K. Bedane (in press) Enabling Equitable Collective Action and Policy Change for Poverty Reduction and Improved Natural Resource Management in the Eastern African Highlands. In: Mwangi, E.,

H. Markelova and R. Meinzen-Dick (eds), *Collective Action and Property Rights for Poverty Reduction*. Johns Hopkins and IFPRI, Baltimore and Washington, D.C.

Masuki, K.F.G., J.G. Mowo, S. Rao, R. Kamugisha, C. Opondo, and J. Tanui (2008) Improving Smallholder Farmers' Access to Information for Enhanced Decision Making in Natural Resource Management: Experiences from South Western Uganda. In: Bationo, A., B.S. Waswa, J. Okeyo, and F. Maina (eds), Innovations as Key to the Green Revolution in Africa: Exploring the Scientific Facts. Springer, Dordrecht.

Meliyo, J.L., K.F.G. Masuki, and J.G. Mowo (2007) Integrated Natural Resources Managament: A Strategy for Food Security and Poverty Alleviation in Kwalei Village, Lushoto District, Tanzania. In: Bationo, A., B. Waswa, J. Kihara, and J. Kimetu (eds), *Advances in Integrated Soil Fertility Management in Sub Saharan Africa: Challenges and Opportunities*, pp. 781–785. Springer, Dordrecht.

Mowo, J.G., S.R. Shemdoe, and A. Stroud (2007) Interdisciplinary Research and Management in the Highlands of Eastern Africa: AHI Experiences in the Usambara Mountains, Tanzania. In: M.F. Price (ed.), *Mountain Area Research and Management: Integrated Approaches*, pp. 118–130. London: Earthscan.

Working papers

Amede, T., Bekele, A., and Opondo, C. (2006) Creating Niches for Integration of Green Manures and Risk Management through Growing Maize Cultivar Mixtures in the Southern Ethiopian Highlands. *AHI Working Papers* No. 14.

Amede, T., T. Belachew, and E. Geta (2001) Reversing the Degradation of Arable Land in the Ethiopian Highlands. *Managing Africa's Soils* No. 23. IIED, London.

Amede, T., T. Belachew, and E. Geta (2006) Reversing Degradation of Arable Lands in Southern Ethiopia. *AHI Working Papers* No. 1.

Amede, T. and R. Delve (2006) Improved Decision-Making for Achieving Triple Benefits of Food Security, Income and Environmental Services through Modeling Cropping Systems in the Ethiopian Highlands. *AHI Working Papers* No. 20.

Amede, T. and R. Kirkby (2006) Guidelines for Integration of Legumes into the Farming Systems of the East African Highlands. *AHI Working Papers* No. 7.

Amede, T. and E. Taboge (2006) Optimizing Soil Fertility Gradients in the Enset (Ensete ventricosum) Systems of the Ethiopian Highlands: Trade-offs and Local Innovations. *AHI Working Papers* No. 15.

Amede, T., A. Stroud, and J. Aune (2006) Advancing Human Nutrition without Degrading Land Resources through Modeling Cropping Systems in the Ethiopian Highlands. *AHI Working Papers* No. 8.

Beyene, H. and T. Mulatu (1999) A selection of diagnostic and characterisation studies conducted between 1986 and 1994 at Ginchi and Nzaret, Ethiopia. *AHI Technical Report Series* No. 8. Nairobi, Kenya: African Highlands Initiative.

Braun, A.R., E.M.A. Smaling, E.I. Muchugu, K.D. Shephherd, and J.D. Corbett, (eds) (1997) Maintenance and Improvement of Soil Productivity in the Highlands of Ethiopia, Kenya, Madagascar and Uganda: An Inventory of Spatial and Non-spatial, Survey and Research Data on Natural Resources and Land Productivity. *AHI Technical Report Series* No. 6. Nairobi, Kenya: African Highlands Initiative.

David, S. (ed.) (2000) Planning for Farmers' Seed Requirements: Proceedings of Workshops at AHI Benchmark Sites in Eastern Africa. *AHI Technical Report Series* No. 12. Nairobi, Kenya: African Highlands Initiative.

Esilaba, A.O., T. Mulatu, F. Reda, J.K. Ransom, G. Woldewahid, A. Tesfaye, I. Fitwy, and G. Abate (1999) A Diagnostic Survey on Striga in the Northern Ethiopian Highlands. *AHI Technical Report Series* No. 5. Nairobi, Kenya: African Highlands Initiative.

Gachene, C.K.K., C.A. Palm, and J.G. Mureithi (2000) Legume Cover Crop for Soil Fertility Improvement in the Eastern Africa Region: Report of an AHI Workshop, 18–19 February, 1999. *AHI Technical Report Series* No. 11. Nairobi, Kenya: African Highlands Initiative.

German, L. (2006) Approaches for Mountain Regions: Operationalizing Systems Integration at Farm and Landscape Scales. *AHI Working Papers* No. 21.

German, L. (2006) Social and Environmental Trade-Offs in Tree Species Selection: A Methodology for Identifying Niche Incompatibilities in Agroforestry. *AHI Working Papers* No. 9.

German, L., S. Charamila, and T. Tolera (2006) Managing Trade-Offs in Agroforestry: From Conflict to Collaboration in Natural Resource Management. *AHI Working Papers* No. 10.

German, L., B. Kidane, and K. Mekonnen (2005) Watershed Management to Counter Farming Systems Decline: Toward a Demand-Driven, Systems-Oriented Research Agenda. *AgREN Network Paper* No. 145.

German, L., B. Kidane, and K. Mekonnen (2006) Watershed Management to Counter Farming Systems Decline: Toward a Demand-Driven, Systems-Oriented Research Agenda. *AHI Working Papers* No. 16.

German, L., H. Mansoor, G. Alemu, W. Mazengia, T. Amede, and A. Stroud (2006) Participatory Integrated Watershed Management: Evolution of Concepts and Methods. *AHI Working Papers* No. 11.

German, L., K. Masuki, Y. Gojjam, J. Odenya, and E. Geta (2006) Beyond the Farm: A New Look at Livelihood Constraints in the Eastern African Highlands. *AHI Working Papers* No. 12.

German, L., W. Mazengia, S. Ayele, W. Tirwomwe, J. Tanui, H. Taye, L. Begashaw, S. Nyangas, A. Chemangeni, W. Cheptegei, M. Tsegaye, Z. Adimassu, F. Alinyo, A. Mekonnen, K. Aberra, T. Tolera, Z. Jotte, and K. Bedane (2007) Enabling Equitable Collective Action & Policy Change for Poverty Reduction and Improved Natural Resource Management in Ethiopia and Uganda. *AHI Working Papers* No. 25.

German, L., W. Mazengia, W. Tirwomwe, S. Ayele, J. Tanui, S. Nyangas, L. Begashaw, H. Taye, Z. Adimassu, M. Tsegaye, F. Alinyo, A. Mekonnen, K. Aberra, A. Chemangeni, W. Cheptegei, T. Tolera, Z. Jotte, and K. Bedane (2008) Enabling Equitable Collective Action and Policy Change for Poverty Reduction and Improved Natural Resource Management in the Eastern African Highlands. *CAPRi Working Paper* No. 86. IFPRI, Washington, D.C.

German, L., A. Stroud, C. Opondo, and B. Mwebesa (2004) Linking Farmers and Policy makers: Experiences from Kabale District, Uganda. *UPWARD Participatory R&D Sourcebook*. CIP, Manila.

German, L., H. Taye, S. Charamila, T. Tolera, and J. Tanui (2006) The Many Meanings of Collective Action: Lessons on Enhancing Gender Inclusion and Equity in Watershed Management. *AHI Working Papers* No. 17 and *CAPRi Working Paper* No. 52.

Himmelfarb, D. (2007) Moving People, Moving Boundaries: The Socio-economic Effects of Protectionist Conservation, Involuntary Resettlement and Tenure Insecurity on the Edge of Mt. Elgon National Park, Uganda. *AHI Working Papers* No. 24.

Mowo, J., B. Janssen, O. Oenema, L. German, P. Mrema, and R. Shemdoe (2006) Soil Fertility Evaluation and Management by Smallholder Farmer Communities in Northern Tanzania. *AHI Working Papers* No. 18.

Nderitu, J. H., R.A. Buruchara, and J.K.O. Ampofo (1997) Relationship Between Bean Stem Maggot, Bean Root Rots and Soil Fertility: Literature Review with Emphasis on Research in Eastern and Central Africa. *AHI Technical Report Series* No. 4. Nairobi, Kenya: African Highlands Initiative.

Opondo, C., L. German, A. Stroud, and E. Obin (2006) Lessons from Using Participatory Action Research to Enhance Farmer-Led Research and Extension in Southwestern Uganda. *AHI Working Papers* No. 3.

Opondo, C., P. Sanginga, and A. Stroud (2006) Monitoring the Outcomes of Participatory Research in Natural Resources Management: Experiences of the African Highlands Initiative. *AHI Working Papers* No. 2.

Opondo, C., A. Stroud, L. German, and J. Hagmann (2003) Institutionalising Participation in East African Research Institutes, Ch. 11. *PLA Notes* 48. IIED, London.

Salasya, B.D.S and S. Ajanga (1999) A Selection of Diagnostic and Characterisation Studies Conducted Between 1986 and 1995 in Western and Central Highland Areas of Kenya. *AHI Technical Report Series* No. 9. Nairobi, Kenya: African Highlands Initiative.

Stroud, A. (2006) Transforming Institutions to Achieve Innovation in Research and Development. *AHI Working Papers* No. 4.

Stroud, A. (2006) Understanding People, Their Livelihood Systems and the Demands and Impact of Innovations. *AHI Working Papers* No. 13.

Stroud, A. and L. German (2006) A Framework for the Integration of Diverse Learning Approaches: Operationalizing Agricultural Research and Development (R&D) Linkages in Eastern Africa. *AHI Working Papers* No. 23.

Stroud, A. and J. Hagmann (2006) Shared Experiences of an ASARECA Programme: Key Challenges for Institutions to Operationalise INRM from Ecoregional and NARS Perspectives. *AHI Working Papers* No. 5.

Stroud, A. and R. Khandelwal (2006) In Search of Substance: "State of the Art" of Approaches, Strategies and Methods for Improving Natural Resource Management and Livelihoods. *AHI Working Papers* No. 6.

Stroud, A., E. Obin, R. Kandelwahl, F. Byekwaso, C. Opondo, L. German, J. Tanui, O. Kyampaire, B. Mbwesa, A. Ariho, Africare, and Kabale District Farmers' Association (2006) Managing Change: Institutional Development under NAADS: A Field Study on Farmer Institutions Working with NAADS. *AHI Working Papers* No. 22.

Tanui, J. (2006) Incorporating a Landcare Approach into Community Land Management Efforts in Africa: A Case Study of the Mount Kenya Region. *AHI Working Papers* No. 19.

Tukahirwa, J.M. (1999) Diagnostic and characterisation studies conducted from 1945 to 1995 in southwestern Uganda. *AHI Technical Report Series* No. 10. Nairobi, Kenya: African Highlands Initiative.

Wickama, J.M. and J.G. Mowo (2001) Indigenous nutrient resources in Tanzania. *Managing Africa's Soils* No. 21.

AHI Working Papers available online at:
www.worldagroforestry.org/projects/african-highlands/archives.html#wps

Methods guides

Barrios, E., M. Bekunda, R. Delve, A. Esilaba, and J. Mowo (2000) Methodologies for Decision Making in Natural Resource Management: Identifying and Classifying Local Indicators of Soil Quality. Eastern Africa Version. Online at: www.ciat.cgiar.org/downloads/pdf/isq_contents.pdf.

German, L., B. Kidane and S. Charamila with W. Mazengia, S. Ayele, and T. Tolera (2007) Niche-Compatible Agroforestry: A Methodology for Understanding and Managing Trade-Offs in Tree Species Selection at Landscape Level. *AHI Methods Guides* C1.

German, L., W. Mazengia, S. Charamila, H. Taye, S. Nyangas, J. Tanui, S. Ayele, and A. Stroud (2007) Action Research: An Approach for Generating Methodological Innovations for Improved Impact from Agricultural Development and Natural Resource Management. *AHI Methods Guide* E1.

German, L., K. Mekonnen, J.G. Mowo, E. Geta, and T. Amede (2006) A Socially-Optimal Approach to Participatory Watershed Diagnosis. *AHI Methods Guide* B2.

German, L., J. Mowo, M. Kingamkono, and J. Nuñez (2006) Technology Spillover: A Methodology for Understanding Patterns and Limits to Adoption of Farm-Level Innovations. *AHI Methods Guide* A1.

German, L., A. Stroud, G. Alemu, Y. Gojjam, B. Kidane, B. Bekele, D. Bekele, G. Woldegiorgis, T. Tolera, and M. Haile (2006) Creating an Integrated Research Agenda from Prioritized Watershed Issues. *AHI Methods Guide* B4.

Mekonnen, K. (2009) Watershed Management. In: M. Nigussie, A. Girma, C. Anchala, and A. Kirub (eds), *Improved Technologies and Resource Management for Ethiopian Agriculture: A Training Manual*, pp. 289–296. RCBP-MoARD, Addis Ababa, Ethiopia.

AHI Methods Guides available online at:
www.worldagroforestry.org/projects/african-highlands/archives.html#mgs

AHI briefs

Stroud, A. (2003) Program Brief.

Theme A – strategies for systems intensification

Amede, T. (2003) Restoring Soil Fertility in the Highlands of East Africa through Participatory Research. *AHI Brief* A1.

Amede, T. (2003) Differential Entry Points to Address Complex Natural Resource Constraints in the Highlands of Eastern Africa. *AHI Brief* A2.

Stroud, A. (2003) Linked Technologies for Increasing Adoption and Impact. *AHI Brief* A3.

Amede, T. (2004) Boosting Human Nutrition through Land Use Modelling: An Alternative to Biofortification. *AHI Brief* A4.

Amede, T. (2004) Soil Fertility Decision Guide Formulation: Assisting Farmers with Varying Objectives to Integrate Legume Cover Crops. *AHI Brief* A5.

Amede, T. (2007) Soil and Water Conservation through Attitude Change and Negotiation. *AHI Brief* A6.

Theme B – institutional innovations for R&D

Chemengei, A., S. Nyangas, W. Cheptegei, J. Tanui, F. Alinyo, and L. German (2007) Co-Management is About Cultivating Relationships. *AHI Brief* B7.
German, L. and A. Stroud (2004) Integrating Learning Approaches for Agricultural R&D. *AHI Brief* B4.
German, L., A. Stroud, and E. Obin (2003) A Coalition for Enabling Demand-Driven Development in Kabale District, Uganda. *AHI Brief* B1.
German, L., W. Tirwomwe, J. Tanui, S. Nyangas, and A. Chemengei (2007) Searching for Solutions: Technology-Policy Synergies in Participatory NRM. *AHI Brief* B6.
Opondo, C., L. German, S. Charamila, A. Stroud, and R.K Khandelwal (2005) Process Monitoring and Documentation for R&D Team Learning: Concepts and Approaches. *AHI Brief* B5.
Tanui, J., A. Chemengei, S. Nyangas, and W. Cheptegei (2007) Rural Development and Conservation: The Future Lies with Multi-Stakeholder Collective Action. *AHI Brief* B8.
Stroud, A. (2003) Self-Management of Institutional Change for Improving Approaches to Integrated NRM. *AHI Brief* B2.
Stroud, A. (2003) Combining Science with Participation: Learning Locally and Generalizing Regionally. *AHI Brief* B3.

Theme C – integrated watershed management

German, L. (2003) Beyond the Farm: A New Look at Livelihood Constraints in the Highlands of Eastern Africa. *AHI Brief* C1.
German, L. (2006) Environmental Service Rewards in ECA: "Environmental Signatures" and Scenario Analysis to Minimize Trade-Offs. *AHI Brief* C3.
German, L., B. Kidane, R. Shemdoe, and M. Sellungato (2005) A Methodology for Understanding Niche Incompatibilities in Agroforestry. *AHI Brief* C2.

Theme D – advancing impact

German, L. (2004) Adding Nuance: The Role of Constructivist Inquiry in Agricultural R&D. *AHI Brief* D3.
German, L. and A. Stroud (2004) Social Learning in Regional R&D Programs. *AHI Brief* D2.
Mowo, J., L. Nabahungu, and L. Dusengemengu (2007) The Integrated Watershed Management Approach for Livelihoods and Natural Resource Management in Rwanda: Moving Beyond AHI Pilot Sites. *AHI Brief* D5.
Tanui, J. (2003) What about the Land User? An African Grassroots Innovation for Livelihood and Environment (AGILE) Approach. *AHI Brief* D1.
Tanui, J. (2005) Revitalizing Grassroots Knowledge Systems: Farmer Learning Cycles in AGILE. *AHI Brief* D4.

Theme E – strengthening local institutions and equity

Ayele, S., A. Ghizaw, Z. Adimassu, M. Tsegaye, G. Alemu, T. Tolera, and L. German (2007) Enhancing Collective Action in Spring "Development" and Management through Negotiation Support and By-Law Reforms. *AHI Brief* E5.

Begashaw, L., W. Mazengia, and L. German (2007) Mobilizing Collective Action for Vertebrate Pest Control. *AHI Brief* E3.
German, L. (2003) Watershed Entry: A Socially-Optimal Approach. *AHI Brief* E1.
German, L., S. Charamila, and T. Tolera (2005) Negotiation Support in Watershed Management: A Case for Decision-Making beyond the Farm Level. *AHI Brief* E2.
Mazengia, W., A. Tenaye, L. Begashaw, L. German, and Y. Rezene (2007) Enhancing Equitable Technology Access for Socially and Economically Constrained Farmers. *AHI Brief* E4.

Theme T – training briefs

AHI (2005) Training Course on System Optimization Based on Demand, Markets and the Resource Base. *AHI Brief* T1.
AHI (2005) Training Course on Participatory Integrated Watershed Management. *AHI Brief* T2.
AHI (2005) Training Course on Tracking Technology "Spillover." *AHI Brief* T3.
AHI (2005) Facilitators' Course on District Institutional Collaboration for Integrated Livelihoods and Conservation. *AHI Brief* T4.
AHI (2005) Workshop on Self-Led Institutional Learning and Change for NARS and NPPs. *AHI Brief* T5.

AHI Briefs available online at:
www.worldagroforestry.org/projects/african-highlands/archives.html#briefs

Conference proceedings

Adimassu, Z., K. Mekonnen, and B. Gorfu (2011) Understanding and Managing Complexities in Integrated Natural Resources Management at Watershed Scale: Lessons from the Central Highland of Ethiopia. Proceedings of the "International Conference on Ecosystem Conservation and Sustainable Development," February 10–12, 2011, Ambo University, Ethiopia.
Amede, T., H. Assefa, and A. Stroud (eds) (2004) Participatory Research in Action: Ethiopian Experience. Proceedings of a Participatory Research Workshop, June 12–17, 2002. Addis Ababa, Ethiopia: Ethiopian Agricultural Research Organisation and African Highlands Initiative. 144 p.
Masuki, K.F.G, R. Kamugisha, J.G. Mowo, J. Tanui, J. Tukahirwa, J. Mogoi, and E.O. Adera (2010) Role of Mobile Phones in Improving Communication and Information Delivery for Agricultural Development: Lessons from South Western Uganda. Paper presented in "ICT and Development – Research Voices from Africa," March 22–23, 2010, Makerere University, Uganda.
Masuki, K.F.G., H.A. Mansoor, J.G. Mowo, A.J. Tenge, J.M. Wickama, J. Mogoi, and J. Tanui (2009) Institutional Approach to Grassroots Communities' Engagement in Natural Resource Management in the Northeastern Tanzania: Entry Point to Forging Partnership in SLM. Paper presented at the "25th Soil Science Society of East Africa," December 7–11, 2009, Moshi, Tanzania. Proceedings in preparation.
Masuki, K., J. Mowo, R. Kamugisha, A. Tibingana, E. Adera, J. Tanui, J. Tukahirwa, and J. Mogoi (2009) Rural Information and Communication System and its Implication to Landcare Movement in Eastern Africa. Presented at the "Fourth Biennial Landcare Conference," July 12–16, 2009, Limpopo, South Africa. Proceedings in preparation.

Mekonnen, K. and Z. Adimassu (2009) Watershed Management Approach for NRM: Experiences and Lessons from Galessa, Central Ethiopia. In: Alemu, A., S. Yifredew, A. Mekonnen and M. Kassie (eds), *Sustainable Land Management and Poverty Alleviation*, pp. 182–190. Proceedings of a workshop held from May 18–19, 2009, CRDA Training Center, Addis Ababa, Ethiopia.

Mogoi, J., J. Tanui, D. Catacutan, R. Kamugisha, and K. Masuki (2010) A Report on the "National Landcare Workshop," Hotel Des Mille Collines, Kigali, Rwanda, June 2010. Nairobi: African Highlands Initiative.

Mowo, J.G., R.S. Kurothe, M.N. Shem, N.L. Kanuya, and L. Dusengemungu (2007) Adopting the Integrated Watershed Management Approach in Rwanda. In: R.W. Njeru, D.M. Kagabo, T. Ndabamenye, D. Kayiranga, P. Ragama, P.Y.K. Sallah, D. Nkeabahzi, A. Ndayiragije, L. Ndiramiye, G. Night, S.O.S. Akinyemi, N. Kanuya, M.C. Bagabe, and J. Mugabe (2007), *Sustainable Agricultural Productivity for Improved Food Security and Livelihoods*, pp. 497–506. Proceedings of the "National Conference on Agricultural Research Outputs," March 26–27, 2007, Kigali, Rwanda.

Mowo, J.G., H. Masoor, K.F.G. Masuki, and J. Meliyo (2009) Influence of Eucalypts on selected soil properties and maize growth in the highlands of northeastern Tanzania. Paper presented at the "25th Soil Science Society of East Africa," 7–11 December, 2009, Moshi, Tanzania. Proceedings in preparation.

Mowo, J.G., L.N. Nabahungu, L. Dusengemungu, and S. Sylveri (2007) Opportunities for Overcoming Soil Fertility Constraints to Agricultural Production in Gasharu Watershed, Southern Province, Rwanda. In: R.W. Njeru, D.M. Kagabo, T. Ndabamenye, D. Kayiranga, P. Ragama, P.Y.K. Sallah, D. Nkeabahzi, A. Ndayiragije, L. Ndiramiye, G. Night, S.O.S. Akinyemi, N. Kanuya, M.C. Bagabe, and J. Mugabe (2007), *Sustainable Agricultural Productivity for Improved Food Security and Livelihoods*, pp. 506–514. Proceedings of the "National Conference on Agricultural Research Outputs," March 26–27, 2007, Kigali, Rwanda.

Tanui, J., A. Chemangei, and J.G. Mowo (2007) Enabling Negotiation and Conflict Resolution for Area Wide Planning: The Case of Collective Action for Watershed Management. In: R.W. Njeru, D.M. Kagabo, T. Ndabamenye, D. Kayiranga, P. Ragama, P.Y.K. Sallah, D. Nkeabahzi, A. Ndayiragije, L. Ndiramiye, G. Night, S.O.S. Akinyemi, N. Kanuya, M.C. Bagabe, and J. Mugabe (eds), *Sustainable Agricultural Productivity for Improved Food Security and Livelihoods*, pp. 515–524. Proceedings of the "National Conference on Agricultural Research Outputs," 26–27 March, 2007, Kigali, Rwanda.

Note

1 This list includes only those chapters not otherwise included in AHI books.

INDEX

For Product Safety Concerns and Information please contact our EU
representative GPSR@taylorandfrancis.com
Taylor & Francis Verlag GmbH, Kaufingerstraße 24, 80331 München, Germany

www.ingramcontent.com/pod-product-compliance
Ingram Content Group UK Ltd.
Pitfield, Milton Keynes, MK11 3LW, UK
UKHW022216190926
475702UK00003B/381

* 9 7 8 1 8 4 9 7 1 4 2 4 2 *